PHYSICS FOUNDATIONS SOCIETY

The
Dynamic Universe

Toward a unified picture of physical reality

Fourth, complemented edition

Tuomo Suntola

Published by

PHYSICS FOUNDATIONS SOCIETY, Finland

www.physicsfoundations.org

Printed by Kindle Direct Publishing, USA

Copyright © 2018 by *Tuomo Suntola*. All rights reserved. No part of this publication may be reproduced, stored in a retrieval system, or transmitted, in any form or by any means without written permission from the copyright owner.

ISBN-13: 978-1725863415
ISBN-10: 1725863413

Contents

Preface   11

1. Introduction to the Dynamic Universe   17
  1.1 Basic concepts   18
    1.1.1 Space as a spherically closed entity   18
      The zero-energy principle   *19*
    1.1.2 Mass objects and the two-fold expression of energy   20
    1.1.3 Linkage between GR space and DU space   21
      The balance of the rest energy and gravitational energy   *21*
      Zero-energy balance and the critical mass density   *22*
    1.1.4 Definitions and notations   23
    1.1.5 Reinterpretation of Planck's equation   25
      The wave nature of mass   *26*
      The "antenna solution" of blackbody radiation   *27*
      The unified expression of energy   *27*
      Mass objects as resonant mass wave structures   *29*
  1.2 Buildup of energy in space   30
    1.2.1 The primary energy buildup in space   30
    1.2.2 Buildup of kinetic energy in space   32
      Kinetic energy at constant gravitational potential   *32*
      Kinetic energy in free fall in a gravitational field   *33*
      Tilting of local space   *35*
    1.2.3 Energy structures in space   36
      The system of nested energy frames   *36*
      The topography of the fourth dimension   *38*
      The local velocity of light   *39*
    1.2.4 The frequency of atomic clocks   40
      The quantum mechanical solution   *40*
      The effect of motion and gravitation   *41*
      Clocks in the Earth's gravitational frame   *43*
      Experiments on the effects of motion and gravitation on atomic clocks   *43*
    1.2.5 Propagation of light   46
      The Michelson–Morley experiment   *47*
      M-M experiment in the DU framework   *47*
      Sagnac effect   *48*
      Slow transport of clocks   *48*
    1.2.6 Observables in a local gravitational frame   50
      Perihelion advance   *50*

Black hole, critical radius	*52*
Orbital decay	*54*
Shapiro delay	*55*
Deflection of light, gravitational lens	*57*

1.3 Cosmological considerations 58
   1.3.1 The linkage of local to the whole 58
   1.3.2 Distances in FLRW cosmology 59
   1.3.3 Distances in DU space 62
      Angular size of cosmological objects *63*
      The magnitude of the standard candle *64*
   1.3.4 The length of a day and a year 66
   1.3.5 Timekeeping and near-space distances 67
      SI Second and meter *67*
      Annual variation of the Earth to Moon distance *68*

1.4 Summary 70
   1.4.1 Hierarchy of physical quantities and theory structures 70
      The postulates *70*
      The force-based versus energy-based perspective *70*
   1.4.2 Some fundamental equations 73
      The rest energy of matter *73*
      The total energy of motion *73*
      Kinetic energy *74*
      The laws of motion *74*
      The Planck equation *75*
      Physical and optical distance in space (cosmology) *75*
   1.4.3 Dynamic Universe and contemporary physics 76
      Linkage of local and global *78*
      The buildup of local structures *78*
      The destiny of the universe *78*

2. Basic concepts, definitions, and notations 79

2.1 Closed spherical space and the universal coordinate system 79
   2.1.1 Space as a spherically closed entity 79
   2.1.2 Time and distance 80
   2.1.3 Absolute reference at rest, the initial condition 81
   2.1.4 Notation of complex quantities 81

2.2 Base quantities 84
   2.2.1 Mass 84
   2.2.2 Energy and the conservation laws 85
      Gravitational energy in homogeneous space *85*
      The energy of motion in homogeneous space *85*
      Conservation of total energy *86*
   2.2.3 Force, inertia, and gravitational potential 86

3. Energy buildup in spherical space — 87
3.1 Volume of spherical space — 87
3.2 Gravitation in spherical space — 88
3.2.1 Mass in spherical space — 88
3.2.2 Gravitational energy in spherical space — 89
3.3 Primary energy buildup of space — 91
3.3.1 Contraction and expansion of space — 91
3.3.2 Mass and energy of space — 93
3.3.3 Development of space with time — 95
3.3.4 The state of rest and the recession of distant objects — 98
3.3.5 From mass to matter — 100

4. Energy structures in space — 103
4.1 The zero-energy balance — 104
4.1.1 Conservation of energy in mass center buildup — 104
Mass center buildup in homogeneous space — 104
Mass center buildup in real space — 108
4.1.2 Kinetic energy — 112
Kinetic energy obtained in free fall — 113
Kinetic energy obtained via insertion of mass — 114
Kinetic energy obtained in free fall and via the insertion of mass — 116
4.1.3 Inertial work and a local state of rest — 117
Energy as a complex function — 117
The concept of internal energy — 117
Reduction of rest mass as a dynamic effect — 120
4.1.4 The system of nested energy frames — 122
4.1.5 Effect of location and local motion in a gravitational frame — 124
Local rest energy of orbiting bodies — 124
Energy object — 127
4.1.6 Free fall and escape in a gravitational frame — 128
4.1.7 Inertial force of motion in space — 132
4.1.8 Inertial force in the imaginary direction — 134
4.1.9 Topography of space in a local gravitational frame — 137
4.1.10 Local velocity of light — 139
4.2 Celestial mechanics — 142
4.2.1 The cylinder coordinate system — 142
4.2.2 The equation of motion — 142
4.2.3 Perihelion direction on the flat space plane — 144
4.2.4 Kepler's energy integral — 148
4.2.5 The fourth dimension — 151
4.2.6 Effect of the expansion of space — 153

4.2.7 Effect of the gravitational state in the parent frame	154
4.2.8 Local singularity in space	155
4.2.9 Orbital decay	158
The effect of orbit plane rotation on the angular momentum of the orbit	*159*
Keplerian orbit	*160*
Dependence of dP on dL in Keplerian orbit	*161*
GR prediction for the orbital decay	*162*

5. Mass, mass objects, and electromagnetic radiation — 165

5.1 The mass equivalence of radiation	166
5.1.1 Quantum of radiation	166
The Planck equation	*166*
Maxwell's equations: solution of one cycle of radiation	*167*
The intrinsic Planck constant	*169*
Physical meaning of a quantum	*170*
The intensity factor	*171*
5.1.2 The fine structure constant and the Coulomb energy	172
The fine structure constant	*172*
The Coulomb energy	*172*
Energy carried by electric and magnetic fields	*174*
5.1.3 Wavelength equivalence of mass	174
The Compton wavelength	*174*
Wave presentation of the energy four vector	*175*
Resonant mass wave in a potential well	*177*
5.1.4 Hydrogen-like atoms	178
Principal energy states	*178*
The effects of gravitation and motion	*180*
Characteristic absorption and emission frequencies	*180*
5.2 Effect of gravitation and motion on clocks and radiation	183
5.2.1 Effect of gravitation and motion on clocks and radiation	183
5.2.2 Gravitational shift of electromagnetic radiation	185
5.2.3 The Doppler effect of electromagnetic radiation	187
Doppler effect in a local gravitational frame	*187*
Doppler effect in nested energy frames	*190*
5.3 Localized energy objects	194
5.3.1 Momentum of radiation from a moving emitter	193
Emission from a point source	*194*
Emission from a plane emitter	*194*
5.3.2 Resonator as an energy object	197
5.3.3 Momentum of spherical emitter	201
5.3.4 Mass object as a standing wave structure	202
5.3.5 The double slit experiment	204
5.3.6 Planck units in the DU framework	205

5.4 Propagation of electromagnetic radiation in local frames	207
5.4.1 Shapiro delay in a local gravitational frame	207
5.4.2 Shapiro delay in general relativity and in the DU	213
5.4.3 Bending of light	213
5.4.4 Measurement of the Shapiro delay	214
5.4.5 Effects of moving receiver and moving source	216
5.4.6 The effect of a dielectric propagation medium	218
5.5 Propagation of light from stellar objects	222
5.5.1 Frame to frame transmission	222
5.5.2 Gravitational lensing and momentum of radiation	223
5.5.3 Transversal velocity of the source and receiver	224
5.6 The development of the lengths of a year, month and day	227
5.6.1 Earth to Moon distance	227
Effect of the expansion of space on the Earth to Moon distance	*227*
Annual perturbation of the Earth to Moon distance	*228*
5.6.2 Development of rotational and orbital velocities	230
5.6.3 Days in a year based on coral fossil data	231
5.7 Timekeeping in the Dynamic Universe	234
5.7.1 Periodic phenomena and timescales	234
Characteristic wavelength and frequency of atomic objects	*234*
Natural periodic phenomena	*235*
Coordinated Universal Time	*236*
5.7.2 Units of time and distance, the frames of reference	237
The Earth second	*237*
The meter	*240*
The Earth geoid	*241*
5.7.3 Periodic fluctuations in Earth clocks	243
The effect of the eccentricity of the Earth-Moon barycenter orbit	*243*
Rotation and the inclination angle of the Earth	*244*
5.7.4 Galactic and extragalactic effects	245
Solar system in Milky Way frame	*245*
Milky Way galaxy in Extragalactic space	*245*
5.7.5 Summary of timekeeping	246
Average frequency of the SI-second standard	*246*
6. The dynamic cosmology	**250**
6.1 Redshift and the Hubble law	250
6.1.1 Expanding and non-expanding objects	250
6.1.2 Redshift and Hubble law	252
Optical distance and redshift in DU space	*252*
Classical Hubble law	*253*
Redshift in standard cosmology model	*254*

Recession velocity of cosmological objects	*255*
Effects of local motion and gravitation on redshift	*255*
6.1.3 Light propagation time in expanding space	257
The effect of the local structure of space	*258*
6.2 Angular sizes of a standard rod and expanding objects	260
6.2.1 Angular size of a standard rod in FLRW space	260
6.2.2 Angular size of a standard rod in DU space	260
6.2.3 Angular size of expanding objects in DU space	261
6.3 Magnitude and surface brightness	264
6.3.1 Luminosity distance and magnitude in FLRW space	264
6.3.2 Magnitude of standard candle in DU space	265
6.3.3 Bolometric magnitudes in multi-bandpass detection	267
6.3.4 K-corrected magnitudes	269
6.3.5 Time delay of bursts	274
6.3.6 Surface brightness of expanding objects	275
6.4 Observations in distant space	277
6.4.1 Microwave background radiation	277
6.4.2 Double image of an object	278
6.4.3 Radiometric dating	279

7. Summary   281

7.1 The picture of reality behind theory and experiments	281
The relativistic reality	*281*
The velocity of light	*281*
Discontinuity and discreteness of physical systems	*281*
Wavenumber, mass, and energy	*283*
7.2 Changes in paradigm	284
7.2.1 The basic postulates	284
7.2.2 Natural constants	285
Gravitational constant	*285*
Total mass in space	*285*
The velocity of light	*286*
Planck's constant	*287*
The fine structure constant	*287*
The Bohr radius	*287*
Vacuum permeability	*288*
Summary of natural constants	*288*
7.2.3 Energy and force	288
Unified expression of energy	*290*
7.3 Comparison of DU, SR, GR, QM, and FLRW cosmology	293
Philosophical basis	*293*
Physics	*294*
Cosmology	*295*

7.4 Conclusions — 297

8. Index — 299

Appendix 1, Blackbody radiation — 303
Energy density of radiation in a blackbody cavity — *303*
Radiation emittance — *304*
Spectral distribution of blackbody radiation — *304*

References — 307

Preface

The modern view of physical reality is based on the theory of relativity, the related standard cosmology model, and quantum mechanics. The development of these theories was triggered by observations on the velocity and emission/absorption properties of light in the late 19th and early 20th centuries. Theories have attained a high degree of perfection during the last 100 years. When assessed in the context of the huge progress in the 20th century, they have been exceedingly successful, not only in increasing our knowledge and understanding of nature but also in bringing the knowledge into practice in technological achievements, in applications ranging from nanostructures to nuclear energy and space travel.

In spite of their significant successes, there has also been continuing criticism of the theories since their introduction. The theory of relativity raised a lot of confusion, not least by redefining the concepts of time and distance, the basic coordinate quantities for human conception. This was quite a shock to the safe and well-ordered Newtonian world, which had governed scientific thinking for more than two hundred years. Another shock came with the abstraction related to quantum mechanics — particles and waves were interrelated, deterministic precision was challenged by stochasticity and probabilities, and continuity was replaced by discrete states. As a consequence, nature was no longer expected to be consistent with human logic; it is not unusual that a lecturer in physics starts his talk by advising the audience not to try to "understand" nature.

In a philosophical sense, neglecting the demand of human comprehension is somewhat alarming, since it is a primary challenge and purpose of a scientific theory to make nature understandable. It is easier to verify the merits of a scientific theory through its capability of describing and predicting observable phenomena, and that is what the present theories do well in most cases.

As a mathematical description of observable physical phenomena, a scientific theory need not be based on physical assumptions. The Ptolemy sky was based on a direct description of observations as seen from the Earth. It related the motions of planets to the motion of the Sun across the sky without any physical law, other than continuity, behind the motions. Kepler's laws, which still form the basis of celestial mechanics, were originally purely mathematical formulations of the observations made by the Danish astronomer Tycho Brahe. Several decades later, Newton's laws of motion and the formulation of gravitational force revealed the physical meaning of Kepler's laws, which formed the basis of celestial mechanics for the succeeding centuries.

Newtonian space does not recognize limits to physical quantities. Newtonian space is Euclidean to infinity, and velocities in space grow linearly as long as there is a constant force acting on an object. As realized in the late 19th century, the velocity of accelerated objects does not grow linearly but saturates at the velocity of light. The theory of relativity describes the finiteness of velocities by linking time to space in four-dimensional spacetime and by postulating the velocity of light to be a natural constant and invariant to all observers. An observer in relativistic space sees a time interval in an object in relative motion approach infinity so that the velocity of light is never exceeded. The dilated flow of time is also used as the explanation of the observed slower frequency of clocks moving relative to the observer or at a lower gravitational potential than the observer.

Authorized by the relativity principle, the theory of relativity ignores the effects of the space around the observer; an observer studying a particle accelerated in a laboratory on the Earth is subject to the rotation of the Earth and the orbital motion around the Sun, at a periodically changing velocity and gravitational potential due to the eccentricity of the orbit. Further, the whole solar system is in motion and in gravitational interaction in the Milky Way galaxy, which interacts with neighboring galaxies as a part of the cosmological structure of the universe.

The Dynamic Universe theory presented in this book is a holistic approach to the universe and interactions in space. The energy structure of space is described as a system of nested energy frames starting from the hypothetical homogeneous space as the universal frame of reference to all local frames in space. In DU space, everything is interconnected; the energy available for running a physical process on the Earth is not only affected by the local motion and gravitation but also the motion and gravitational state of the Earth in its parent frames. Relativity in the Dynamic Universe theory is primarily the relativity of the local to the whole rather than relativity between the observer and an object.

The whole is not composed as a sum of elementary units, but the multiplicity of elementary units emerges as a diversification of the whole. There are no independent objects in space — everything is linked to the rest of space and thereby to each other.

Although the Dynamic Universe theory means a full replacement of special and general relativity, it had not been found without the relativity theory. All key elements in the DU can be found in special and general relativity, once we replace the time-like fourth dimension with the fourth dimension of a metric nature and adopt a holistic perspective to space as a whole.

In the DU, space is postulated as a three-dimensional structure closed through a fourth dimension, like the 3-dimensional surface of a 4-dimensional sphere – following Einstein's original view of the cosmological structure of space in general relativity. Unlike the time-like fourth dimension of the relativity theory, the metric fourth dimension of the DU, the direction of the 4-radius of the structure, allows contraction and expansion of space like a spherical pendulum in the fourth dimension. Spherically closed space does not need an energizing quantum jump or Big Bang; space has gained its energy of motion against the release of its gravitational energy in a contraction phase and pays it back to gravitational energy in the ongoing expansion phase. As observers in space, we observe the energy of motion of space in the fourth dimension as the rest energy of matter. Any motion in 3D space is associated with the motion of space in the fourth dimension. As shown by the detailed energy bookkeeping, any momentum built up in a spatial direction reduces the momentum in the fourth dimension. The relativistic mass increase taught by special relativity is not a consequence of velocity, but the energy input needed to build up the kinetic energy. In free fall in a local gravitational frame, the kinetic energy is built up against the reduced rest energy via tilting of local space; there is no mass increase associated with the velocity of free fall, which means the cancellation of the equivalence principle behind general relativity.

Due to the kinematic approach, special relativity discloses the increase of the inertial mass in motion but is blind to the associated decrease of the rest mass which is observed, e.g., as the reduced frequency of atomic clocks in motion; general relativity discloses the tilting of space near mass centers but is blind to the associated reduction of the rest energy resulting, e.g., in the reduced frequency of atomic clocks near mass centers.

In DU space, the velocity of light is not constant but fixed to the velocity of space in the local fourth dimension. All local structures in the DU space are linked to the rest of space. Unlike in GR space, gravitationally bound local structures like galaxies and planetary systems expand in direct proportion to the expansion of space. About 2.8 cm of the 3.8 cm annual increase in the Earth-to-Moon distance comes from the expansion of space, and only one centimeter comes from tidal interactions. Four billion years ago, the solar luminosity was about 25% lower than it is today, which makes it very difficult to explain the geological history of the Earth and the free water on early Mars if the planets had been at their present distances from the Sun as taught by general relativity. According to the DU, 4 billion years ago, planets were about 30% closer to the Sun, which overcompensates for the lower luminosity of the Sun and offers a natural explanation for the ancient warm oceans on Earth and liquid water on Mars.

The expansion of DU space occurs with the energies of motion and gravitation in balance. Such a condition corresponds to the "flat space" condition in the GR framework. For matching cosmological predictions with observations, the flat space condition in GR space is associated with a remarkable amount of "dark energy" with gravitational push instead of attraction. In the DU space, a precise match between predictions and observations is obtained without dark energy or any other additional parameters.

The zero-energy approach of the Dynamic Universe allows the derivation of local and cosmological predictions with a minimum number of postulates – by honoring universal time and distance as the natural coordinate quantities for human comprehension.

Philosophically, the relativity of observations is an indication that something in space is finite. In the kinematic approach of the relativity theory, finiteness is fixed to the velocity of light – in the dynamic approach of the DU, the finiteness of the velocity of light is a consequence of the conservation of energy, or more fundamentally, the balance of the energies of motion and gravitation in space. The velocity of light and several "natural constants" are observed as constant because the measuring instruments are subject to the same energy balance as the quantities measured. The late Finnish professor Raimo Lehti called this "the conspiracy of the laws of nature".

As a basic principle of scientific thinking, the reality behind natural phenomena is independent of the models by which we describe them. The best a scientific model can give is a description that makes the reality understandable. The model should rely on sound basic assumptions and inherently coherent logic, and, specifically in physics and cosmology, give precise predictions of phenomena observed and to be observed.

We are not free to choose the laws of nature, but we have considerable freedom in choosing the coordinate quantities used in the models. Time and distance are the most fundamental coordinate quantities. For human perception and logic, time and distance should be universal for all physical phenomena described.

The origin of the Dynamic Universe concept lies in the continuing interest I have had in the basic laws of nature and human comprehension of reality since my student time in the 1960s. I recognize my friend and former colleague, Heikki Kanerva, as an important early inspirer in the thinking that paved the way for the Dynamic Universe theory. After many years of maturing, the active development of the theory was triggered by a stimulus from my late colleague Jaakko Kajamaa in the mid-90s. I express my sincere gratitude to my early inspirers.

The breakthrough in the development of the Dynamic Universe concept occurred in 1995, once I replaced the time-like fourth dimension with the fourth dimension of a metric nature – thereby revealing the physical meaning of the quantity mc, the rest momentum, the momentum of mass m in a fourth dimension, perpendicular to the three spatial directions. Momentum and the related energy of motion against the energy of gravitation in spherically closed space showed the dynamics of space as that of a spherical pendulum in the fourth dimension, showing the buildup and release of the rest energy of matter as a continuous process in contraction and expansion periods of the structure. In the DU framework, energizing of space did not happen in a mysterious quantum fluctuation or Big Bang. The energy buildup and release of space is described as a continuous contraction–expansion process. Mass can be understood as a wavelike substance for the expression of energy. By assuming conservation of the total energy in interactions in space, the overall energy structure of space can be described as a system of nested energy frames, proceeding from large-scale gravitational structures down to atoms and elementary particles.

The development of the Dynamic Universe model has been documented in annually updated monographs titled *"The Dynamic Universe"* in 1996-99, *"The Dynamic Universe, A New Perspective on Space and Relativity"* in 2000-2003, *"Theoretical Bases of the Dynamic Universe"* in 2004, and *"The Dynamic Universe, Toward a Unified Picture of Physical Reality"*, editions 1, 2, and 3 in 2009-2012. The first peer-reviewed papers on the Dynamic Universe were published in Apeiron in 2001. For several years, the main channel for scientific discussions and publications was the PIRT (Physical Interpretations of Relativity Theory) conference, biannually organized in London and occasionally in Moscow, Calcutta, and Budapest. I would like to express my respect to the organizers of PIRT for keeping up critical discussion on the basis of physics and pass my sincere gratitude to Michael Duffy, Peter Rowlands, and many conference participants. At the national level, the Finnish Society for Natural Philosophy has organized seminars and lectures on the Dynamic Universe concept. I express my gratitude to the Society and many members of the Society for the encouragement and inspiring discussions. I am exceedingly grateful to the co-founders of the Physics Foundations Society, Ari Lehto, Heikki Sipilä, and Tarja Kallio-Tamminen, for their initiatives in promoting the search for the fundamentals of physics and the essence of the philosophy of science, complemented by Avril Styrman with his doctoral thesis *Economical Unification as a Method of Philosophical Analysis*, presented at the University of Helsinki in 2016. I also like to express my sincere thanks to Robert Day for his early analysis of the DU supernova predictions and his assistance with my publications, and Mervi Hyvönen-Dabek and Jan Dabek for polishing my English language. Many good friends and colleagues are thanked for their encouragement during the years of my treatise. The unfailing support of my wife, Soilikki, and my daughter Silja and her family has been of special importance, and I am deeply grateful to them.

The 4th edition is restructured; Chapter 1, Introduction, gives an overview of the theory and presents important theoretical outcomes and experimental results with minimal mathematics. Chapters 2 to 6 document the formal derivation of the theory. As an addition to the earlier editions, Section 4.2.9 *Orbital decay* has been added.

In the Dynamic Universe, several physical quantities get meanings and notations different from those in traditional theories. For example, mass in the DU is not a form or expression of energy like in the theory of relativity, but the wavelike *substance* for the expression of energy. Mass, momentum, and energy are described as complex quantities. For example, the real component of the complex momentum is the momentum we observe in space; the imaginary part of the momentum is the rest momentum due to the expansion of space in the fourth dimension. The imaginary part of momentum is not recognized in current theories because of the time-like fourth dimension and the postulated constancy of the velocity of light and rest mass. The modulus of the complex energy is equal to the concept of energy as a scalar quantity used in the traditional formalism.

1. Introduction to the Dynamic Universe

A new theory is necessary when existing theories grow in complexity, fail in producing predictions matching observations or fail in producing an understandable picture of reality. The theory of relativity has succeeded well in producing mathematical descriptions for the observations, but failed in creating a comprehensive picture of reality.

The theory of relativity continues the Galilean–Newtonian tradition; it is built on kinematics and metrics on a local basis and relies on the relativity and equivalence principles. As additional bases, the theory of relativity needs the assumption of the constancy of the velocity of light and coordinate transformations for moving from one frame of reference to another. The Friedmann-Lemaître-Robertson-Walker (FLRW) cosmology relies on the general theory of relativity and the cosmological principle. Quantum mechanics completes the theory of relativity in the description of phenomena on the micro-scale. QM breaks the deterministic nature of the classical theories and brings up the concept of discrete states and probabilities.

The Dynamic Universe theory means a major change in the paradigm. DU replaces the relativity principle and the associated concept of inertial frames of reference with a system of nested energy frames that relates any energy state in space to the state of rest in hypothetical homogeneous space. Instead of expressing relativity in terms of coordinate transformations, relativity is expressed in terms of locally available energy in the DU. The concept of time-like fourth dimension is replaced with the metric fourth dimension. Time is a universal scalar allowing motion equally in the three space directions – and in the fourth dimension, as the expansion of the spherically closed space.

The Introduction to the Dynamic Universe is presented in four chapters:

1.1 *"Basic concepts" introduces the basic structure of the theory and the central definitions and notations needed.*

1.2 *"Buildup of energy in space" introduces the contraction-expansion process, building up the rest energy of matter as the primary energy available for the buildup of local energy structures, the system of nested energy frames in space. Starting from the quantum mechanical solution of atomic structures, the frequency of atomic clocks is linked to the local state of gravitation and motion. The properties of the velocity of light are derived from the overall energy balance in space.*

1.3 *"Cosmological considerations" gives an overview of the basis of the Friedmann-Lemaître-Robertson-Walker (FLRW) cosmology and introduces the reconsiderations needed in the DU framework. DU predictions for key observables are introduced with a comparison to observations. The linkage between GR space and DU space is discussed.*

1.4 *"The summary" compares the theory structure and the hierarchy of physical quantities in the DU and contemporary physics.*

1.1 Basic concepts

The first Chapter of Copernicus' De Revolutionibus (1543) is titled **The Universe is Spherical**: *"First of all, we must note that the universe is spherical. The reason is either that, of all forms, the sphere is the most perfect, needing no joint and being a complete whole, which can be neither increased nor diminished; or that it is the most capacious of figures, best suited to enclose and retain all things; or even that all the separate parts of the universe, I mean the sun, moon, planets, and stars, are seen to be of this shape; or that wholes strive to be circumscribed by this boundary, as is apparent in drops of water and other fluid bodies when they seek to be self-contained. Hence, no one will question the attribution of this form to the divine bodies".*

The Copernican system allowed dynamic analyses of physical interactions in the planetary system and created the basis for mathematical physics. The Dynamic Universe takes the next step: Not only the planetary system but the whole three-dimensional space is described as a spherically closed entity, allowing a dynamic analysis linking all local structures and phenomena to space as a whole.

1.1.1 Space as a spherically closed entity

In the DU, space is studied as a closed energy system, the three-dimensional "surface" of a four-dimensional sphere [a]. Space as the 3D surface of a 4D sphere is quite an old concept for describing space as a closed but endless entity. The concept of a 4D sphere is based on differential geometry developed in the 19th century by Ludwig Schläfli, Arthur Cayley, and Bernhard Riemann. Space as the 3D surface of a 4D sphere was Einstein's original view of the cosmological picture of general relativity in 1917 [1]. Gravitation in spherically closed space tends to shrink the structure, leading to dynamic space; dynamic space requires a metric fourth dimension, which did not fit the concept of four-dimensional spacetime of the theory of relativity. To prevent the dynamics of spherically closed space, Einstein completed the theory with the famous cosmological constant, which was recently reawakened as the "dark energy" needed to match cosmological predictions to observations.

In his lectures on gravitation in the early 1960s, Richard Feynman [2] returned to the idea of spherically closed space:

"…One intriguing suggestion is that the universe has a structure analogous to that of a spherical surface. If we move in any direction on such a surface, we never meet a boundary or end, yet the surface is bounded and finite. It might be that our three-dimensional space is such a thing, a tridimensional surface of a four sphere. The arrangement and distribution of galaxies in the world that we see would then be something analogous to a distribution of spots on a spherical ball."

In the same lectures,[3] Feynman also pondered the equality of the rest energy and gravitational energy in space:

---

[a] In mathematics, the 3-dimensional surface of a 4-dimensional sphere is referred to as 3-sphere. The terms 3D surface of a 4D sphere are used to avoid the confusion.

"If now we compare the total gravitational energy $E_g = GM^2_{tot}/R$ to the total rest energy of the universe, $E_{rest} = M_{tot} c^2$, lo and behold, we get the amazing result that $GM^2_{tot}/R = M_{tot} c^2$, so that the total energy of the universe is zero. — It is exciting to think that it costs nothing to create a new particle, since we can create it at the center of the universe where it will have a negative gravitational energy equal to $M_{tot} c^2$. — Why this should be so is one of the great mysteries — and therefore one of the important questions of physics. After all, what would be the use of studying physics if the mysteries were not the most important things to investigate?"

The Dynamic Universe can be seen as a detailed analysis of combining Feynman's "great mystery" of zero-energy space with the "intriguing suggestion of spherically closed space" — by the dynamics of space as a spherically closed structure.

Such a solution does not work in the framework of the relativity theory, which is based on the constant velocity of light, and time as the fourth dimension. A dynamic solution requires universal time and a metric fourth dimension that allows velocity and momentum equally in the three space dimensions and the fourth dimension.

Relativity in the Dynamic Universe means relativity of local to the whole. Local velocities in space become related to the velocity of space in the fourth dimension, and local gravitation becomes related to the total gravitational energy in space. Local gravitational systems expand in direct proportion to the expansion of the whole space. It means that, unlike in GR-based cosmology, galaxies and planetary systems expand in direct proportion to the expansion of the whole space. Everything in space is interrelated. The velocity of light is linked to the velocity of space in the local fourth dimension, and the frequency of atomic clocks, as well as the rates of most physical processes, are related to the local velocity of light.

The dynamic approach does not allow the relativity principle and the associated freedom of choosing the state of rest. The local state of motion is related to the state preceding the buildup of the kinetic energy. Any motion in space has a history that links it to the system energizing the motion. The energy structure of space is described as a system of nested energy frames starting from a hypothetical homogeneous space as the universal frame of reference and proceeding down to local frames in space.

The zero-energy principle

The zero-energy principle has its roots in Aristotle's entelechy, the actualization of potentiality. Gottfried Leibniz referred to Aristotle's entelechy in his Essays in Dynamics[4]: *"There is neither more nor less power in an effect than in its cause."* The concept of energy was fully recognized first as a part of the development of thermodynamics in the late 19th century. The possibility of the overall zero-energy balance in space was stated by Dennis Sciama in his lectures on inertia in 1953[5] and Richard Feynman in his lectures on gravitation cited above.

In the Dynamic Universe, the primary energy buildup and release of matter in space are described as a zero-energy process of the spherical structure; in the contraction phase, the energy of motion is obtained against the release of gravitational energy, and in the expansion, the energy of motion is released back to the energy of gravitation. The energy of motion obtained in the contraction is observed as the rest energy of matter, as the energy of motion in the fourth dimension. The rest energy is balanced by global gravitational energy due to the rest of space. The buildup of local structures in space conserves the total energy; kinetic energy in 3D space reduces the rest energy of the object in motion. Relativity in the DU is a direct consequence of the conservation of energy; relativity gets its expression in terms of the locally available energy, e.g., the rates of physical processes become a function of the local energy state. Atomic clocks in motion or at a high gravitational field run slow due to the reduced rest energy of the oscillating electrons.

1.1.2 Mass objects and the two-fold expression of energy

The Dynamic Universe theory means a major change in the paradigm. We need to go back to the Greek philosophers to reawaken the discussion of the essence of mass as a substance. Mass as a wavelike substance for the expression of energy in the DU has something in common with the Greek *Apeiron* as the indefinite substance for material forms, originally introduced by Anaximander in the 6th century BC. *Apeiron* was not defined precisely; the descriptions given by different philosophers deviate substantially from each other but comprise the basic feature of *Apeiron* as the primary source for all visible forms in the cosmos.

The DU shows "unity via duality"; mass is the substance in common for the energies of motion and gravitation that emerge and then vanish in a dynamic zero-energy process, giving existence to observable physical reality. As a philosophical concept, the primary energy buildup process in the DU is related to the Chinese yin-yang concept, where the two inseparable opposites are thought to arise from emptiness and end up in emptiness. In Greek philosophy, perhaps the ideas closest to the yin-yang concept are expressed by Heraclitus, contemporary to Anaximander.

Mathematically, the abstract role of mass as the substance for the expression of the complementary energies of motion and gravitation is seen in the equation

$$E_m + E_g = c_0 m c_0 - \frac{GM''}{R_4} m = 0, \tag{1.1.2:1}$$

with mass m as a first-order factor, both in the energy of motion and gravitation. In (1.1.2:1), written for the balance of the energies of gravitation and motion in hypothetical homogeneous space, where the 4-velocity of space is c_0, and the 4-radius, the distance to the mass equivalence M'' at the barycenter of space, is R_4.

The energy of motion expressed by mass m is local by its nature. The counterbalancing energy of gravitation is due to all the rest of mass in space. Equation (1.1.2:1) does not only mean complementarity of the two types of energies but also complementarity of the local and the whole. We may say that the antibody of a local mass object is the rest of space, or that the localized expression of the energy of a mass object is its rest energy, and the non-localized expression of its energy is the global gravitational energy arising from the rest of space, Figure 1.1.2-1. It looks like the complementary nature of local and the rest of space in the Dynamic Universe reflects the idea of Leibniz's monads as "perpetual, living mirrors of the universe".

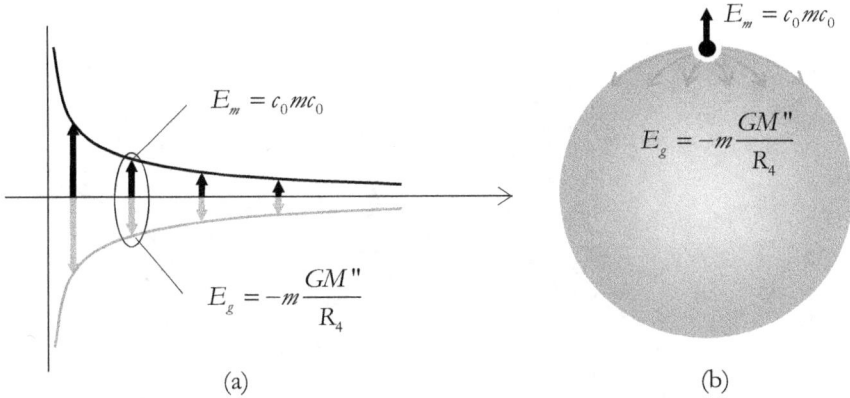

Figure 1.1.2-1(a) The twofold nature of matter at rest in space is manifested by the energies of motion and gravitation. The intensity of the energies of motion and gravitation declines as space expands along the 4-radius. (b). Complementarity of local and whole can be seen in the complementarity of the local rest energy and the global gravitational energy arising from all the rest of mass in space. *The antibody of a local mass object is the rest of space.*

1.1.3 Linkage between GR space and DU space

The balance of the rest energy and gravitational energy

One of the characteristic features of the DU is the balance between the global gravitational energy and the rest energy of any mass object in space, which, in the complex quantity presentation, appears as the balance of complementary energies in the fourth dimension. For making sense with velocity, momentum, and the corresponding energy of motion in the fourth dimension, the fourth dimension shall be studied as a metric dimension.

In fact, the stress-energy tensor in general relativity has the same message when interpreted in the light of Gauss's divergence theory or simply as the physical linkage of pressure and energy content. On the cosmological scale, in homogeneous space, the stress-energy tensor can be expressed in the form

$$(T^{\mu\nu})_{\mu,\nu=0,1,2,3} = \begin{pmatrix} mc^2/dV & 0 & 0 & 0 \\ 0 & F_{11}/dA & 0 & 0 \\ 0 & 0 & F_{22}/dA & 0 \\ 0 & 0 & 0 & F_{33}/dA \end{pmatrix}, \qquad (1.1.3:1)$$

where the energy density mc^2/dV is constant in the whole space, and the locally observed net force densities F_{11}/dA, F_{22}/dA, and F_{33}/dA in the space directions are equal to zero. The energy content of volume dV is equal to the pressure uniformly from all space directions, which can be interpreted as the integrated gravitational force from the whole space. Once the global gravitation on element mc^2/dV appears in the fourth dimension, the center of gravity must be in the fourth dimension at equal distance from all space locations.

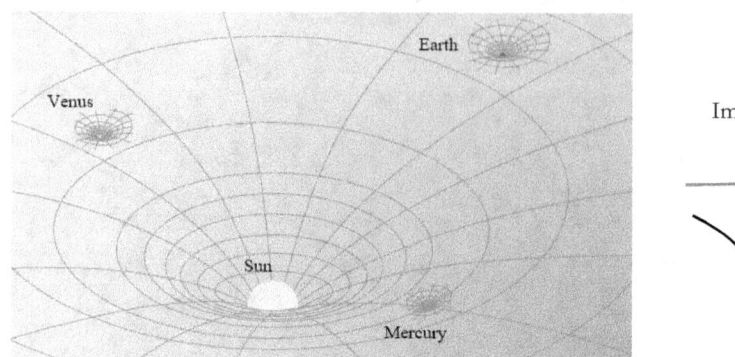 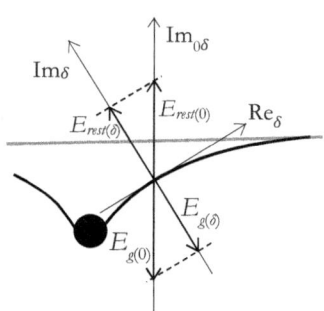

Figure 1.1.3-1. The overall energy balance in space is conserved via tilting of space in local mass center buildup creating the kinetic energy of free fall and the local gravitational energy. Due to the tilting, the velocity of space in the local fourth dimension is reduced compared to the 4-velocity of the surrounding non-tilted space. The buildup of dents in space occurs in several steps; dents around planets are dents in the larger dent around the Sun – which is a local dent in the much larger Milky Way dent.

Einstein drew a similar conclusion in his Berlin Writings in 1914–1917 [6]: *"... If we are to have in the universe an average density of matter which differs from zero, however small may be that difference, then the universe cannot be quasi-Euclidean. On the contrary, the results of calculation indicate that if matter be distributed uniformly, the universe would necessarily be spherical (or elliptical)."*

To save the equality of all locations in space, elliptic solutions must be excluded, and we enter the DU equation

$$mc^2 = \frac{GM''}{R_4} m . \tag{1.1.3:2}$$

In real space, to conserve the balance of the energies in the local mass center buildup, the total gravitational energy is divided, via the tilting of local space, into orthogonal components with the local gravitational energy in a space direction and the reduced global gravitational energy in the fourth dimension. This also means a reduction of the local rest energy of objects and consequently, e.g., reduction of the characteristic frequencies of atomic oscillators in tilted space, Figure 1.1.3-1.

Zero-energy balance and the critical mass density

Based on measurements of microwave background radiation by the Wilkinson Microwave Anisotropy Probe (WMAP), the mass density in space is concluded to be essentially equal to Friedmann's critical mass density

$$\varrho_c = \frac{3H_0^2}{8\pi G} \approx 9.2 \cdot 10^{-27} \ [\text{kg/m}^3], \tag{1.1.3:3}$$

where G ($\approx 6.67 \cdot 10^{-11}$ [Nm²/kg²]) is the gravitational constant and H_0 the Hubble constant [≈ 70 (km/s)/Mpc]. In FLRW cosmology, such a condition means "flat space" expanding with the energy of motion and gravitation in balance. Assuming the volume of space as the volume of a 3D sphere with radius $R_H = c/H_0$, equation (1.1.3:3), the total mass in space and the velocity of light can be expressed as

$$M = \varrho_c \frac{4\pi R_H^3}{3} = \frac{3c^2}{R_H^2} \frac{4\pi R_H^3}{3 \cdot 8\pi G} = \frac{c^2 R_H}{2G} \quad \Rightarrow \quad c^2 = \frac{2GM}{R_H}. \tag{1.1.3:4}$$

Solved from the Friedmann's critical mass density, the rest energy of mass m, and the total mass $M = \Sigma m$ in GR space are

$$mc^2 = \frac{2GMm}{R_H} \quad ; \quad \tfrac{1}{2}Mc^2 = \frac{GM^2}{R_H} \quad c = \sqrt{\frac{2GM}{R_H}}. \tag{1.1.3:5}$$

Formally, the last form of (1.1.3:5) describes c as the Newtonian velocity of free fall or the escape velocity at distance R_H from mass M at the barycenter representing the total mass in space. This means that the rest energy, as the Newtonian kinetic energy of mass m, is counterbalanced with the global gravitational energy arising from hypothetical mass M at a distance R_H from mass m anywhere in space. Such a solution is possible only in 3D space as the surface of a 4D sphere with radius R_H. A detailed study of (1.1.3:5) shows that the factor ½ in the rest energy Mc^2 comes from the numerical factors used in Einstein's field equations to make them consistent with Newtonian gravitation at a low gravitational field in 3D space.

1.1.4 Definitions and notations

In comparison with classical mechanics and the theory of relativity, the most significant differences in the Dynamic Universe approach come from the holistic perspective and the use of dynamics instead of kinematics and metrics. In the Dynamic Universe, space is described as the 3D surface of a 4D sphere. The properties of local structures are derived from the whole by conserving the zero-energy balance initially assumed in homogeneous space with all mass uniformly distributed in the volume. Dynamic Universe honors absolute time and distance as coordinate quantities; in DU, relativistic effects appear as consequences of the conservation of total energy in all energy conversions in space.

In homogeneous space, the direction of the fourth dimension is the direction of the 4-radius of space. In locally curved space near mass centers, the fourth dimension is the direction perpendicular to the three spatial directions.

It is useful to denote the fourth dimension as the imaginary direction. Phenomena that act both in the fourth dimension and in a spatial direction are expressed in the form of complex functions. For example, the **rest energy** that in the DU framework is the energy of motion an object has due to the motion of space in the fourth dimension appears as the imaginary component of the total energy of motion. Correspondingly, the **global gravitational energy** appears as the imaginary component of the gravitational energy arising from the total mass in space. The total mass is represented by the mass equivalence at the barycenter of the 4D sphere. As the inherent form of the energy of gravitation, Newtonian gravitational energy is assumed in hypothetical homogeneous space. The global gravitational energy of mass m in hypothetical homogeneous space is

$$E^{\scriptscriptstyle\square}_{g(0)} = E''_{g(0)} \equiv -mG \int_V \frac{\varrho dV(r)}{r} = -\frac{GM''m}{R_4}, \qquad (1.1.4:1)$$

where m is a test mass, G the gravitational constant, ϱ the mass density in space, R_4 the 4-radius of space, $V = 2\pi^2 R_4^3$ the volume of the 3D space and $M'' = 0.776 \cdot M_\Sigma$ the mass equivalence of the total mass M_Σ in space [see Chapter 2 for formal derivation of (1.1.4:1)]. The subscript "$_{(0)}$" in the global gravitational energy in (1.1.4:1) refers to hypothetical homogeneous space. Superscript ($^{\scriptscriptstyle\square}$) is used to denote a complex function. A single apostrophe ('), [or no apostrophe to meet traditional notations], denotes the real part of the complex function, and a double apostrophe (") the imaginary part.

In the four-dimensional manifold allowing motion of space and motion in space, a mass particle moving in space has the momentum both in a space direction and in the fourth dimension. As a consequence of the zero-energy balance of motion and gravitation, the velocity of light in space is equal to the velocity of space in the fourth dimension. In hypothetical homogeneous space, the velocity of light, c_0, is equal to the expansion velocity of space in the direction of the 4-radius. In locally tilted space, the velocity of light is equal to the velocity of space in the local fourth dimension, c. In the vicinity of the Earth, the local velocity of light, c, is estimated as $c \approx 0.999\,999\,c_0$.

In the complex quantity presentation, the momentum in space is the real component, and the momentum in the fourth dimension, the rest momentum, the imaginary component of the complex total momentum

$$\mathbf{p}^{\scriptscriptstyle\square} \equiv \mathbf{p} + \mathbf{i}\,p'' = \mathbf{p} + \mathbf{i}mc. \qquad (1.1.4:2)$$

The energy of motion is expressed as the product of the system velocity, the expansion velocity of the 4-sphere c_0 as a scalar quantity, and the complex momentum in the system as a vector quantity.

$$E^{\scriptscriptstyle\square} \equiv c_0 \mathbf{p}^{\scriptscriptstyle\square} = c_0(\mathbf{p} + \mathbf{i}mc), \qquad (1.1.4:3)$$

Traditionally, energy is used as a scalar quantity. Equation (1.1.4:3) allows the study of energy as a complex quantity where the imaginary component represents the global contribution, and the real component the local contribution to the total energy. The absolute value of the total energy of motion, the modulus of the complex energy, is

$$\mathrm{Mod}\{E^{\scriptscriptstyle\square}\} = c_0\sqrt{p^2 + (mc)^2}, \qquad (1.1.4:4)$$

which is essentially the same as the total energy in special relativity.

Any motion **in** space is associated with the motion **of** space, which brings the imaginary components to the momentum and the energy of motion. In the DU framework, energy is the primary postulated quantity, and force is a derived quantity as the gradient of potential energy or the time derivative of momentum. Force, as the gradient of potential energy, is considered the trend towards minimum energy, the driving force of Aristotle's *entelechy*, the actualization of a potentiality.

The complex quantity presentation of energy unifies the expression of the energy of motion of mass objects and electromagnetic radiation. A mass object at rest in a local energy frame has the imaginary component only; a mass object moving in a local energy frame in space has both imaginary and real components. Electromagnetic radiation propagating

Introduction

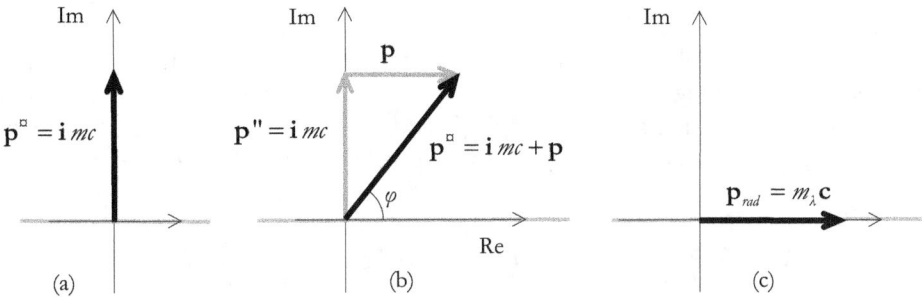

Figure 1.1.4-1. The complex presentation of momentum. (a) Mass object at rest in space, (b) mass object with momentum **p** in space, (c) electromagnetic radiation with momentum **p**$_{rad}$ in space.

in space has only the real component of the energy. Applying the mass equivalence of electromagnetic radiation, $m_\lambda = h_0/\lambda$, the expressions for the rest momentum and energy of electromagnetic radiation are formally identical with the expressions of the imaginary rest momentum and rest energy of mass m, Figure 1.1.4-1.

1.1.5 Reinterpretation of Planck's equation

For understanding the wave nature of mass and the essence of quantum, it is necessary to take a look at the physical messages of Planck's equation.

The Planck equation originates from the need to solve the wavelength spectrum of blackbody radiation. Around 1900, Max Planck realized that the atoms emitting and absorbing radiation at the walls of a blackbody cavity could be considered as harmonic oscillators able to interact with radiation at the resonant frequency of the oscillator only [7]. As an intuitive view, he proposed that the energy, which each oscillator emits or absorbs in a single emission/absorption process, is proportional to the frequency of the oscillator.

He described the energy of such a single interaction with the equation $E = hf$, where h is a constant. According to Planck's interpretation, electromagnetic radiation is emitted or absorbed only in energy quanta proportional to the frequency of the radiation. Planck saw this to contradict the classical electromagnetism as expressed by Maxwell's equations. Once we solve Maxwell's equations for the energy emitted into one cycle of radiation by a single electron transition in an antenna related to the wavelength, we obtain Planck's equation.

The physical interpretation of Planck's equation as the energy emitted into a cycle of radiation can be seen directly by writing Planck's equation in the form

$$E = hf = h/dt , \qquad (1.1.5:1)$$

where h [Js] is the Planck constant and $dt = 1/f$ means the cycle time. Planck related the energy of a quantum to the average kinetic energy of a particle, $hf = kT$, which, in this connection, can be interpreted as the kinetic energy of an electron.

The Maxwellian solution of a quantum does not require the DU theory or any other new assumptions. The required interpretation of a point source as "one-wavelength dipole" in the fourth dimension can be equally concluded as the line element cdt in the fourth dimension in the SR/GR framework.

We have to apply vacuum permeability, μ_0, instead of vacuum permittivity, ε_0, as the vacuum electric constant. The solution relates the dipole characteristics (number of oscillating electrons, dipole length/emitted wavelength, radiation geometry) to the Planck constant and the frequency of the radiated electromagnetic wave. Also, it relates the Planck constant to the fundamental electromagnetic constants, the unit charge, e, the vacuum permeability, μ_0, and the velocity of light, which appears as a hidden factor in the Planck constant

$$E_\lambda = hf = 1.1049 \cdot 2\pi^3 e^2 \mu_0 c_0 \cdot f = h_0 \cdot c_0 \cdot f = \frac{h_0}{\lambda} c_0 c = c_0 m_\lambda c = c_0 |\mathbf{p}|, \quad (1.1.5:2)$$

where h is the Planck constant, f and λ are the frequency and wavelength of the radiation emitted. Factor 1.1049 is related to the fine structure constant and regarded as the geometry constant of a *Planckian* antenna. For disclosing the physical meaning of Planck's equation, it is important to remove the velocity of light from the Planck constant by defining "the intrinsic Planck constant" $h_0 = h/c$. Applying the intrinsic Planck constant, Planck's equation for a cycle or quantum of electromagnetic radiation obtains the form

$$E_\lambda = h_0 c \cdot f = \frac{h_0}{\lambda} c^2 = m_\lambda c^2, \quad (1.1.5:3)$$

where λ is the wavelength of the radiation cycle emitted. The quantity $h_0/\lambda = m_\lambda$ has the dimension of kilograms [kg] that, physically, can be interpreted as *the mass equivalence of a cycle of radiation* generated by a unit charge oscillation in the emitter. For N electrons oscillating in the emitter, the energy emitted into a cycle of radiation is

$$E_{\lambda(N)} = N^2 \frac{h_0}{\lambda} c^2 = m_{\lambda(N)} c^2 = c_0 |\mathbf{p}_{(N)}|, \quad (1.1.5:4)$$

where N^2 is the intensity factor.

A further important result of the breakdown of Planck's constant into primary electrical constants is the disclosure of the physical nature of the fine structure constant a. Substituting equation (1.1.5:2) into the expression of the fine structure constant shows the fine structure constant as a pure numerical or geometrical constant without any connection to other natural constants

$$a \equiv \frac{e^2 \mu_0 c}{2 h_0 c} = \frac{e^2 \mu_0}{2 \cdot 1.1049 \cdot 2\pi^3 e^2 \mu_0} = \frac{1}{4 \cdot 1.1049 \cdot \pi^3} \simeq \frac{1}{137.0360}. \quad (1.1.5:5)$$

The wave nature of mass

Applying the intrinsic Planck constant, the Compton wavelength of mass m obtains the form

$$\lambda_{Compton} \equiv \frac{h}{mc} = \frac{h_0}{m} = \lambda_m \quad ; \quad k_{Compton} \equiv \frac{mc}{h} = \frac{m}{h_0} = k_m, \quad (1.1.5:6)$$

where the last equation uses the wave number $k = 2\pi/\lambda$ and the reduced intrinsic Planck constant $\hbar_0 = h_0/2\pi$. The Compton wavelength of a mass object is the wavelength equivalence of mass, like the inverse of the mass equivalence of electromagnetic radiation

$$m = h_0/\lambda_m \quad ; \quad m = h_0 k_m, \qquad (1.1.5:7)$$

which illustrates the wave nature of mass.

The concept of mass as the substance for the expression of energy can be extended to Coulomb energy E,

$$E_e = N^2 \frac{e^2 \mu_0}{4\pi r} c^2 = N^2 a \frac{h_0}{2\pi r} c^2 = m_e c^2, \qquad (1.1.5:8)$$

where m_e is the mass equivalence of Coulomb energy. The final form of (1.1.5:8) indicates that the acceleration of a charged mass object in an accelerator can be expressed as a transfer of mass to the accelerated object. Transfer of mass gives a physical explanation for the mass increase of accelerated objects. In the framework of special relativity, the increase of "relativistic mass" is introduced as a consequence of the velocity and described in terms of coordinate transformations.

The solution of Planck's equation from Maxwell's equations confirms the interpretation of Planck's equation as the energy conversion occurring at the emission or absorption of electromagnetic radiation.

Planck's equation does not describe an inherent property of radiation. Radiation propagating in expanding space loses power and energy density with the expanding wavelength but conserves the energy carried by a cycle of radiation. This is exceedingly important, e.g., in the interpretation of the luminosity of cosmological objects and the microwave background radiation.

The "antenna solution" of blackbody radiation

Planck's radiation law can be directly concluded from the antenna theory as a combination of two limiting mechanisms: At the long wavelength part of the spectrum, the thermal energy, $kT>hf$, is enough to activate all the emitters, *"the surface antennas"*, and the power density is limited by the emitter density. At the short wavelength part of the spectrum, where $kT<hf$, the limitation comes from the thermal energy available for activation of the emitters, Figure 1.1.5-1.

The unified expression of energy

The breakdown of Planck's constant allows a unified expression for the rest energy of mass, the quantum of electromagnetic radiation, and the Coulombian energy. In a precise analysis, we make a distinction between the local velocity of light, c, and the velocity of light in hypothetical homogeneous space c_0, which is equal to the expansion velocity of space in the direction of the 4-radius, Figure 1.1.5-2.

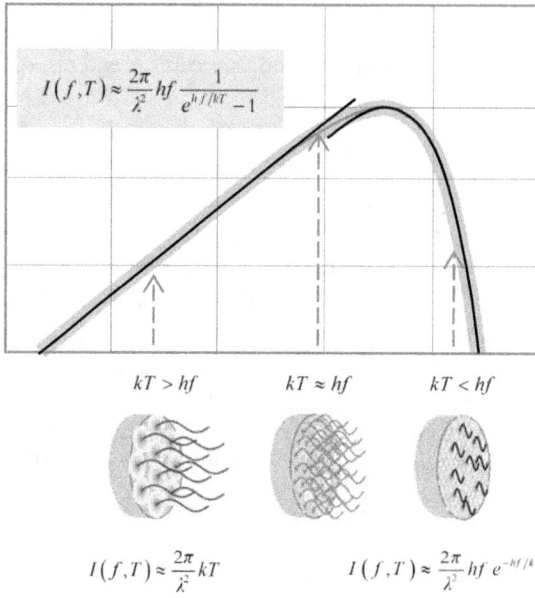

Figure 1.1.5-1. Blackbody surface as an antenna field. Antenna active area is related to the wavelength as $A_\lambda = \lambda^2/4\pi$, with emission intensity to half-space as $I_\lambda = 1/A_\lambda = 2\pi/\lambda^2$, which leads to the Rayleigh-Jeans formula applying when all antennas are activated by the thermal energy, $kT > hf$. When only a part of the antennas are activated as described by the Maxwell-Boltzmann distribution of the thermal energy, $kT << hf$, the emission intensity follows Wilhelm Wien's radiation law. Max Planck's radiation law combines the two, plus covers the transition region where $kT \approx hf$.

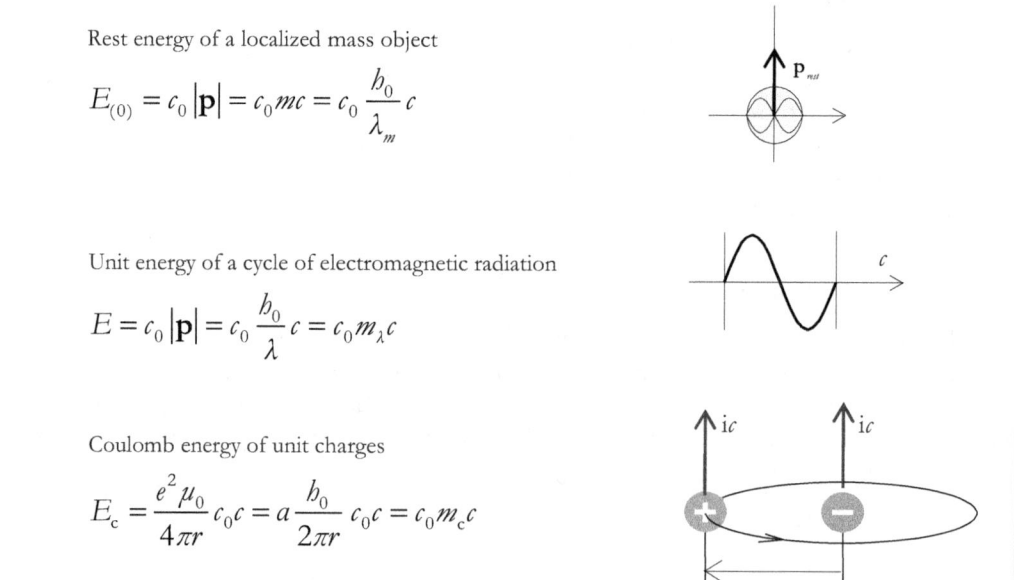

Figure 1.1.5-2. Unified expressions for the rest energy of a localized mass object, the unit energy of a cycle of electromagnetic radiation, and the Coulomb energy.

Mass objects as resonant mass wave structures

In the DU framework, a mass object (particle) is described as a spherically symmetric resonator in 3-space. The momentum of a resonator with standing waves built up of opposing waves and momenta in a space direction appears as the imaginary momentum, the rest momentum of the resonator. When a mass object is put into motion in space, the imaginary rest momentum will be accompanied by a real component equal to the net mass wave of the Doppler shifted front and back waves of the resonator, thus building up the de Broglie momentum

$$\mathbf{P}_{\rightleftarrows} = \mathbf{P}_{\rightarrow} + \mathbf{P}_{\leftarrow} = \frac{h_0}{\lambda_m} \frac{v/c}{\sqrt{1-(v/c)^2}} \mathbf{c} = \frac{h_0/\lambda_m}{\sqrt{1-(v/c)^2}} \mathbf{v} = \frac{h_0}{\lambda_{dB}} \mathbf{c}. \qquad (1.1.5:9)$$

As illustrated by equation (1.1.3:10), the net momentum wave of a mass object can be interpreted as a mass wave corresponding to the wave equivalence of the "relativistic mass" $m_0 / \sqrt{1-(v/c)^2}$ propagating at the velocity v of the moving mass object, or the de Broglie wave propagating at the velocity of light.

In fact, the removal of the velocity of light from the Planck constant discloses the matter wave de Broglie was looking for: a wave does not disperse but follows the moving object.

The intrinsic Planck constant and the mass wave concept disclose the close connection between the Compton wave and the de Broglie wave.

The momentum wave is the sum of the momenta of the Doppler-shifted back and front waves of a moving resonator. Passing a double-slit, the momentum wave explains the interference pattern observed on the screen (Section 5.3.5), Figure 1.1.5-3.

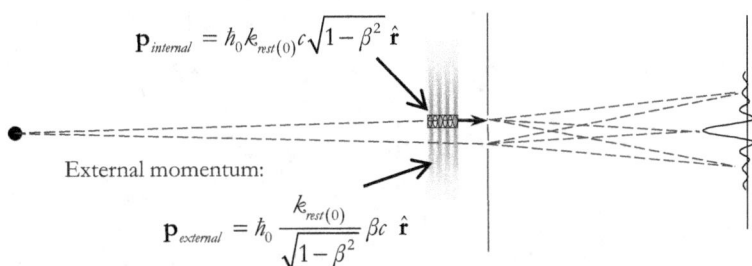

Figure 1.1.5-3. A mass object as a standing wave structure (drawn in the direction of the real axis). The momentum of the object moving at velocity βc is the external momentum as the sum of the Doppler-shifted front and back waves, which can be described as the momentum of a wavefront propagating in the local frame in parallel with the propagating mass object. The interference pattern observed in the double slit experiment demonstrates the momentum as a wavefront, resulting in deflection of the propagation path observed as an interference pattern between wavefronts passing the two different slits.

1.2 Buildup of energy in space

1.2.1 The primary energy buildup in space

In Chapter 3, the buildup of the rest energy of matter is described as a contraction–expansion process of spherically closed space. Starting from the state of rest in homogeneous space with essentially infinite radius means an initial condition where both the energy of motion is zero, and the energy of gravitation is zero, due to very high distances. A trend to minimum potential energy in a spherically closed space converts gravitational energy into the energy of motion in a contraction phase. Space gains motion from gravitation in a contraction phase and pays it back in an expansion phase after passing a singularity. The dynamics of spherically closed space works like that of a spherical pendulum in the fourth dimension, as illustrated in Figure 1.2.1-1.

Homogeneous space has dynamics in the fourth dimension only, in the direction of the 4-radius of the spherically closed space. Applying the inherent energies of motion and gravitation to the zero-energy balance of motion and gravitation, we get the equation for the zero-energy balance of homogeneous space

$$E_m + E_g = c_0 \cdot M_\Sigma c_0 - \frac{GM_\Sigma M''}{R_4} = 0, \qquad (1.2.1{:}1)$$

where M_Σ is the total mass in space, and $M'' = 0.776 \cdot M_\Sigma$ is the mass equivalence of the whole space in the center of the spherical structure.

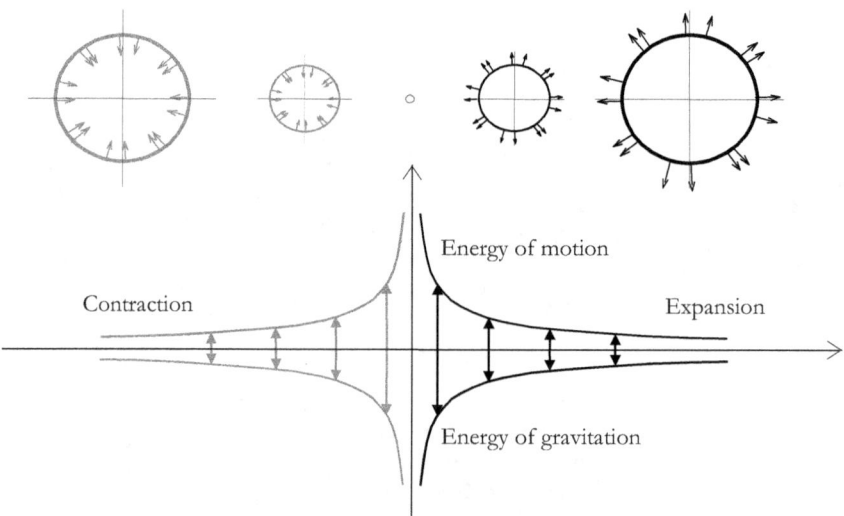

Figure 1.2.1-1. Energy buildup and release in spherical space. In the contraction phase, the velocity of motion increases due to the energy gained from the release of gravitational energy. In the expansion phase, the velocity of motion gradually decreases, while the energy of motion gained in contraction is returned to the energy of gravitation.

The contraction-expansion cycle creating the motion of space is referred to as the primary energy buildup process of space.

Using today's estimates for the mass density in space, and the 4-radius, which corresponds to the Hubble radius, $R_4 = R_H \approx 14$ billion light years, the present velocity of the expansion, c_0, in (1.2.1:1) is

$$c_0 = \pm \sqrt{\frac{GM''}{R_4}} \approx 300\,000 \quad [\text{km/s}], \qquad (1.2.1:2)$$

which is equal to the present velocity of light. It can be shown that the velocity of the expansion of space in the direction of the 4-radius determines the maximum velocity in space and the velocity of light.

Due to the dynamic nature of the zero-energy balance in space, the velocity of space in the fourth dimension and, accordingly, the velocity of light slows down in the course of the expansion of space. The present annual increase of the R_4 radius of space is $dR_4/R_4 \approx 7.2 \cdot 10^{-11}/y$. The deceleration rate of the expansion is $dc_0/c_0 \approx -3.6 \cdot 10^{-11}/y$, which also means that the velocity of light slows down as $dc/c \approx -3.6 \cdot 10^{-11}/\text{year}$. In principle, the change is large enough to be detected. However, the change is reflected in the ticking frequencies of atomic clocks via the degradation of the rest momentum, i.e., the frequencies of clocks slow down at the same rate as the velocity of light, thus disabling the detection.

The velocity of light in the Dynamic Universe is not a natural constant but is determined by the velocity of space in the fourth dimension — the velocity of space in the fourth dimension is determined by the zero-energy balance in equation (1.2.1:1).

An important conclusion from the primary energy buildup process is that the rest energy is not a property of mass or matter but has the nature of the energy of motion, not due to motion in space but due to the motion of space. In expanding space, the velocity of space decreases due to the work the expansion does against the gravitation of the structure. It means that also the rest energy of mass in space diminishes, although the amount of mass in space is conserved.

In the prevailing Friedman-Lemaître-Robertson-Walker (FLRW) cosmology, or "Big Bang cosmology", all energy and the flow of time in space were triggered by a sudden event or quantum jump about 14 billion years ago. A major difference between the primary energy buildup in the DU and the energy buildup in the Big Bang cosmology is that the energy of matter in the DU has developed against the reduction of the gravitational energy in a continuous process in a contraction phase preceding the ongoing expansion phase. Space has lost volume and gained velocity in the contraction phase and pays the velocity back to volume in the ongoing expansion phase.

The basis of the zero-energy concept was first expressed, at least implicitly, by Gottfried Leibniz, contemporary with Isaac Newton. Although the concept of energy was not yet mature, the idea of the zero-energy principle can be recognized in Leibniz's *vis viva*, the living force mv^2 (kinetic energy) that is obtained against the release of *vis mortua*, the dead force (potential energy) – or vice versa. *"Action is always equal to reaction, and the complete effect is always equivalent to the total cause"* [8,9].

1.2.2 Buildup of kinetic energy in space

Buildup of velocity, momentum, and kinetic energy in space requires an input of energy. The energy input may come from an accelerating system at constant gravitational potential or via the release of gravitational energy in free fall.

Kinetic energy at constant gravitational potential

In the framework of special relativity, the kinetic energy is expressed in the form

$$E_{kin} = E_{total} - E_{rest} = mc^2 / \sqrt{1-(v/c)^2} - mc^2 = \Delta mc^2, \tag{1.2.2:1}$$

where Δm is the relativistic mass increase. In the SR framework, following the relativity principle and the kinematic explanation, Δm is a consequence of the velocity and the associated coordinate transformations.

Equation (1.2.2:1) also holds in the DU framework; following the conservation of energy, Δmc^2 is the energy released by the accelerating system to build up the kinetic energy and the associated momentum and velocity. The total energy of a moving object is

$$\left|E_{total}^{\square}\right| = c_0 \left|p_{total}^{\square}\right| = c_0 \left|(m+\Delta m)v + \mathrm{i}mc\right| = c_0(m+\Delta m)c, \tag{1.2.2:2}$$

which relates the total energy to the total momentum consisting of the rest momentum as the imaginary part imc, and the real part $p=(m+\Delta m)v$, the momentum in space. The last form of (1.2.2:2) is the modulus of the complex function, Figure 1.2.2-1.

The complex function formalism allows a detailed study of the energy structure and the buildup of momentum in space. The momentum in space consists of two parts: momentum mv as the real component of mc in the total momentum, and Δmv as the real component of Δmc in the total momentum.

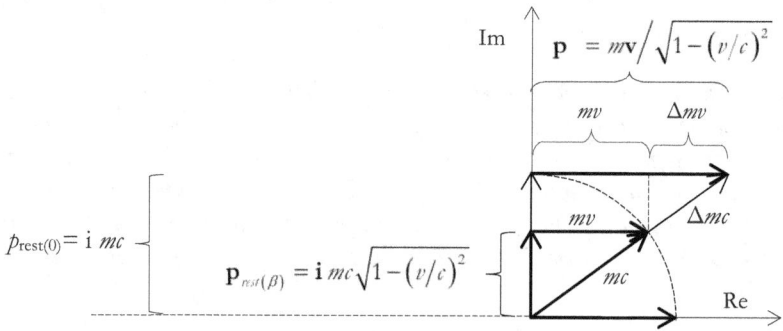

Figure 1.2.2-1. The momentum of an object moving in space at velocity v consists of the contribution mv as the real component of momentum mc contributing to the total energy, and component Δmv as the real component of Δmc contributing to the total energy. The imaginary component of mc is the rest momentum of the moving object, $p_{rest(v)} = \mathrm{i}mc\sqrt{1-(v/c)^2}$.

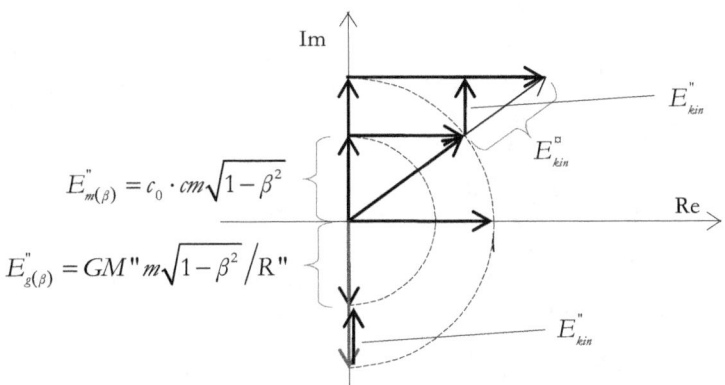

Figure 1.2.2-2. The reduction of the rest mass means an equal reduction in the rest energy E_m and the global gravitational energy E_g of the moving object. The imaginary component E''_{kin} of the complex kinetic energy $E^\square_{kin}$ is the work done against the global gravitational energy by the buildup of motion as central motion in spherically closed space. This means a quantitative expression of Mach's principle.

The rest momentum of the moving object is reduced. Part mc of the total momentum of the moving object can be divided into orthogonal components: momentum mv in a space direction and $mc\sqrt{1-\beta^2}$ in the imaginary direction, which is the reduced rest momentum of the moving object. While the local 4-velocity of space is c, the reduction of the rest momentum shall be understood as the reduction of the rest mass. The reduced rest mass $m_{rest(\beta)}$ is the counterpart of the increased mass contributing to the momentum in space, m_β (the relativistic mass in the SR framework)

$$m_{rest(\beta)} = m\sqrt{1-\beta^2} \quad ; \quad m_\beta = m/\sqrt{1-\beta^2}, \qquad (1.2.2:3)$$

where $\beta=v/c$ and m is the rest mass of the object at rest in the local frame of reference.

The reduction of the rest mass means a corresponding reduction in the rest energy and the global gravitational energy of the moving object, which completes the conservation of the overall balance of the energies of motion and gravitation. The imaginary component E''_{kin} of the complex kinetic energy $E^\square_{kin}$ is the work done against the global gravitational energy by the buildup of the motion as central motion. **This means a quantitative expression of Mach's principle**, Figure 1.2.2-2.

Kinetic energy in free fall in a gravitational field

The buildup of mass centers in space requires the buildup of motion and kinetic energy via the release of global gravitational energy. For conserving the total energy of motion and gravitation in such an energy conversion, space as the zero-energy "surface" is tilted, which turns part of the 4-velocity of space into velocity in space by rotating the local fourth dimension relative to the fourth dimension in the non-tilted space, Figure 1.2.2-3. In free fall towards a mass center, there is no source for increased mass, like in the case of kinetic energy buildup by local energy release. The velocity and momentum are obtained against

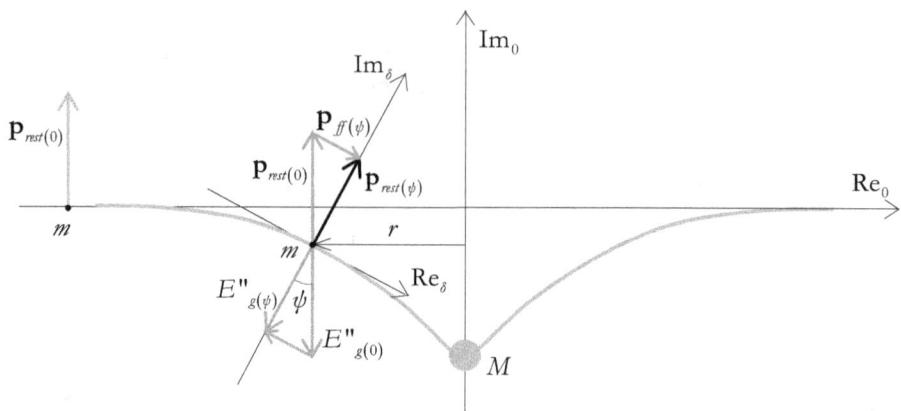

Figure 1.2.2-3. Free fall of mass *m* towards mass center *M* in space. The momentum of free fall is obtained against a reduction of the rest momentum in the local fourth dimension.

the reduction of the velocity and momentum due to the motion of space in the local fourth dimension. It means that, unlike in acceleration at constant gravitational potential, there is no insert of mass in free fall. Kinetic energy is obtained against a reduction of the rest energy of the falling object,

$$E_{kin} = E_{rest(0)} - E_{rest(\psi)} = E_{rest(0)}\left(1-\cos\psi\right) = c_0 m\Delta c , \qquad (1.2.2:4)$$

where Δc is the reduction of the local velocity of light associated with the local tilting of space. The buildup of kinetic energy, combining inertial acceleration and gravitational acceleration, can be written as

$$E_{kin} = c_0 \left(c\Delta m + m\Delta c\right), \qquad (1.2.2:5)$$

where the first term shows the kinetic energy obtained at constant gravitational potential by the insertion of energy $c_0\Delta mc$, and the second term, $c_0 m\Delta c$, the kinetic energy obtained in free fall against reduction of the rest energy due to the reduction of the local velocity of light.

The zero-energy principle behind the Dynamic Universe cancels the equivalence principle behind the general theory of relativity.

The reduction of the velocity of space in the local fourth dimension also means a reduction of the local velocity of light and the local rest momentum.

The reduced rest momentum in tilted space results, e.g., in reduced ticking frequency of atomic clocks in the vicinity of mass centers in space.

The reason for the instability of orbits near black holes predicted by general relativity can be traced to the equivalence principle increasing the "relativistic mass" in free fall. DU predicts stable orbits at any distance from local singularities (Sections 1.2.6 and 4.2.8).

Tilting of local space

The reduced local rest energy in tilted space is balanced by the reduced global gravitational energy $E''_{g(\delta)}$. The tilting angle of space near mass centers can be solved from the release of global gravitational energy. The global gravitational energy of a mass object in homogeneous space builds up from gravitation affecting uniformly in all space directions. Due to the spherical geometry, the integrated effect of gravitation is seen in the fourth dimension, as if affected by the mass equivalence of space at the 4-center of the structure. Removal of a certain amount of mass from the symmetry for creating a local mass center M means a reduction in the global gravitation in the vicinity of the center. At distance r from the local center, the reduction in global gravitational energy of test mass m is equal to the gravitational energy created by mass M uniformly distributed around at distance r from m. Accordingly, the locally observed global gravitational energy is

$$E_{g(\delta)} = E_{g(0)} \cos\psi = E_{g(0)}\left(1 - GM/rc_0^2\right) = E_{g(0)}\left(1-\delta\right), \qquad (1.2.2{:}6)$$

where δ is the gravitational factor at distance r from the local mass center M

$$\delta \equiv GM/rc_0^2 = 1 - \cos\psi. \qquad (1.2.2{:}7)$$

The 4-velocity of space, determining the local velocity of light at distance r from M, is

$$c = c_0 \cos\psi = c_0\left(1-\delta\right). \qquad (1.2.2{:}8)$$

Equations (1.2.2:6) and (1.2.7) are closely related to Schwarzschild space. In Schwarzschild space, the fourth dimension is the time-like dimension, and the tilting angle is related to the reduced proper time

$$dt' = dt\cos\psi = dt\sqrt{1-2\delta} \approx dt\left(1-\delta\right). \qquad (1.2.2{:}9)$$

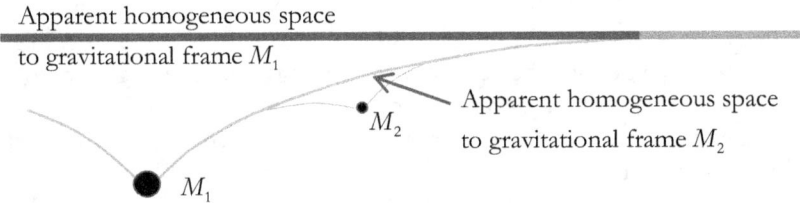

Figure 1.2.2-4. Space around a local gravitational frame, as it would be without the mass center, is referred to as apparent homogeneous space to the local gravitational frame. Accumulation of mass into mass centers occurs in several steps. Starting from hypothetical homogeneous space, the "first-order" gravitational frames, like the frame around mass M_1 in the figure, have hypothetical homogeneous space as the apparent homogeneous space. In subsequent steps, smaller mass centers may be formed within the tilted space in the "first order" frame. For those frames, like M_2 in the figure, space in frame M_1, as it would be without the mass center M_2, serves as the apparent homogeneous space to frame M_2.

In real space, the buildup of mass centers occurs in several steps, Figure 1.2.2-4. In each step, the local dent forming a local gravitational frame is related to its parent frame. The global gravitational energy in the n:th mass center is

$$E''_{g(n)} = -\frac{GM''m}{R''_0}\prod_{i=1}^{n}(1-\delta_i) = -\frac{GM''m}{R''}, \qquad (1.2.2:10)$$

where $R''_0 = R_4$ is the 4-radius of homogeneous space. The local apparent 4-radius R'' is

$$R'' = R''_n = R''_0 \bigg/ \prod_{i=1}^{n}(1-\delta_i). \qquad (1.2.2:11)$$

The local velocity of light at gravitational state δ_n in the n:th frame is

$$c = c_n = c_0 \prod_{i=1}^{n}(1-\delta_i), \qquad (1.2.2:12)$$

where δ_i is the gravitational factor in the i:th gravitational frame.

Combining the effects of motion and gravitation, the rest energy $E_{rest(\delta,\beta)}$ of a mass object moving at gravitational state $\delta=\delta_n$, at velocity $\beta=\beta_n$ in the n:th frame is

$$E_{rest(\delta,\beta)} = E''_{m(n)} = c_0 mc = E_{rest(0,0)}\prod_{i=1}^{n}(1-\delta_i)\sqrt{1-\beta_i^2}. \qquad (1.2.2:13)$$

1.2.3 Energy structures in space

The system of nested energy frames

The system of nested energy frames is a central feature of the Dynamic Universe. It means full replacement of the observer-centered frame of reference of special and general relativity.

The buildup of the system of nested energy frames follows the conservation of energy. Any local frame is obtained via the release of rest energy and global gravitational energy in the parent frame. A state of motion in a local frame is related to the state of rest in the frame; the kinetic energy is obtained against the release of potential energy in the frame. A state of gravitation in a local frame is related to the gravitational state as it were, without the local frame (or mass center) in a specific location in space.

In the Earth's gravitational frame, referred to as the Earth Centered Inertial frame (ECI-frame) in the SR/GR framework, any gravitational state in the Earth's gravitational frame is related to the gravitational state in the Sun's gravitational frame at a specific location in space. Any state of motion in the Earth's gravitational frame is related to the state of rest in the Earth frame, which means a state excluding the rotation of the Earth and motion relative to the surface of the Earth.

The Earth's gravitational frame moves at the orbital velocity of the Earth in the Solar frame. The state of motion and gravitation of the Earth's gravitational frame in the Solar gravitational frame is determined by the orbital velocity and radius, respectively. Due to the eccentricity of Earth's orbit, the state of motion and gravitation change periodically during the year.

Introduction

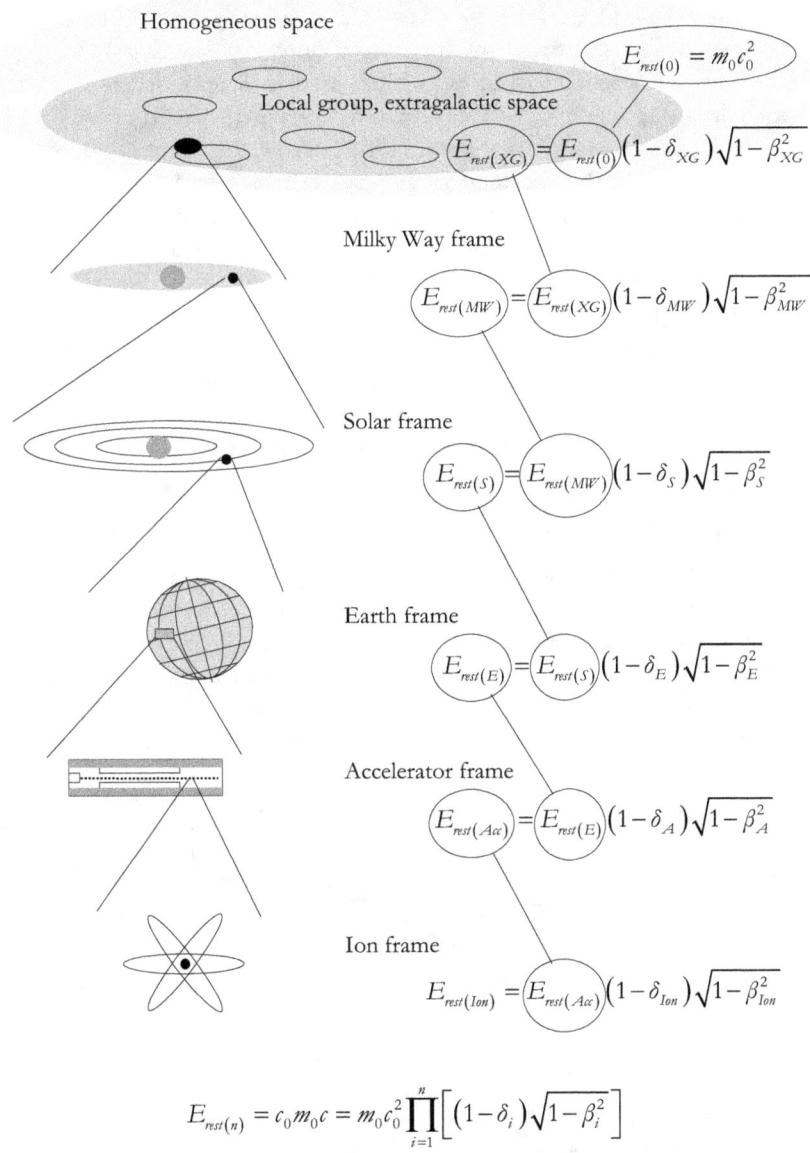

Figure 1.2.3-1. The rest energy of an object in a local frame is linked to the rest energy of the local frame in its parent frame. The system of nested energy frames relates the rest energy of an object in a local frame to the rest energy the object would have at rest in homogeneous space.

Any local frame on the Earth is subject to Earth's rotation, affecting the rest mass of objects in the frame, and the local distance to the barycenter of the Earth, which determines the local gravitational state and the local velocity of light. An electron or any other charged particle accelerated in an accelerator obtains its kinetic energy from the electric field in the accelerator, which thus forms a frame as a subframe in the Earth's gravitational frame. In an accelerator, the accelerated particle moves at a constant gravitational potential, which means that its rest energy is affected by the motion only.

The system of nested energy frames is illustrated in Figure 1.2.3-1. Starting from a charged particle in an accelerator on the Earth, the chain of nested energy frames can be listed as follows: The local frame to the ion is the accelerator frame; the rest energy of an ion moving at velocity v in the accelerator is related to the rest energy of an ion at rest in the accelerator frame or the laboratory frame housing the accelerator. The parent frame to the accelerator/laboratory frame is the Earth's gravitational frame, where the accelerator/laboratory frame has the rotational velocity and gravitational factor characteristic to the location and altitude of the laboratory on the Earth.

The Earth's gravitational frame orbits the Sun in the Solar gravitational frame; the Solar gravitational frame orbits the center of the Milky Way in the Milky Way gravitational frame, which is a subframe of the Local group moving in hypothetical homogeneous space, which serves as the universal frame of reference to the chain of nested frames.

The topography of the fourth dimension

The curvature of space near local mass centers is a consequence of the conservation of the energy balance created in the primary energy buildup of space. Because the fourth dimension is a geometrical dimension, the shape of space can be expressed in distance units, including the topography in the fourth dimension. Dents in space are associated with a reduced velocity of light. Figure 1.2.3-2 illustrates the "depth" profile of the planetary system in the vicinity of the Earth.

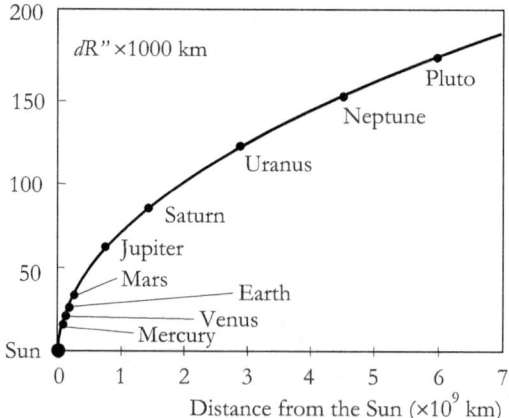

Figure 1.2.3-2. The topography of the Solar System in the fourth dimension. Earth is about 26 000 km higher than the Sun; Pluto is about 180 000 km higher than the Sun in the fourth dimension.

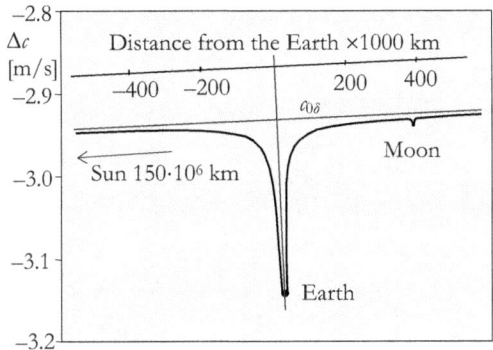

Figure 1.2.3-3. Effect of the gravitation of the Sun, Earth, and Moon on the velocity of light. The tilted baseline at the top shows the effect of the Sun on the velocity of light, which is the apparent homogeneous space velocity of light for the Earth, $c_{0\delta(Earth)}$. The Moon is shown in its "Full Moon" position, opposite to the Sun. The curves in the figure are based on equation (4.1.10:2) as separately applied to the Earth and the Sun. The effect of the mass of the Milky Way on the velocity of light in our planetary system is about $\Delta c \approx -300$ m/s.

The velocity of light in the dents around mass centers in space is reduced, resulting in the Shapiro delay, and the path of light is bent. The Shapiro delay is affected both by the lengthening of the path and the reduction of the velocity of light near mass centers.

The effect of the difference in the GR and DU predictions for the Shapiro delay is not detectable in the experiments that have been performed (see Section 7.3.4).

The local velocity of light

The local velocity of light is determined by the velocity of space in the local fourth dimension. The local velocity light is a function of the distance from mass centers in space. On the surface of the Earth, the velocity of light is reduced by about 20 cm/s compared to the velocity of light at the distance of the Moon from the Earth. The velocity of light at the Earth's distance from the Sun is about 3 m/s lower than the velocity of light far from the Sun, Figure 1.2.3-3.

The velocity of light is linked to the local 4-velocity of space, which is a property of the gravitational state in the local gravitational frame. The motion of an object at a constant gravitational potential in the local frame, as well as the motion of the local gravitational frame at a constant gravitational potential in its parent frame, does not change the local velocity of light. This means that the velocity of light observed in a space location is not summed to the velocity of the observer as long as the observer moves at constant gravitational potential.

In the DU framework, e.g., an interferometer or a resonator is a local energy frame where the opposite waves have the velocity of light equal to the local 4-velocity of space in the particular space location. The motion of a resonator in its parent frame creates a net Doppler wave observable in the parent frame but conserves the resonance condition in the resonator frame, see Section 5.3.2. for a detailed analysis.

As a simplified picture, the linkage of the local velocity of light to the local 4-velocity of space means that electromagnetic radiation in space propagates at a "satellite orbit" relative to the mass equivalence of space seen in the local fourth dimension. Moving at the local 4-velocity, radiation as an energy object (with mass equivalence) creates a centrifugal force opposite to the gravitational force arising from the mass equivalence of space in the fourth dimension, thus making electromagnetic radiation or photons look like "massless objects", Figure 1.2.3-4.

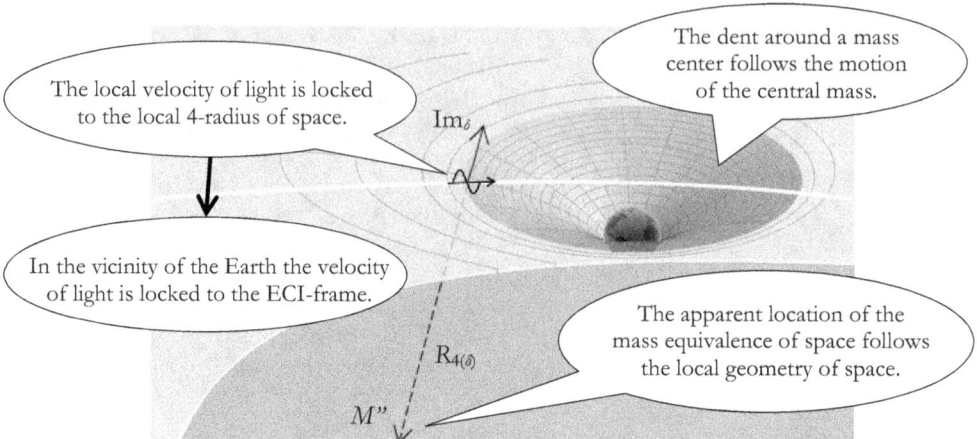

Figure 1.2.3-4. Electromagnetic radiation propagating in space is balanced by the gravitation due to the mass equivalence in the fourth dimension. The mass equivalence appears at distance R'' in the direction of the local fourth dimension following the motion of the local frame in its parent frame in space.

The linkage of the local velocity of light to the local 4-velocity of space explains the constancy of the velocity of light observed in interferometric measurements, like, e.g., the famous Michelson-Morley experiment, which was one of the key triggers for the development of coordinate transformations behind the theory of relativity. Also, it explains the necessity of expressing the frequencies of atomic clocks on the Earth and in Earth satellites in terms of motion and gravitational state in the Earth's gravitational frame (the Earth Centered Inertial (ECI) frame in the framework of the theory of relativity).

1.2.4 The frequency of atomic clocks

The quantum mechanical solution

At the time special and general theories of relativity were formulated, there was no theory on the emission and absorption frequencies of atomic objects that determine the ticking frequency of atomic clocks. The solution to the characteristic frequencies was obtained first by applying the quantum mechanical analyses of atomic structures. In a general form, the emission/absorption frequency can be expressed in terms of the rest energy of the oscillating electrons, the Planck constant, and the quantum numbers characterizing the energy states related to the oscillation

$$f_{(n1,n2)} = \frac{\Delta E_{(n1,n2)}}{h} = \frac{m_e c^2}{h} \Delta F(a,n,l,m_l,m_s), \qquad (1.2.4{:}1)$$

where $\Delta E_{(n1,n2)}$ is the difference of the rest energy of an electron in the two energy states relevant to the emission/absorption process, h is the Planck constant, m_e is the rest mass of the electron of the atom in the local energy frame, and c is the local velocity of light. The function $\Delta F(a,n,l,m_l,m_s)$ is determined by the fine structure constant a and the quantum

Introduction

numbers characterizing the energy states in question. Applying the intrinsic Planck constant h_0, defined in equation (1.1.5:3), equation (1.2.4:1) obtains the form

$$f_{(n1,n2)} = \frac{\Delta E_{(n1,n2)}}{h} = \frac{m_e c}{h_0} \Delta F(a, n, l, m_l, m_s), \tag{1.2.4:2}$$

where ΔF consists of numerical factors only; the quantum numbers and the fine structure constant, which is also a numerical constant, see equation (1.1.5:5).

The effect of motion and gravitation

As shown by equation (1.2.4:2), the characteristic frequency of atomic oscillators is directly proportional to the rest momentum of the oscillating electrons – and thereby affected by the velocity and gravitational state of the oscillator in a local frame – and through the velocity and gravitational state of the local frame in its parent frames

$$f_{n(\beta,\delta)} = f_{(0,0)} \prod_{i=1}^{n} (1-\delta_i)\sqrt{1-\beta_i^2}, \tag{1.2.4:3}$$

where $f_{(0,0)}$ is the frequency of the oscillator at rest in hypothetical homogeneous space.

The frequency of an atomic emitter in an accelerator is related to the frequency f_{lab} of a similar atom at rest in the accelerator or the laboratory housing the accelerator

$$f_{acc} = f_{lab}\sqrt{1-\beta_{acc}^2}. \tag{1.2.4:4}$$

The frequency of an atomic clock in a laboratory with the velocity and gravitational factors β_{lab} and δ_{lab}, respectively, is related to the frequency of a hypothetical clock with the gravitational factor $\delta_{Earth}=0$, and the velocity factor $\beta_{Earth}=0$.

$$f_{lab} = f_{Earth(0,0)}(1-\delta_{lab})\sqrt{1-\beta_{lab}^2}. \tag{1.2.4:5}$$

The frequency of an atomic clock in an Earth satellite is related to the hypothetical clock with the gravitational factor $\delta_{Earth}=0$, and the velocity factor $\beta_{Earth}=0$, just like the laboratory clock

$$f_{satellite} = f_{Earth(0,0)}(1-\delta_{satellite(Earth)})\sqrt{1-\beta_{satellite(Earth)}^2}. \tag{1.2.4:6}$$

The frequency of an atomic clock in a spacecraft flying in the solar system outside the gravitational interaction of the Earth is related to a hypothetical clock with the gravitational factor $\delta_{Sun}=0$, and the velocity factor $\beta_{Sun}=0$.

$$f_{spacecraft} = f_{Sun(0,0)}(1-\delta_{spacecraft(Sun)})\sqrt{1-\beta_{spacecraft(Sun)}^2}. \tag{1.2.4:7}$$

For comparing clocks outside the near space region, a hypothetical clock at rest in the Sun's gravitational frame is needed.

The velocity terms in equations (1.2.4:3-7) are related to the kinetic energy obtained/released in an exchange with the potential energy in the frame. For example, spacecrafts leaving the gravitational interaction with the Earth, like the Pioneer spacecrafts, continue to have a *kinematic velocity* relative to the Earth but not a velocity related to the exchange of gravitational energy due to the Earth.

BOX 1.2.4-I

In the DU framework, all experiments on the effects of motion and gravitation on the frequency of atomic clocks can be explained by equation (1.2.4:3). On the Earth and in near space, the frequency f_A of a clock A is

$$f_{A(\beta,\delta)} = f_{(0,0)Earth}\left(1-\delta_{A,Earth}\right)\sqrt{1-\beta^2_{A,Earth}}\sqrt{1-\beta^2_{A,Equipment}} \qquad (1.2.4:10)$$

where $f_{(0,0)Earth}$ is a hypothetical reference clock ($\delta_{0,Earth}=\beta_{0,Earth}=0$) in the Earth gravitational frame. Factors $\delta_{A,Earth}$ and $\beta_{A,Earth}$ are the gravitational and velocity factors of clock A in the Earth gravitational frame, respectively, and $\beta_{A,Equipment}$, the velocity factor of the clock in a possible equipment frame. An equipment frame may be an accelerator or centrifuge, which are closed systems where the clock is given the local kinetic energy. The frequency of a reference clock B in the same equipment as clock A is

$$f_{B(\beta,\delta)} = f_{(0,0)Earth}\left(1-\delta_{B,Earth}\right)\sqrt{1-\beta^2_{B,Earth}}\sqrt{1-\beta^2_{B,Equipment}} \qquad (1.2.4:11)$$

The frequency of clock A is related to the frequency of the reference clock as

$$f_{A(\beta,\delta)} = f_{B(\beta,\delta)} \frac{\left(1-\delta_{A,Earth}\right)\sqrt{1-\beta^2_{A,Earth}}\sqrt{1-\beta^2_{A,Equipment}}}{\left(1-\delta_{B,Earth}\right)\sqrt{1-\beta^2_{B,Earth}}\sqrt{1-\beta^2_{B,Equipment}}} \qquad (1.2.4:12)$$

The frequency $f_{B(\delta,\beta)}$ of the reference clock B can be related to the frequency consistent with International Atomic Time (TAI) and the SI second by relating clock B to the frequency f_{SI} of a reference clock at the Earth geoid (see Section 5.7.2) (at the poles the rotational velocity of the Earth is zero; the frequency of a clock at the geoid is determined by the gravitational factor only)

$$f_{B(\delta,\beta)} = f_{SI} \frac{\left(1-\delta_{B,Earth}\right)\sqrt{1-\beta^2_{B,Earth}}\sqrt{1-\beta^2_{B,Equipment}}}{\left(1-\delta_{Polar}\right)} \qquad (1.2.4:13)$$

where

$$f_{SI} = f_{(0,0)Earth}\left(1-\delta_{Polar}\right) = f_{(0,0)Earth}\cdot\left(1-6.977\cdot10^{-10}\right) \qquad (1.2.4:14)$$

All velocities $\beta=v/c$ in the equations in BOX 1.2.4-I apply equally to linear velocities and rotational velocities. Acceleration does not affect the ticking frequency of an atomic clock. In the DU framework, this is obvious because the characteristic frequency of atomic emitters is a function of the energy states of the oscillating electrons – not a function of an external force on the atom.

Introduction

Clocks in the Earth's gravitational frame

Most "clock experiments" for testing general relativity are performed in the Earth's gravitational frame, which allows the use of the Schwarzschild solution of the GR field equations. In the GR framework, the rest energy of a mass object is constant anywhere in space; the effects of local gravitation and motion are explained as consequences of the different flow of time in different frames of reference. According to Schwarzschild's solution, the proper time $d\tau$ experienced by a clock at a state characterized by gravitational factor δ $(=GM/rc^2)$ and velocity factor β $(=v/c)$ is

$$d\tau_{(\beta,\delta)} = dt_{(0,0)}\sqrt{1-2\delta-\beta^2} \approx dt_{0,0}\left(1-\delta-\frac{1}{2}\beta^2-\frac{1}{8}\beta^4-\frac{1}{2}\delta\beta^2-\frac{1}{2}\delta^2+\ldots\right), \quad (1.2.4{:}8)$$

where dt is the coordinate time measured by a hypothetical clock at a state characterized by the gravitational factor $\delta=0$ and the velocity factor $v/c=0$ in the Earth's gravitational frame. Schwarzschild solution is derived for the proper time around a single mass center in otherwise empty space, but it works well enough in the Earth's gravitational frame up to typical satellite altitudes.

Equation (1.2.4:8) is the GR counterpart of the DU equation (1.2.4:3), which in a single gravitational frame reduces to the form

$$f_{(\beta,\delta)} = f_{(0,0)}(1-\delta)\sqrt{1-\beta^2} \approx f_{0,0}\left(1-\delta-\frac{1}{2}\beta^2-\frac{1}{8}\beta^4+\frac{1}{2}\delta\beta^2+\ldots\right). \quad (1.2.4{:}9)$$

In the Earth's near space, the difference between the GR and DU predictions in equations (1.2.4:8) and (1.2.4:9) appears only in the 18th decimal (assuming $\beta^2=\delta$ like in the case of circular satellite orbits), which is too small a difference to be detected.

At extreme conditions like close to black holes, the difference between the predictions, however, is dramatic; GR clock stops when $2\delta+\beta^2 \to 1$, whereas the frequency of a DU clock approaches gradually zero when δ or $\beta^2 \to 1$, which means achieving the critical radius of a black hole or a velocity approaching the local velocity of light.

Experiments on the effects of motion and gravitation on atomic clocks

Before the 1970's, experiments on the time dilation of clocks or characteristic frequency of atomic oscillators were based on observation of electromagnetic radiation transmitted from an object moving relative to the observer. In the late 1930's, Ives and Stillwell, and in the early 1960's, Mandelberg and Witten measured the effect of motion on the wavelength of radiation emitted by fast-moving hydrogen ions [10,11,12]. In the 1960's, several groups studied time dilation in experiments based on the Mössbauer effect in absorbers rotated in a centrifuge [13,14,15,16]. In 1971, atomic clocks were taken around the world in airplanes [17,18]. In 1976, a hydrogen maser was sent to an altitude of 10 000 km to test the combined effects of motion and gravitation on the frequency of the maser [19,20].

Figure 1.2.4-1 summarizes some well-known laboratory and near-Earth experiments on the effects of motion and gravitation on atomic emitters and clocks. In laboratory experiments, the kinetic energy and velocity of the clock are supplied by the accelerator or centrifuge, which serve as a local energy frame for the clocks, cases A [10,11,12], and B [13,14,15,16] in Figure 1.2.4-1.

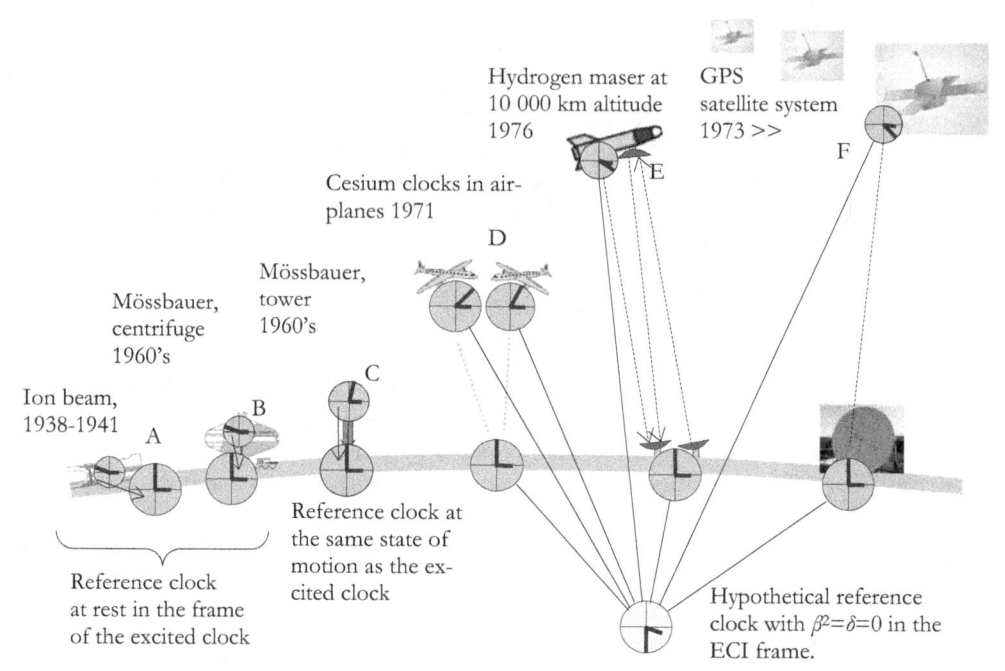

Figure 1.2.4-1. Laboratory and near space experiments for testing the effects of motion and gravitation on atomic emitters and clocks.

A Experiments with hydrogen canal rays emitting blue-green 4861 Å H_β spectral line. Increase of wavelength by factor $\frac{1}{2}(v/c)^2$ with increasing velocity v of the emitting ions was confirmed.

B Experiments with Co-57 γ-ray source at the center of a rotating disk and a resonant Fe-57 absorber at the periphery of the disk. The observed change in the absorption with the rotation speed suggested a change in the peak absorption frequency by factor $\frac{1}{2}(v/c)^2$ with the increasing velocity v of the absorber.

C Experiment with Co-57 γ-ray source at the top and Fe-57 absorber at the bottom of a 75 ft high tower. The observed gravitational shift corresponded to the difference in the gravitational factor between the top and bottom of the tower in the Earth gravitational frame.

D Experiment with cesium clocks flown eastward and westward around the world on commercial airplanes. The experiment confirmed that the hypothetical clock with $\beta^2=\delta=0$ in the Earth gravitational frame shall be used as the reference for both the airplane clocks and the Earth station clock according to equation (1.2.4:6).

E The frequency of a hydrogen maser in a spacecraft sent up to 10 000 km altitude was monitored via a microwave link to Earth station. The effect of gravitation confirmed the GR/DU prediction. The effect of the velocities of the spacecraft and the Earth station was reported as a confirmation of special relativity. A detailed analysis showed that the apparent match with the special relativity prediction was due to an extra term resulting from the two-way Doppler cancellation signal used in the experiment. Corrected analysis showed full match with the GR/DU predictions, in accordance with equation (1.2.4:6).

F The Global Positioning System serves as a modem high accuracy test for the effects of motion and gravitation on atomic clocks. The prediction of equation (1.2.4:6) is confirmed.

Laboratories housing a local frame move at the same velocity at the same gravitational state as the local frames, which means that the state of rest in the laboratory is equal to the state of rest in an accelerator or centrifuge frame in the laboratory. The parent frame to each accelerator or centrifuge frame is the Earth's gravitational frame. The state of motion and gravitation of each accelerator or centrifuge in the Earth's gravitational frame is determined by the altitude and location of the laboratory.

An accelerator on the 5th floor of a laboratory is at a higher gravitational potential than an accelerator on the basement floor. Accordingly, the frequencies of the reference clocks of the accelerators are different, but the effects of the motion in the accelerators relative to the local reference of each accelerator are the same.

The effect of gravitation on the energy of γ-ray radiation was first demonstrated over a 75-foot vertical path in the Jefferson Laboratory, case C [21,22,23] in Figure 1.2.4-1. In the GR framework, the gravitational potential does not affect the emission energy of the radiation source. The increase in the energy of the radiation propagating down from the top of the measuring tower, the gravitational blueshift, was explained as an effect of gravitational acceleration on the radiation.

In the DU framework, the velocity of light and the rest energy, and consequently the emission energy and the emission frequency of the source S_1 at a higher gravitational potential at the top of the measuring tower, are higher than the velocity of light and the emission energy of a reference source S_0 at the basement level. The wavelength of the emitted radiation is independent of the gravitational factor δ because the emission frequency and the velocity of light depend on the gravitational potential in the same way $\lambda = c(1-\delta)/f(1-\delta) = c/f$. On the way down, the frequency of the radiation is conserved, but the velocity of light is decreased, which means that also the wavelength of the radiation is decreased $\lambda = c(1-\delta)/f$, which is observed as a gravitational blueshift.

Gravitational potential affects the rest energy and the characteristic frequency of atomic emitters via the local velocity of light. Radiation propagating from one gravitational potential to another conserves its frequency and the energy carried by a cycle. The wavelength of radiation propagating in a changing gravitational potential changes in direct proportion to the local velocity of light.

The effect of gravitational potential on the characteristic frequency of atomic oscillators can be confirmed with modern atomic clocks, showing the cumulative reading of elapsed time.

Airplanes and satellites move in the Earth's gravitational frame, where the effect of motion and gravitation on atomic clocks is compared to a hypothetical clock with $\beta_2 = \delta = 0$ in the Earth's gravitational frame. This was first confirmed with cesium clocks flown around the world on commercial airplanes in 1971, case D [17,18,24] in Figure 1.2.4-1.

The experiment confirmed that after the round trip, the reading of the clock flown eastward was lower than the reading of the clock flown westward. The difference was due to the rotational velocity of the Earth, which, in the Earth's gravitational frame, added the velocity of the airplane flying to the east and reduced the velocity of the airplane flying to the west, respectively.

Case E [19,20] in Figure 1.2.4-1 refers to the test with a hydrogen maser launched to an altitude of 10 000 km in a nearly vertical trajectory. The test was reported to confirm that the gravitational shift of the maser frequency follows the prediction of the general theory of relativity at the level of $70 \cdot 10^{-6}$. The effect of velocity on the frequency shift was reported as the second-order Doppler effect of special relativity, based on the relative velocity between the spacecraft and the receiver at the Earth station

$$\frac{\Delta f}{f_e} = \frac{GM_e}{c^2}\left(\frac{1}{r_e} - \frac{1}{r_s}\right) - \tfrac{1}{2}|\boldsymbol{\beta}_s - \boldsymbol{\beta}_e|^2, \tag{1.2.4:15}$$

where r_e and r_s are the distances to the barycenter of the Earth from the Earth station and the spacecraft, respectively. Velocities $\boldsymbol{\beta}_s$ and $\boldsymbol{\beta}_e$ are the velocities of the Earth station and the spacecraft in the Earth's gravitational frame (ECI frame), respectively.

The theory given as the reference suggested the prediction based on the velocities of both the spacecraft and the Earth station in the Earth's gravitational frame, like in case D of Figure 1.2.4-1. A detailed analysis [25] shows that the apparent match with the special relativity prediction is due to an extra term resulting from the two-way Doppler cancellation signal used in the experiment. The corrected analysis shows full agreement with the general DU prediction in equation (1.2.4:6), approximated as

$$\frac{\Delta f}{f_e} = \frac{GM_e}{c^2}\left(\frac{1}{r_e} - \frac{1}{r_s}\right) - \tfrac{1}{2}\left(\beta_s^2 - \beta_e^2\right), \tag{1.2.4:16}$$

in agreement with the corresponding approximation of the GR prediction (1.2.4:8). Equation (1.2.4:6) is confirmed in all tests between Earth satellites and Earth stations like the Global Positioning System, which serves as a modern high-accuracy test for the effects of motion and gravitation on atomic clocks, Case F.

The system of nested energy frames is a central feature of the DU. It describes the energy structure of space and produces a quantitative expression for the locally available rest energy. Motion in a local energy frame is related to the state of rest in the frame where the motion and the related kinetic energy are obtained. The motion of the local frame in its parent frame is related to the state of rest in the parent frame, etc. The gravitational state in a local frame is related to the state of gravitation in the particular location in the parent frame, as it were, without the local gravitational frame.

The system of energy frames conveys the relativity of observations in terms of locally available energy that determines the frequencies of atomic oscillators and the rate of essentially all physical processes. The system of nested energy frames means a full replacement of observers' inertial frames of reference applied in the theory of relativity. Also, it means the cancellation of the principle of relativity.

1.2.5 Propagation of light

The apparent constancy and the independence of the observer's motion of the velocity of light in the late 19th century were one of the early signs of imperfections in Newtonian physics. James Clerk Maxwell proposed the world ether[26] as the medium for the propagation of electromagnetic radiation. The world ether was not recognized in spite of several attempts in the 1880s [27,28]; on the contrary, the experiments indicated that the velocity of light is independent of the observer's velocity; the orbital velocity of the Earth did not sum up to the velocity of light as suggested by the world ether theory.

The Michelson–Morley experiment

The best known and historically most important attempt to determine the velocity of the Earth in an assumed "world ether", the Michelson–Morley experiment, was based on a comparison of phase angles in light beams parallel and perpendicular to the assumed velocity of the Earth in the world ether [29]. Neither the size of the Milky Way nor its relation to other galaxies nor the velocity of the Sun in the Milky Way was known at the time of the experiment. Accordingly, it was supposed that the velocity of the interferometer frame relative to the ether would be about 30 km/s, the velocity of the Earth in its planetary orbit around the Sun.

Given the limits of accuracy of the classical instruments, the zero result in the original Michelson–Morley experiments meant that the velocity of the interferometer frame was zero or at least less than 5 km/s. The rotational velocity of the Earth, which is below 400 m/s at European latitudes, was thus more than an order of magnitude too small to be detected. The sensitivity obtained with a classical Michelson–Morley interferometer by Georg Joos in 1930, a phase resolution corresponding to a frame velocity of 1.5 km/s, was still not good enough for the detection of a possible effect of the rotational velocity of the Earth [30]. Many variations of the M–M interferometer were developed to provide an improved sensitivity. One approach was the elimination of the rotation of the interferometer table. The lack of rotation excludes, however, the possibility of detecting the effect of the rotational motion of the Earth [31,32].

There are several variations of the M–M experiment based on masers, lasers, and microwave cavities in a system on a rotating table [33,34,35]. The higher sensitivity of these systems is based on the measurement of the frequency of resonators in different orientations relative to the frame velocity. The accuracy of laser and maser interferometers is good enough to distinguish the possible effect of the rotational velocity of the Earth; all experiments confirm that the motion of the interferometer with the rotation of the Earth does not create interference patterns. An interferometer behaves like an energy object; motion in a local frame creates a momentum wave in the local frame and reduces the frequency observed in the interferometer frame, which makes it look like a confirmation of the theory of relativity and the relativity principle (Section 5.3.2).

M-M experiment in the DU framework

In the DU framework, the zero result of the interferometer experiments is a consequence of the linkage of the velocity of light to the local gravitational state and the balance with global gravitational energy. The effect is illustrated in Figure 1.2.3-4; the motion of a local frame "draws" the apparent location of the barycenter of space with the motion. The motion does not affect the velocity of light observed in the moving frame, but it reduces the rest mass by the factor $\sqrt{1-\beta^2}$ which, in the case of a resonator, means an increase in the wavelength of the resonating waves in the resonator (see Section 5.3.2). Atomic size (proportional to the Bohr radius) increases in direct proportion to the rest mass, which guarantees that a resonator in motion conserves the resonance condition.

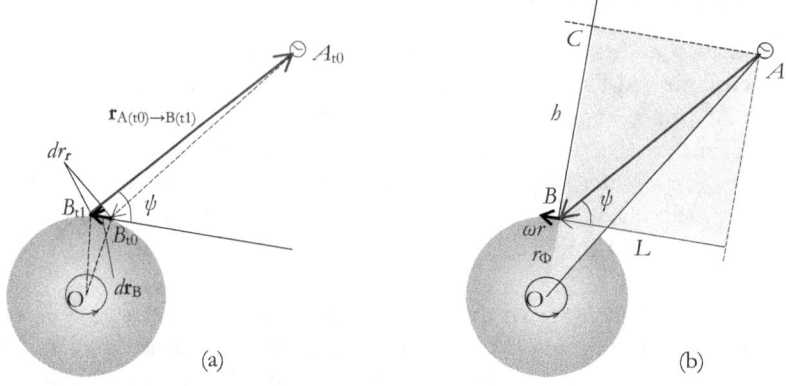

Figure 1.2.5-1. During the signal transmission from a satellite, the rotation of the Earth results in displacement $d\mathbf{r}_B$ relative to a stationary receiver on the Earth. Mathematically the GR expression is identical with the DU expression (see Section 7.3.2).

(a) In DU, the lengthening of the signal path due to the rotation is the component $d\mathbf{r}_r$ in the direction of the signal path

$$\Delta T_{to(Earth)} = \frac{r_{AB(t0)}}{c^2} \omega r_\varphi \cos\psi .$$

(b) The GR expression for the Sagnac correction is related to the area of the equatorial plane projection of triangle O, A_{t0}, B_{t1}

$$\Delta T_{to(Earth)} = \frac{2\omega A_{ABO}}{c^2} .$$

Sagnac effect

The theory of special relativity was challenged by the Sagnac effect, a phase shift due to rotation between opposite beams in an optical loop. The effect was first observed by Harres in 1911 and Sagnac in 1913 [36,37]. In 1925, Albert Michelson and Henry Gale constructed a large optical loop near Chicago and observed the effect of the rotation of the Earth as a phase shift between opposite beams in the loop [38]. Modern versions of optical loops are ring lasers used, e.g., as optical gyroscopes.

In the DU framework, the Sagnac effect is a direct consequence of the motion of the receiver, resulting in an increase or a decrease in the effective length of the signal path and propagation time – from the location and time the signal leaves the source to the location and time it reaches the receiver (see Section 7.3.2). Figure 1.2.5-1 illustrates the Sagnac effect of a satellite signal.

Slow transport of clocks

The term "Sagnac correction" is sometimes used in connection with the slow transport of clocks in the Earth's gravitational frame. In this connection, "slow transport" means that the transport velocity of a clock is slow compared to the rotational velocity of the Earth, i.e., the transportation velocity gives a small increase or decrease to the effective rotational velocity the clock experiences.

In the DU framework, the transport of the clock means that during the transportation, the frequency of the clock is affected by the altered states of motion and gravitation due to the transportation, i.e., the gravitational factor and the velocity factor are functions of time. During the transportation of a clock in the Earth's gravitational frame, the transportation velocity is added to the rotational velocity of the Earth (as a vector sum), and the gravitational factor is corrected with the transportation altitude. The cumulative count of cycles during transportation is

$$\Delta N = f_A \int_{t1}^{t2} \left[1 - \delta(t)\right]\sqrt{1 - \beta(t)^2}\, dt \,. \tag{1.2.5:1}$$

The change of the frequency of a clock during slow transportation of a clock at a constant altitude and constant speed eastwards can be approximated as

$$df_A \approx f_0 \left[1 - \tfrac{1}{2}(v/c)^2\right] \cdot v \cdot dv = -f_A v\, dv = -f_A v_{rot} v_{tr}, \tag{1.2.5:2}$$

where $v = v_{rot}$ is the velocity due to the rotation of the Earth at the local latitude, and $dv = v_{tr}$ the transport velocity $v_{tr} \ll c$ (positive in the east-west direction). The extra reading of the clock is

$$\Delta N = T \cdot df = -T \frac{v_{rot} \cdot v_{tr}}{c^2} f_A \,. \tag{1.2.5:3}$$

By expressing the transportation distance as longitudinal angle and the velocity of Earth's rotation as angular velocity, equation (1.2.5:3) obtains the form

$$\frac{\Delta N}{f_0} = -T \frac{v_{rot} \cdot v_{tr}}{c^2} = -\frac{2\omega_{rot} A_{\psi,\theta}}{c^2}, \tag{1.2.5:4}$$

which gives the cycles "lost" when transporting an atomic clock towards the east by ψ degrees at latitude θ, Figure 1.2.5-2. Equation (1.2.5:4) states the mathematical equality of the slow east-west transport of a clock and the Sagnac delay of light through longitudinal angle ψ at latitude θ over links following the surface of the earth. The physical mechanism of the delay in the reading of the clock, however, is different from the mechanism of the delay in the electromagnetic signal.

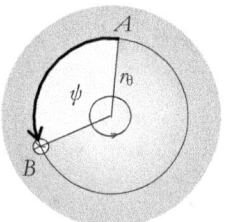

Figure 1.2.5-2. Clock transportation by longitudinal angle ψ from point A to B at latitude θ. The delay due to the additional (slow) velocity is mathematically identical to the Sagnac delay of light transmission from point A to B (through links following the surface of the Earth at the same latitude)

$$\frac{\Delta N}{f_A} = -T \frac{v_{rot} \cdot v_{tr}}{c^2} = -\frac{2\omega_{rot} A_{\psi,\theta}}{c^2}$$

1.2.6 Observables in a local gravitational frame

Perihelion advance

An explanation for the 43 arcsecond/century difference between the observed and predicted perihelion advance of the planet Mercury was a foreseeable proof of general relativity, more than a year before the completion of the theory of general relativity. The missing 43 arc seconds/century was known since the mid-19th century, upon observations and analysis by the French astronomer Urbain Le Verrier, who was able to explain 532 arc seconds out of the observed 575 arc seconds as the gravitational interaction of the other planets.

The GR prediction for the perihelion advance, or main axis rotation, is derived from the Schwarzschild solution of the GR field equations. The solution conserves the diameter of the orbit in the direction of the original main axis used as the initial condition and creates an extended axis serving as the main axis of the rotated orbit, showing the perihelion advance, Figure 1.2.6-1(a).

When solved for a single period, the increase of the main axis is small and omitted as a second-order effect [39,40,41]. However, in the solutions presented, the deformation of the orbit is catastrophic and leads to the escape of the orbiting object when the perihelion advance exceeds 45 degrees.

Applying the GR solution of the main axis rotation to orbits in a high gravitational field near the critical radius, the catastrophic deformation of the orbit occurs in a single period, Figure 1.2.6-2 (a,b). In Schwarzschild space, all orbits, including circular orbits, are unstable at radii smaller than three times the Schwarzschild critical radius, equal to six times the DU critical radius.

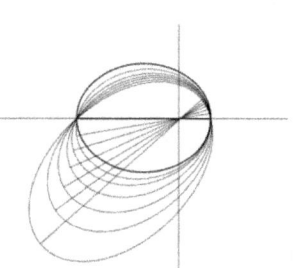
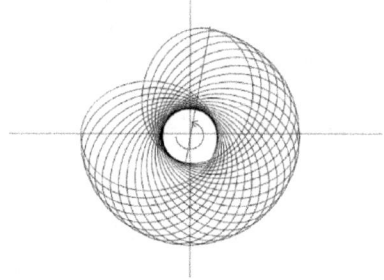

(a) The increase of the orbital radius and the perturbation of the elliptic shape are due to the term 3φ in the denominator [39]

$$r = \frac{a(1-e^2)}{\left\{1+e\sin\varphi - \dfrac{GM}{c^2 a(1-e^2)}\left[e(3\varphi - e\cos\varphi)\cos\varphi + 3 + e^2\right]\right\}}$$

(b) There are no cumulative terms in the DU prediction. The second term results in a small perturbation to the elliptic orbit

$$r_{o,\delta} = \frac{a_{o,\delta}(1-e^2)}{1+e\cos(\varphi - \Delta\psi_{o,\delta})} + \frac{6er[1-\cos(\varphi - \Delta\psi_{o,\delta})]}{(1-e^2)}$$

Figure 1.2.6-1. The development of the orbit of Mercury. (a) The prediction derived from Schwarzschild's solution of general relativity for about 0.3 million years. (b) The prediction given by the Dynamic Universe for about 1.5 million years.

Introduction

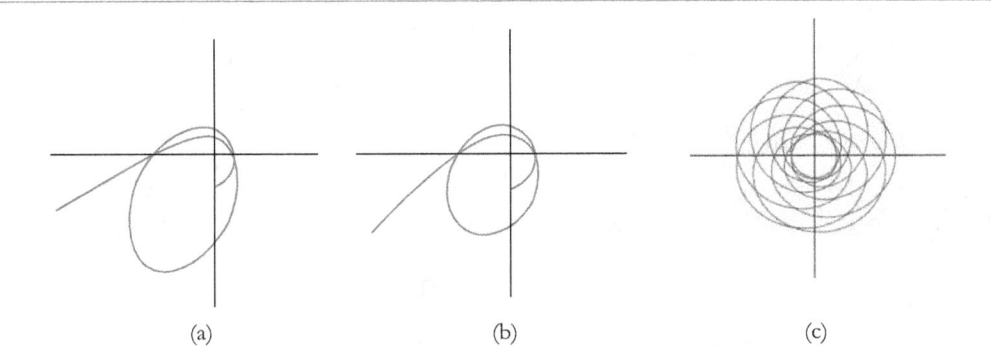

Figure 1.2.6-2. The development of elliptic orbit with gravitational factor $r/r_c = 20$ and eccentricity $e = 0.5$ near a black hole. Like in the case of Mercury, the orbiting object is thrown out of the orbit once the periastron advance reaches about 45 degrees. (a) Schwarzschild's solution by Berry, (b) Schwarzschild's solution by Weber, (c) the Dynamic Universe solution shows stable orbits down to the critical radius.

The reason for the instability can be traced to the equivalence principle that requires the buildup of relativistic mass equally in the buildup of kinetic energy at constant gravitational potential and in free fall in a gravitational field. In the SR/GR framework, relativistic mass is a property of velocity and the associated coordinate transformations, whereas in the DU framework, kinetic energy of free fall is obtained against reduction of the local rest energy and the velocity of light due to tilting of space, which means that the total energy of the object is not changed, Figure 1.2.6-3.

$$E_{m(tot)} = c_0 mc_1 - c_0 m\Delta c + E_{kin} = c_0 mc_1 \quad ; \quad E_{kin} = c_0 m\Delta c \ . \tag{1.2.6:1}$$

The buildup of kinetic energy at constant gravitational potential requires the insertion of extra energy, like Coulombian energy

$$\Delta E_{EM} = \frac{q_1 q_2 \mu_0}{2\pi r} c_0 c = c_0 \Delta m_{EM} c \ , \tag{1.2.6:2}$$

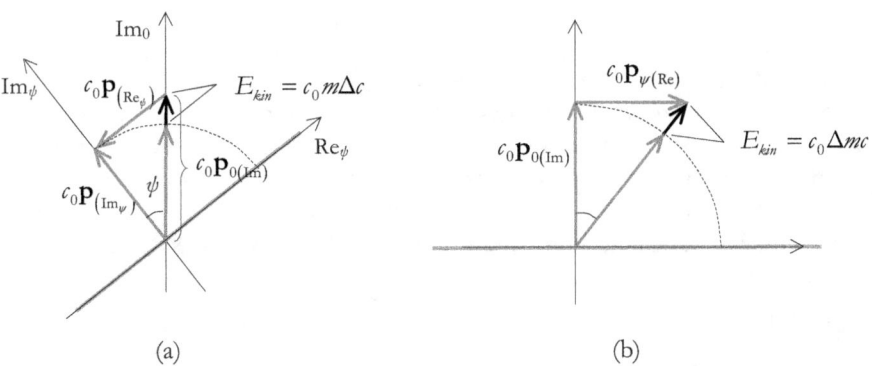

Figure 1.2.6-3. (a) Kinetic energy in free fall is obtained against a reduction of the local rest momentum via tilting of space. (b) At constant gravitational potential kinetic energy is obtained by insertion of excess mass.

which in (1.2.6:2) is written in terms of the mass equivalence Δm_{EM} of Coulomb energy. The total energy of a charged object accelerated in a Coulomb field receives the energy $c_0 \Delta m_{EM} c = c_0 \Delta mc$, resulting in the total energy

$$E_{m(tot)} = E_{rest} + E_{kin} = c_0 (m + \Delta m) c \qquad ; \qquad E_{kin} = c_0 \Delta mc , \qquad (1.2.6:3)$$

which shows the buildup of kinetic energy as the increase of mass, Figure 1.2.6-3(b). In the DU framework, combining the effects of inertial and gravitational acceleration, the kinetic energy is expressed as

$$E_{kin} = c_0 |\Delta \mathbf{p}| = c_0 (|m\Delta c| + |c\Delta m|), \qquad (1.2.6:4)$$

where the first term means kinetic energy obtained in free fall in a gravitational field, and the second term is kinetic energy via the insertion of mass.

The difference in the buildup of kinetic energy in the DU and SR/GR frameworks results in a difference in the predictions for free fall/escape velocity and the orbital velocity, which becomes dramatic in the vicinity of black holes.

Black hole, critical radius

At a low gravitational field, far from a mass center, the velocities of free fall as well as the orbital velocities in Schwarzschild space and in DU space are essentially the same as the corresponding Newtonian velocities. Close to the critical radius, however, the differences become meaningful.

In Schwarzschild space, the critical radius is the radius where Newtonian free fall from infinity achieves the velocity of light

$$r_{c(Schwd)} = \frac{2GM}{c^2} . \qquad (1.2.6:5)$$

The critical radius in DU space is

$$r_{c(DU)} = \frac{GM}{c_0 c_{0\delta}} \approx \frac{GM}{c^2} , \qquad (1.2.6:6)$$

which is half of the critical radius in Schwarzschild space. The two different velocities c_0 and $c_{0\delta}$ in (1.2.6:6) are the velocity of light in hypothetical homogeneous space and the velocity of light in the apparent homogeneous space of the local frame, respectively.

In Schwarzschild space, the predicted orbital velocity at a circular orbit exceeds the velocity of free fall when r is smaller than three times the Schwarzschild critical radius, which makes stable orbits impossible. In DU space, the orbital velocity decreases to zero at $r = r_{c(DU)}$, which means that there are stable slow velocity orbits close to the critical radius, Fig. 1.2.6-4.

The importance of the slow orbits near the critical radius is that they maintain the mass of the black hole.

The instability of orbits in Schwarzschild space can be traced back to the effect of the equivalence principle behind the field equations in general relativity, which assumes the buildup of relativistic mass in free fall in a gravitational field.

Introduction

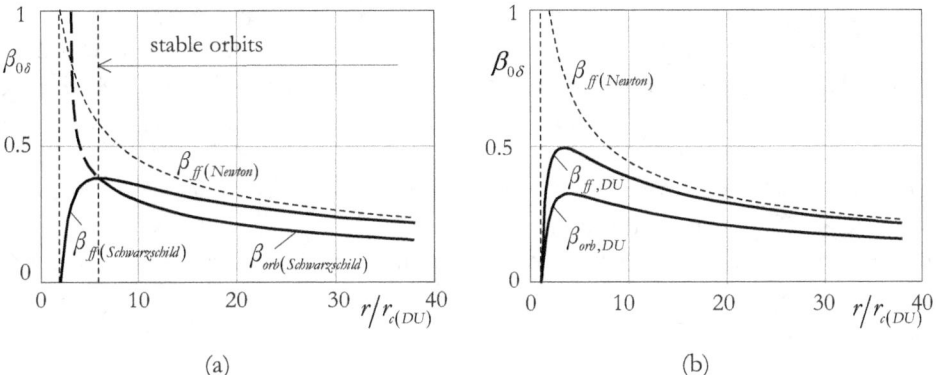

Figure 1.2.6-4. a) The velocity of free fall and the orbital velocity at circular orbits in Schwarzschild space. b) The velocity of free fall and the orbital velocity at circular orbits in DU space. The velocity of free fall in Newtonian space is given as a reference. In DU space, the slow orbits close to the critical radius maintain the mass of the black hole.

According to the DU analysis, there is no source of mass to result in an increase of mass in free fall in a gravitational field – the velocity and momentum of free fall are obtained against a reduction of the local velocity of light and rest momentum. Due to the decreasing orbital velocity close to the critical radius in DU space, the orbital period has a minimum at $r = 2 \cdot r_{c(DU)}$. In Schwarzschild space, the prediction for the orbital period at circular orbits applies only for radii $r > 3 \cdot r_{c(Schwd)}$.

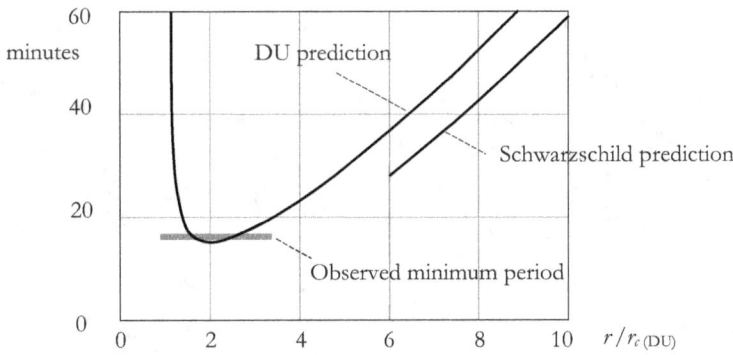

Figure. 1.2.6-5. The predictions in Schwarzschild space and DU space for the period (in minutes) of circular orbits around Sgr A* in the center of Milky Way. The shortest observed period is 16.8 ± 2 min [42] which is close to the minimum period of 14.8 minutes predicted by the DU. The minimum period predicted for orbits for a Schwarzschild black hole is about 28 minutes, which occurs at $r = 3 \cdot r_{c(Schwd)} = 6 \cdot r_{c(DU)}$. A suggested explanation for the observed "too fast" periods is a rotating black hole referred to as the Kerr black hole.

The black hole at the center of the Milky Way, the compact radio source Sgr A*, has an estimated mass of about 3.6 million times the solar mass, which means $M_{black\,hole} \approx 7.2 \cdot 10^{36}$ kg, which in turn means a period of 28 minutes as the minimum for stable orbits in Schwarzschild space. The shortest observed period at Sgr A* is 16.8 ± 2 min, which is very close to the prediction for the minimum period 14.8 min in DU space at $r = 2 \cdot r_{c(DU)}$, Figure 1.2.6-5 [42].

In DU space, the velocity of free fall reaches the local velocity of light when the tilting angle of space is 45°, which occurs at $r_{0\delta} \approx 3.414 \cdot r_c$. We may assume that such a condition is favorable for the matter-to-radiation and elementary particle conversions.

Orbital decay

Gravitational radiation has gained substantial attention after the reported observations of gravitational waves by the LIGO and Virgo Collaboration in 2016. In the GR framework, gravitational radiation is emitted by a changing quadrupole moment of an orbiting system. The energy released from the system results in a decrease in the orbital period.

The GR prediction for the orbital decay of a binary system is [43,65,66]

$$\frac{dP}{dt}_{(GR)} \approx 123 \cdot \frac{G^{5/3}}{c^5} \left(\frac{P}{2\pi}\right)^{-5/3} \cdot \left(\frac{1+(73/24)e^2 + (37/96)e^4}{(1-e^2)^{7/2}}\right) \cdot \frac{m_p m_c}{(m_p + m_c)^2} (m_p + m_c)^{5/3},$$

(1.2.6:7)

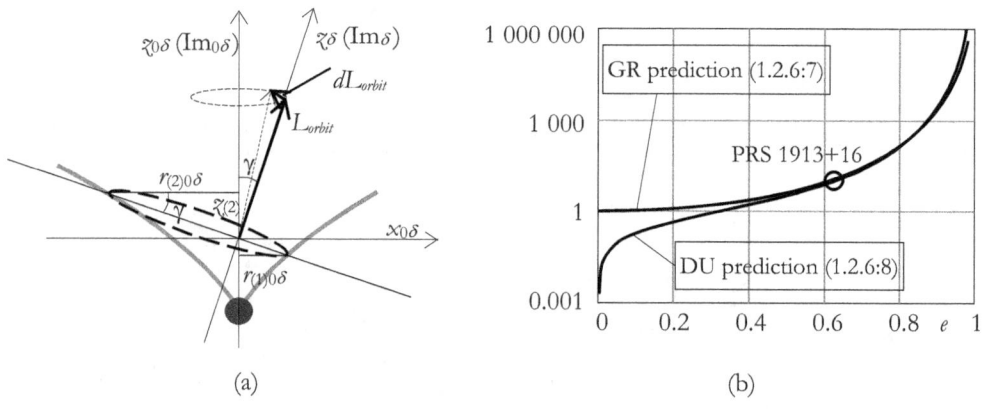

(a)   (b)

Figure 1.2.6-6.(a) The 4D angular momentum L_{orbit} of an eccentric orbit, in the direction of the Im$_\delta$ axis of the orbital plane, rotates with the periastron advance of the obit. The energy released in the decay of the orbital period is assumed to be the energy needed for the rotation of the angular momentum (see Section 4.2.9). b) The eccentricity factor of the decay of the binary star orbit period. At the eccentricity $e = 0.616$ of the PSR 1913+16 orbit, the eccentricity factor in the GR and DU prediction is essentially the same and leads to the same prediction for the decay.

Introduction

In the DU framework, the decay of the period of an elliptic orbit can be seen as a consequence of the periastron rotation and the related rotation of the orbital angular momentum in the fourth dimension, Figure 1.2.6-6(a). Interestingly, the prediction derived from the rotation of the 4D orbital angular momentum (see Section 4.2.9) gives essentially the same prediction as the GR prediction (1.2.6:7)

$$\frac{dP}{dt}_{(DU)} \approx 120 \cdot \frac{G^{5/3}}{c^5} \left(\frac{P}{2\pi}\right)^{-5/3} \left(2 \cdot \frac{\left[\sqrt{1+e_{0\delta}} - \sqrt{1-e_{0\delta}}\right]}{\left(1-e^2\right)^2}\right) \cdot \frac{m_p m_c}{\left(m_p + m_c\right)^2} \left(m_p + m_c\right)^{5/3}. \quad (1.2.6:8)$$

The only difference between the two predictions comes from the eccentricity factor (the factor in parentheses in front of the mass term at the end). In the GR prediction, the eccentricity factor is presented as a serial approximation; in the DU prediction, in algebraic form. As illustrated in Figure 1.2.6-6(b), the eccentricity factor in the DU prediction goes to zero at zero eccentricity but saturates to the value 1 in the GR prediction. At the eccentricity $e = 0.616$ of the PSR 1913+16 orbit, the eccentricity factor in the GR and DU prediction is essentially the same and leads to the same prediction for the decay, Figure 1.2.6-7. The possible energy radiation by the rotating 4D angular momentum in the DU has not been analyzed.

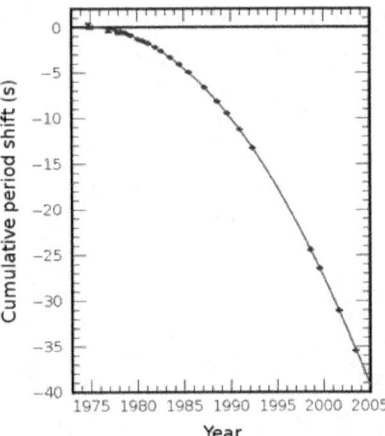

Figure 1.2.6-7. The observed orbital decay (dots) of PSR B1913+16 binary pulsar. Predictions (1.2.6:7) and (1.2.6:8) fit equally the observed decay (solid curve). Picture: *Wikimedia Commons*.

Shapiro delay

In the DU framework, the propagation of light near mass centers is affected both by the reduced velocity of light and the lengthening of the propagation path due to the geometry of the fourth dimension. The resulting delay in the propagation time is referred to as the Shapiro delay. In the GR framework, Shapiro delay is explained in terms of delayed time and lengthened space distance in tilted space-time. The comparison of GR and predictions in Section 5.4.2 shows that the GR prediction can be expressed as the sum of two identical terms, one due to the lengthening of the propagation path and the other due to the delay of time

$$\Delta T_{A,B} = \frac{GM}{c^3} \ln\left[\frac{x_B + r_B}{x_A + r_A}\right]_{\Delta path} + \frac{GM}{c^3} \ln\left[\frac{x_B + r_B}{x_A + r_A}\right]_{\Delta time}. \qquad (1.2.6:9)$$

The corresponding DU equation is

$$\Delta T_{A,B(path)} = \frac{GM}{c_{0\delta}^3}\left\{\ln\left[\frac{x_B + r_B}{x_A + r_A}\right] - \left[\frac{x_B}{r_B} - \frac{x_A}{r_A}\right]\right\}_{\Delta path} + \frac{GM}{c_{0\delta}^3}\ln\left[\frac{x_B + r_B}{x_A + r_A}\right]_{\Delta c}, \qquad (1.2.6:10)$$

which shows that the DU term Δ*time* is equal to the Δ*c* term in the GR prediction, but the Δ*path* terms differ by a subtracting factor in the DU prediction. The subtracting factor in the DU prediction is due to the fact that tilting of space does not lengthen the tangential component of the propagation path, but only the radial component pointing to or from the mass center.

The difference between the GR prediction and DU prediction disappears when there is no tangential component in the propagation path, i.e., propagation occurs in the radial direction — in a direction towards or from the local mass center. Such a situation is due, e.g., in experiments, where the Shapiro delay is measured with a two-way radar signal to a celestial object.

A pioneering test of the effect of the curvature of space in the solar system was the measurement of the delay of the radio signals to and from the Mariner 6 and Mariner 7 spacecraft on their way to Mars in 1970 [44]. The experiments were performed when the spacecrafts were far behind the Sun relative to the Earth, Figure 1.2.6-8. The path of the radio signal passes the Sun at a distance which is small compared the distances from the Earth and planet Mars to the Sun. In such a case, the GR prediction of Shapiro delay (1.2.5:5) reduces to the form

$$\Delta T_{D1,D2} = \frac{2GM}{c^3} \ln\left[\frac{4D_A D_B}{d^2}\right], \qquad (1.2.6:11)$$

and the DU prediction (1.2.5:6) into form

$$\Delta T_{D1,D2} = \frac{2GM}{c^3} \left\{\ln\left[\frac{4D_A D_B}{d^2}\right] - 1\right\}. \qquad (1.2.6:12)$$

In the Mariner experiments, the Shapiro delay at different conjunction distances was compared to the theoretical delay at exact conjunction when the signal passes the Sun at a distance equal to the radius of the Sun, $d = r_{Sun}$.

Such a relative study eliminates the effect of the constant term 1 in the DU prediction and makes predictions of GR and DU identical. The Shapiro delay of the two-way signal to and from the spacecraft at 14.2 light minutes distance from the Sun to the Earth station at 8.3 light minutes distance from the Sun is 250 μs and 230 μs, according to the GR and DU predictions, respectively. The effect of the actual conjunction distances $3.5 \cdot r_{Sun}$ and $5.9 \cdot r_{Sun}$, of Mariner 6 and Mariner 7 spacecrafts, respectively, was 50 μs and 70 μs shorter than the calculated delay at exact conjunction – both according to the GR and DU predictions, Figure 1.2.6-9.

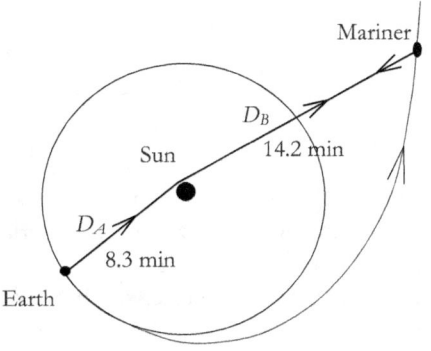

Figure 1.2.6-8. Measurement of the delay in a radio signal due to the gravitation of the Sun. The delay results from two factors: the lengthening of the path due to the topography of space in the fourth dimension and the reduction of the velocity of light near the Sun.

Deflection of light, gravitational lens

The reduced propagation velocity and increased propagation distance of light and radio signals close to a mass center mean not only an increase in the propagation time but also a bending of the propagation path. The prediction for the bending of the light path in the DU framework is the same as it is in the GR framework, $\psi=4GM/c^2d$, where ψ is the bending angle, and M is the mass center the light beam passes at distance d.

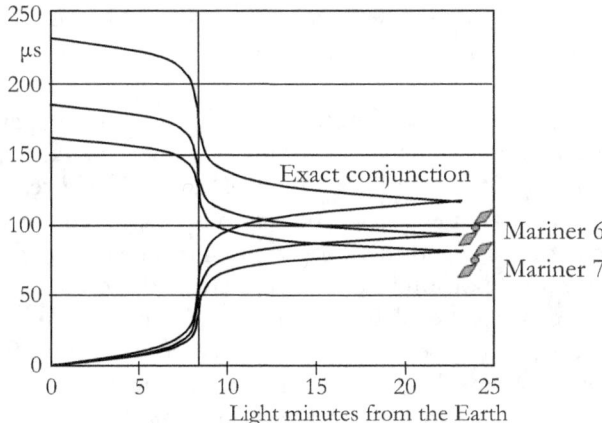

Figure 1.2.6-9. The time delay of a two-way radio signal from Earth to Sun and from Sun to destination, and in reverse direction back to Earth. The curves are based on equation (1.2.6:12).

1.3 Cosmological considerations

1.3.1 The linkage of local to the whole

Space as the 3D surface of a 4D sphere is a basic assumption in the Dynamic Universe. It is a philosophical choice as a natural geometry for closing the 3D space. Spherically closed space allows a precise solution to the dynamics of space – the contraction-expansion process that explains the energy buildup in space and produces precise, mathematically simple predictions to key cosmological observables – as well as an unbroken link between any local phenomenon and space as a whole. Energizing of space does not need new physics or assumptions of unknown quantum fluctuations; mass in space gets its rest energy as the energy of motion gained against the release of gravitational energy in the contraction phase preceding the ongoing expansion phase, releasing the energy of motion back to the energy of gravitation.

The Dynamic Universe means a major change in the cosmological picture of the universe. The overall balance of the energies of gravitation and motion means that all gravitationally bound structures, like galaxies and planetary systems, expand in direct proportion to the expansion of the whole space as the 3D surface of a 4D sphere. For example, 2.8 cm of the 3.8 cm annual increase of the Earth to Moon distance comes from the expansion of space and only 1 cm from the tidal interactions.

The velocity of light is determined by the velocity of expansion in the fourth dimension. The expansion velocity slows down gradually due to the work done against the gravitational energy of the system.

Due to the faster expansion velocity in the past, the age of the expanding space corresponding to the present 14 billion light-years estimate of the 4-radius (equal to the Hubble radius, $R_H = c/H$) is about $2/3*14 \approx 9.3$ billion years. The length of a year increases in direct proportion to the age of expanding space. The length of a day increases in inverse proportion to the 4D velocity of space and the velocity of light. The ticking frequency of atomic clocks decreases in direct proportion to the 4D velocity and the velocity of light.

All the above linkages are in excellent agreement and necessary for a prediction matching the observations on the number of days in a year found in coral fossils and the data calculated from observed solar eclipses for the last 3000 years.

For 4 billion years ago, Earth and Mars were about 30% closer to the Sun than they are today, which gives a natural explanation for the geological development on Earth and the presence of liquid water on Mars, thus providing, e.g., a natural explanation for the "faint Sun paradox [45]".

In DU space, predictions derived for cosmological observables obtain precise mathematical forms due to the exact geometry and dynamics of space. Unlike GR-based cosmology, DU cosmology also produces predictions for the past and future development of the expansion, which allows parameter-free expressions for the physical and optical distances as well as for the redshift and power dilution of light observed from cosmological objects.

Introduction

1.3.2 Distances in FLRW cosmology

The Friedmann-Lemaître-Robertson-Walker (FLRW) cosmology was developed in the 1930s. The 1930s' cosmologists met an immense challenge in putting together the results from special relativity, Friedmann's and Lemaître's solutions of the field equations of general relativity, the Planck equation, the implications of quantum mechanics, and the redshift observations reported by Edwin Hubble. Necessary components for key cosmological predictions were the distance-redshift relation, the dilution of the power density of radiation, and the development of the observed angular size of objects in expanding space. In GR space, not only do material objects from atoms to stars conserve their dimensions in expanding space, but also local systems like galaxies and planetary systems conserve their dimensions as first concluded by de Sitter [46]. The expansion of GR space is considered to occur as "Hubble flow" between galaxies or galaxy groups.

As first proposed by Tolman [47] and later concluded by Hubble and Humason [48,49], de Sitter[50], and Robertson[51], the energy of a quantum is reduced by $(1+z)$ as a consequence of the effect of Planck's equation $E = hf$ as an "energy effect", a reduction of the "intensity of the radiation" due to reduced frequency. When receiving the redshifted radiation at a lowered frequency, a second $(1+z)$ factor was assumed as a "number effect". Hubble considered that the latter is relevant only in the case that the redshift is due to recession velocity [52]. Tolman was careful and added the reservation "evidently" in his conclusions of the distance definition and the effect of Planck's equation in the derivation of the prediction for the observed luminosity of redshifted objects, Figure 1.3.2-1. The double dilution $(1+z)^2$ due to redshift has stayed in the FLRW cosmology since 1930's [53].

Interpretation of Planck's equation as an intrinsic property of radiation means that radiation propagating in expanding space loses energy due to the expansion of wavelength and the associated decrease of frequency. As discussed in Section 5.1, the Planck equation should be understood as the energy injected into one cycle of radiation at the emission; expanding the wavelength does not change the energy carried by a cycle of radiation, but it dilutes the energy received in a unit time. The dilution is observed as the Doppler effect, which means that expansion is associated with one $(1+z)$ dilution factor only.

A primary distance definition in FLRW cosmology is the comoving distance, D_C, which means the physical distance of cosmological objects at the time of the observation. The comoving distance is obtained from Friedmann's solutions of the GR field equations (6.2.1:2)

$$D_C = R_H \int_0^z \frac{1}{\sqrt{(1+z)^2(1+\Omega_m z) - z(2+z)\Omega_\Lambda}} dz, \qquad (1.3.2:1)$$

where z is the redshift, and R_H is the Hubble radius $R_H = c/H_0$ (H_0 is the Hubble constant). Due to the reciprocity demand [76] originating from special relativity, the angle at which an object is seen in GR space is the same as it was when the light from the object was emitted. Accordingly, the angular diameter distance D_A is related to the comoving distance as

$$D_A = \frac{D_C}{1+z}. \qquad (1.3.2:2)$$

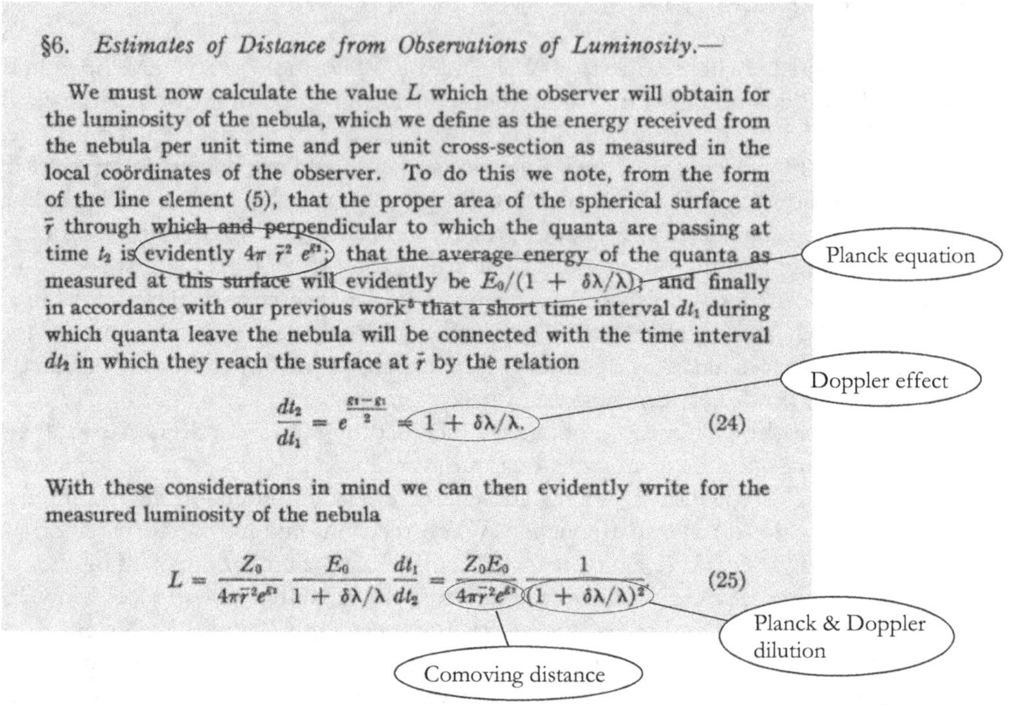

§6. *Estimates of Distance from Observations of Luminosity.*—

We must now calculate the value L which the observer will obtain for the luminosity of the nebula, which we define as the energy received from the nebula per unit time and per unit cross-section as measured in the local coördinates of the observer. To do this we note, from the form of the line element (5), that the proper area of the spherical surface at $\bar{r}$ through which and perpendicular to which the quanta are passing at time t_2 is evidently $4\pi\,\bar{r}^2\,e^{g_1}$; that the average energy of the quanta as measured at this surface will evidently be $E_0/(1+\delta\lambda/\lambda)$; and finally in accordance with our previous work* that a short time interval dt_1 during which quanta leave the nebula will be connected with the time interval dt_2 in which they reach the surface at $\bar{r}$ by the relation

$$\frac{dt_2}{dt_1} = e^{\frac{g_1-g_1}{2}} = 1 + \delta\lambda/\lambda. \qquad (24)$$

(Planck equation — on $E_0/(1+\delta\lambda/\lambda)$)
(Doppler effect — on $1+\delta\lambda/\lambda$)

With these considerations in mind we can then evidently write for the measured luminosity of the nebula

$$L = \frac{Z_0}{4\pi\bar{r}^2 e^{g_1}}\,\frac{E_0}{1+\delta\lambda/\lambda}\,\frac{dt_1}{dt_2} = \frac{Z_0 E_0}{4\pi\bar{r}^2 e^{g_1}}\,\frac{1}{(1+\delta\lambda/\lambda)^2} \qquad (25)$$

(Comoving distance — on $4\pi\bar{r}^2 e^{g_1}$)
(Planck & Doppler dilution — on $(1+\delta\lambda/\lambda)^2$)

Figure 1.3.2-1. Estimates of the luminosity in Tolman[58]. Tolman states that the distance to be used is "evidently" the comoving distance, which is the physical distance at the time the observation. Also, he states that "evidently" the Planck equation results in an energy loss reducing the luminosity observed as $L = L_0/(1+z)$, which together with the Doppler effect results in $L = L_0/(1+z)^2$ attenuation.

As is obvious from (1.3.2:2), the angular size distance increases with the comoving distance at redshifts $z < 1$ but turns into shortening when $z > 1$ and shortens in inverse proportion to z at high redshift $z \gg 1$.

The light-travel distance D_{LT} is expressed

$$D_{LT} = R_H \int_0^z \frac{dz}{(1+z)\sqrt{(1+z)^2(1+\Omega_m z) - z(2+z)\Omega_\Lambda}}. \qquad (1.3.2:3)$$

Luminosity distance, D_L, is the apparent distance corresponding to the luminosity observed from a similar object in static Euclidean space, where luminosity decreases in inverse proportion to the square of the distance. In FLRW cosmology, the observed power density is subject to areal dilution in inverse proportion to the comoving distance squared and dilution due to the expansion of space by the factor $(1+z)^2$ related to the Planck equation and Doppler effect. The luminosity distance is expressed as

$$D_L = (1+z)D_C = (1+z)^2 D_A, \qquad (1.3.2:4)$$

which allows the classical expression of the power density of the radiation flux

$$F_{classical} = \frac{L}{A} = \frac{L}{2\pi D_L^2} \qquad \left[\frac{W}{m^2}\right]. \tag{1.3.2:5}$$

The observed power density is compared to the *K-corrected* radiation flux, which, in addition to correction of instrumental factors, cancels the reciprocity factor by adding an attenuation factor $(1+z)^2$ to the power densities observed in bolometric multi-bandpass photometry (see Section 6.3.3).

Light travel distance or the "lookback" distance, D_{LT}, is the distance light has propagated from the object. In principle, for the redshift approaching infinity, the lookback distance should give the distance from the Big Bang, Figure 1.3.2-2.

As shown in Figure 1.3.2-2, differences in the predictions for the defined distances appear when redshift z approaches 1 and become remarkable at high redshifts. It is concluded that the sum of the density parameters equals one ($\Omega = \Omega_m + \Omega_\Lambda = 1$), which means the "flat space" condition. As the present best estimate, the portion of the visible and dark matter is $\Omega_m = 0.27$ and the portion of the unknown "dark energy" $\Omega_\Lambda = 0.73$. The hypothesis of dark energy comes primarily from recent observations of the magnitude and redshift of supernova explosions. The present estimate for the Hubble constant is about 70 [km/s/Mpc], which corresponds to about 13.7 billion years of age of the expanding *FLRW* universe. Case (b) in Figure 1.3.2-2 corresponds to presently assumed values of the density parameters, which makes the lookback distance D_{LT} approach R_H at high redshifts, $D_{LT} = c/T = R_H$.

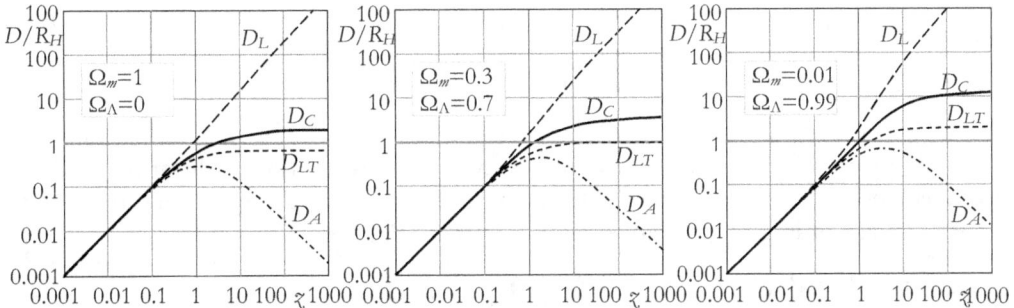

Figure 1.3.2-2. Central definitions of distances in FLRW-space for three combinations of mass density and dark energy. Each case assumes the "flat space" condition which means that the sum of the two density parameters is equal to one, $\Omega_m + \Omega_\Lambda = 1$. The *Co-moving distance* D_C, which means the physical distance at the time of observation, is obtained from Friedmann's solution to the field equations of the general theory of relativity. The *Light-travel distance*, D_{LT} is the length of the light path from the object to the observer in the expanding space. In principle, the *Light-travel distance* approaches the Hubble radius R_H at very high redshifts that occurs in case (b), which has the currently preferred values of the density parameters, $\Omega_m = 0.3$ and $\Omega_\Lambda = 0.7$. The *Angular size distance* D_A is obtained by dividing the *Co-moving distance* D_C with the expansion factor $(1+z)$, which means the distance of the object at the instant the light is left from the object. The *Luminosity distance* D_L is obtained by multiplying the *Co-moving distance* D_C by the expansion factor $(1+z)$, which gives the effect of the increased wavelength on the dilution of the power density. The power density of radiation is proportional to the square of the inverse of the *Luminosity distance* D_L.

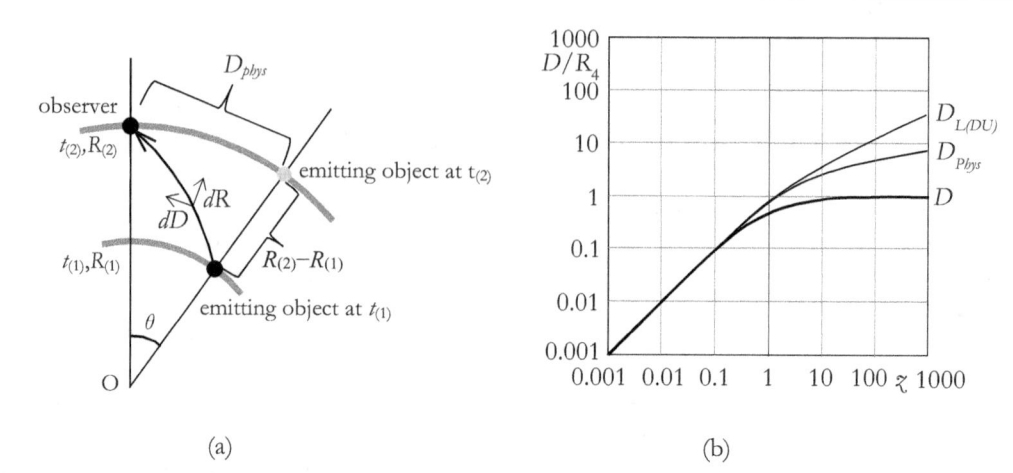

Figure 1.3.3-1. (a) The physical distance D_{phys} is the arc D_{phys}, at the time of the observation. In expanding space, the velocity of light in space (the tangential component of the propagation path) is equal to the velocity of space in the direction of the 4-radius, which means that the optical distance D is equal to the increase of the 4-radius during the propagation of light from the object to the observer, $D = \int_{t(1)}^{t(2)} dD = R_{(2)} - R_{(1)} = R_0 \, z/(1+z)$

(b) The physical distance D_{phys} correspond to the comoving distance in FLRW cosmology. Optical distance D corresponds closest to the light travel distance D_{LT} in FLRW cosmology.

1.3.3 Distances in DU space

In DU space, the physical distance, corresponding to the comoving distance (1.3.2:1) in FLRW cosmology, can be expressed in terms of the separation angle θ seen from the 4-center of space, or in terms of the redshift, Figure 1.3.3-1

$$D_{phys} = \theta \cdot R_4 = R_4 \left(1 - e^{-\theta}\right) = R_H \ln(1+z). \tag{1.3.3:1}$$

The optical distance, referred to as D, is the tangential component of the propagation path from the object to the observer, i.e., the distance light has traveled in the expanding space

$$D = R_4 \, z/(1+z). \tag{1.3.3:2}$$

Luminosity distance is obtained by adding the effect of Doppler dilution to the optical distance

$$D_{L(DU)} = R_4 \, z\sqrt{1+z}/(1+z) = R_4 \, z/\sqrt{1+z}. \tag{1.3.3:3}$$

In DU, the angular diameter of objects is seen at the optical distance.

Introduction

Angular size of cosmological objects

In FLRW cosmology, both stars and gravitationally bound local systems like galaxies and quasars conserve their dimensions in expanding space. In DU space, solid objects like stars conserve their dimensions, but gravitationally bound systems expand in direct proportion to the expansion of space.

For a non-expanding object with a fixed diameter, d_s, the observed angular diameter in DU space is

$$\frac{\psi_{r(s)}}{d_s/R_0} = \frac{z+1}{z}, \qquad (1.3.3:4)$$

and for expanding objects, like galaxies and quasars, with diameter $d = d_0/(1+z)$

$$\frac{\psi_{r(s)}}{d_s/R_4} = \frac{\psi}{\theta_d} = \frac{1}{z}, \qquad (1.3.3:5)$$

where θ_d is the angular diameter of the object, as seen from the 4-center of space. Equation (1.3.3:5) means a Euclidean appearance of galactic objects. In FLRW space, due to the decreasing angular distance at high redshifts, the observed angular size is predicted to turn into an increase at high redshifts[54], Figure 1.3.3-2(a). As illustrated in Figure 1.3.3-2(b), the DU Euclidean prediction (1.3.3:5) for the angular size of galaxies and quasars is in excellent agreement with observations.

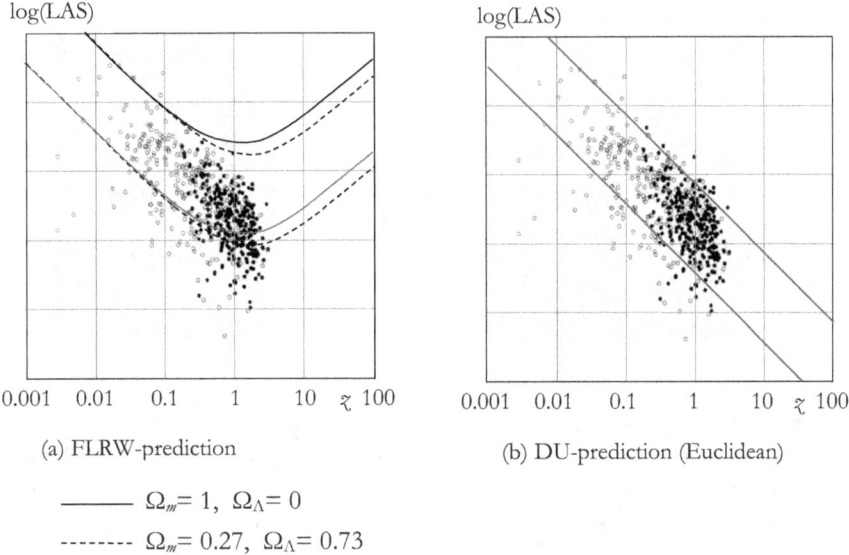

(a) FLRW-prediction   (b) DU-prediction (Euclidean)

——— $\Omega_m = 1$, $\Omega_\Lambda = 0$

- - - - - - $\Omega_m = 0.27$, $\Omega_\Lambda = 0.73$

Figure 1.3.3-2. Dataset of the observed Largest Angular Size (LAS) of quasars and galaxies in the redshift range $0.001 < z < 3$. Open circles are galaxies; filled circles are quasars. In (a) observations are compared to the FLRW prediction (6.2.1:2) with $\Omega_m = 0$ and $\Omega_\Lambda = 0$ (solid curves), and $\Omega_m = 0.27$ and $\Omega_\Lambda = 0.73$ (dashed curves). In (b) observations are compared to the DU prediction (6.2.3:2).

The magnitude of the standard candle

In cosmological observations, absolute magnitude, M, is defined as the logarithm of its luminosity as seen from a distance of 10 parsecs. The magnitude measured by the observer is referred to as the apparent magnitude, m, which is related to the absolute magnitude as

$$m = M + 5\log\frac{R_H}{10\,\text{pc}} + 5 \cdot \log_{10} D_L, \qquad (1.3.3{:}6)$$

where D_L is the luminosity distance of the object. Apparent magnitude grows in inverse proportion to the brightness; the fainter the object observed, the higher the apparent magnitude.

For comparing objects at different distances, one should assume that the absolute magnitudes of the objects are identical. Observations indicate that Type Ia supernovae serve as standard candles when corrected by the shape of the light curve and can be used to test the predictions for the apparent magnitude. The dilution of the power density of radiation from the objects results from areal spreading proportional to the distance squared, and from the dilution of power density due to redshift.

In FLRW cosmology, the spreading distance used in the luminosity distance D_L (1.3.2:4) is the comoving distance D_C. The power dilution affecting the luminosity distance is $(1+z)$, comprising the effects of the Planck equation and Doppler effect.

In DU cosmology, the spreading distance is the optical distance D (1.3.3:2). The power dilution $\sqrt{1+z}$ affecting the luminosity distance (corresponding to the factor $(1+z)$ affecting the luminosity) is due to the Doppler effect.

In FLRW cosmology, the prediction for apparent magnitude is applied to observations corrected to "emitter's rest frame", which means a $5 \cdot \log(1+z)$ increase to the observed bolometric magnitudes (Sections 6.3.3 and 6.3.4)

$$m = M + 5\log\frac{R_H}{10\,\text{pc}} + 5\log\left[(1+z)\int_0^z \frac{1}{\sqrt{(1+z)^2(1+\Omega_m z) - z(2+z)\Omega_\Lambda}}\,dz\right]. \qquad (1.3.3{:}7)$$

In DU cosmology, the apparent magnitude for bolometric observations is

$$m_{DU(bolometric)} = M + 5\log\left(\frac{R_4}{10\,\text{pc}}\right) + 5\log(z) - 2.5\log(1+z). \qquad (1.3.3{:}8)$$

A major difference in the predictions is that the FLRW prediction applies to magnitudes corrected with the K-correction, including instrumental factors + conversion to "emitter's rest frame", which brings a $5 \cdot log(1+z)$ addition to the observed bolometric magnitudes, whereas the DU prediction applies to the observed bolometric magnitudes corrected with the instrumental factors only.

The conversion of the bolometric magnitudes to the emitter's rest frame in FLRW cosmology is related to the reciprocity demand with its origin in the principle of relativity. The effect of the K-correction is analyzed in detail in Section 6.3.4.

In a comparison with the DU prediction based on the DU optical distance and the omission of the "Planck dilution", the extra $5 \cdot \log(1+z)$ term in the FLRW magnitude prediction comes from the areal dimming due to the comoving distance and the "Planck dilution", each resulting in an extra (1+z) dimming factor.

The DU prediction (1.3.3:8) equal to (6.3.3:10) shows an excellent match to direct bolometric observations of Ia supernovae as illustrated in Figure 6.3.3-3 in Section 6.3.3. In Figure 6.3.3-3 (a) and (b), the DU equation (6.3.3:10) has been applied to produce predictions for the magnitudes observed in each filter for hypothetical blackbody sources at 8300 °K and 6600 °K, and compared the predictions to observed magnitudes of Ia supernovae (c)[55].

To make the DU prediction of apparent magnitude comparable to the FLRW prediction and the K-corrected magnitudes, an extra 5·log(1+z) term is added to (1.2.6:13)

$$m_{DU(K-corrected)} = M + 5\log\left(\frac{R_4}{10\text{pc}}\right) + 5\log(z) + 2.5\log(1+z). \qquad (1.3.3:9)$$

Equation (1.2.6:14) gives an excellent match to supernova Ia observations corrected with the K-correction required by the FLRW prediction, Figure 1.3.3-3 (equal to Figure 6.3.4-4). The FLRW prediction in Figure 1.3.3-3 applies Ω_m =0.31 and Ω_Λ=0.69 as the density parameters.

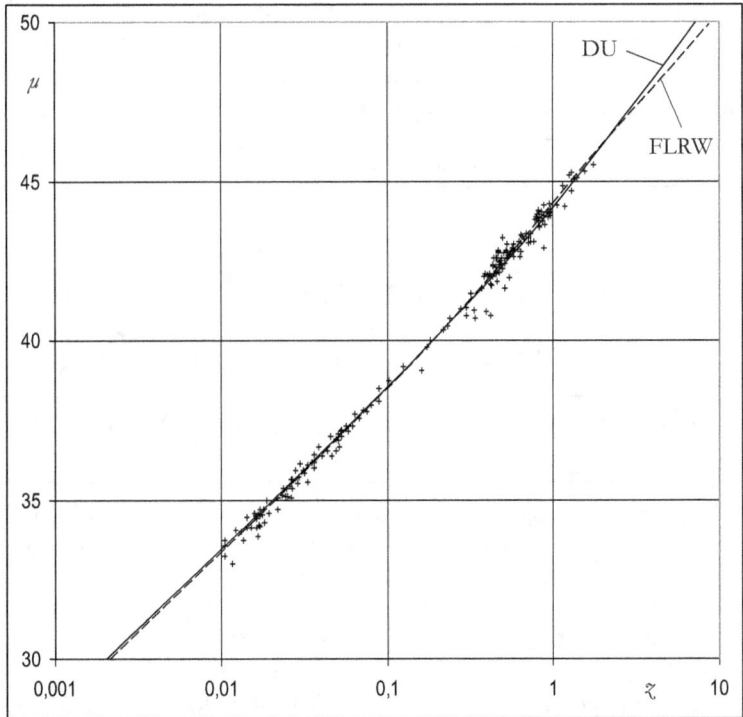

Figure 1.3.3-3. Distance modulus $\mu = m - M$, vs. redshift for Riess et al. "high-confidence" dataset and the data from the HST.

1.3.4 The length of a day and a year

A unique possibility for studying the long-term development of the Earth's rotation comes from paleoanthropological data available back to almost 1000 years in the past. Fossil layers preserve both the daily and annual variations, thus giving the number of days in a year [56,57,58]. At least partly, tidal variations can also be detected, which allows an estimate of the development of the number of days in a lunar month. Reference material from the past 2700 years is available from ancient Babylonian and Chinese eclipse observations [59,60]. The average lengthening of a day based on the eclipse observations is 1.7–1.8 ms/100y, which is about 0.6 ms/100y less than the estimated effect of tidal friction (2.3–2.5 ms/100y). The length of a day has been measured with atomic clocks since 1955. Since 1988, the length of a day has been monitored by the International Earth Rotation and Reference Systems Service (IERS). Monitoring is based on atomic clocks and Very Long Baseline Interferometry (VLBI). In the time interval 1962–2018, the long-term trend is hidden by the short-term variations [61].

According to current theories, planetary systems do not expand with the expansion of space, and atomic clocks conserve their frequencies. It means that the length of a year is assumed unchanged, and the length of a day is affected only by tidal interactions with the Moon and Sun.

In the DU framework, planetary systems expand in direct proportion to the expansion of space and the frequency of atomic clocks slows down in direct proportion to the decrease of the velocity of light. As a consequence, the length of a year, the length of a day, and the frequency of atomic clocks change with the expansion of space, Table 1.3.4-I.

	GR, FLRW cosmology	Dynamic Universe
The length of a year	Constant	Increases with the expansion $y \sim t$
The length of a day	Increases due to tidal friction	Increases due to tidal friction + increases with the expansion $d \sim t^{1/3}$
The frequency of atomic clocks	Constant	Decreases with the expansion as $f \sim t^{-1/3}$

Table 1.3.4-I. The predicted change of the length of a year, the length of a day, and the frequency of an atomic clock with the expansion of space is expressed in terms of the time from the singularity, t.

Figure 1.3.4-1 illustrates the development of the length of the year (in current days) and the number of days in a year during the last 1000 million years. The number of days given by equation (7.4.2:9) in Section (7.4.9 follows well the development of the number of days in a year counted in fossil samples since almost one billion years back. The estimate based on the tidal effect only shows too great a change, but when corrected with the lengthening of a year, a perfect match with the coral fossil data is obtained.

Experimental values, shown as squares in the figure, have been collected from papers comprising coral fossil data [56,57,58] and stromatolite data from the Bitter Springs Formation[62]

Introduction

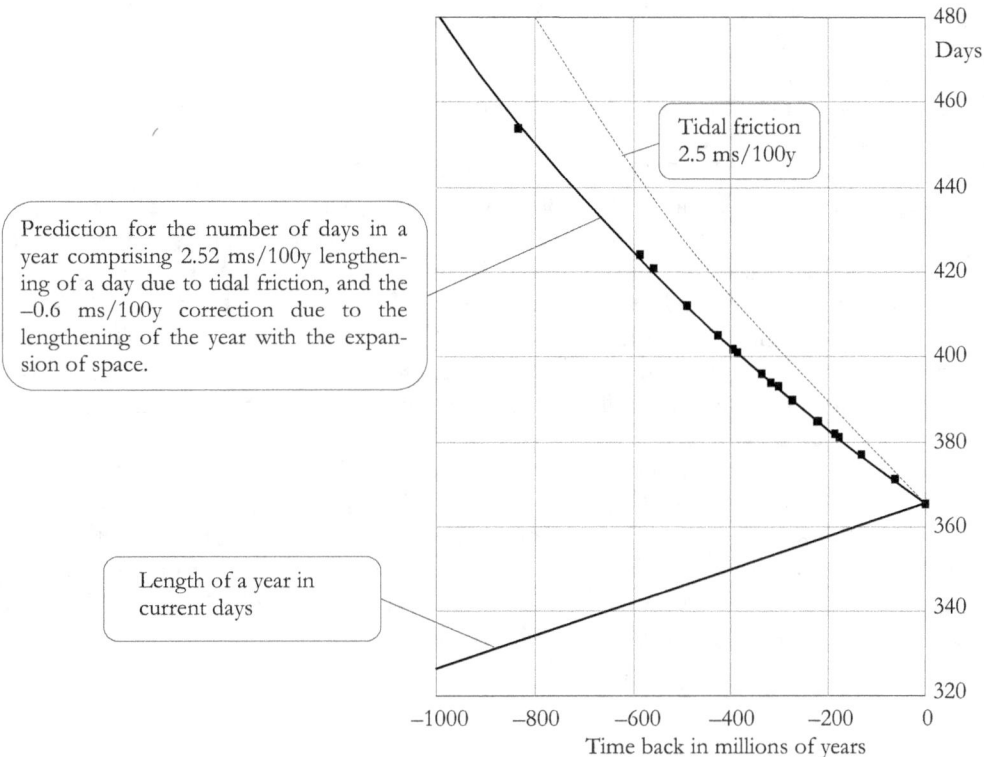

Figure 1.3.4-1. The development of the length of a year in current days, and the number of days in a year according to the DU predictions (5.6.3:5) and (5.6.3:9) respectively. The squares are observed counts of the number of days in a year in fossils [56,57,58,62]. The dashed line is the prediction based on tidal effects of the Moon and the Sun, 2.5 ms/100y. The age data of the fossils are based on radiometric dating, which has been adjusted according to equation (6.4.3:8) for the faster decay rate in the past. In the oldest data, 850 million years by linear decay, the correction is 13 million years, i.e., about 1.5 %.

(the data from the samples going back to more than 800 million years). In all data points in Figure 1.3.4-1, the DU correction in the age estimate is made according to equation (6.4.3:10).

The predicted 1.9 ms/100y lengthening of a day, taking into account both the tidal friction and the lengthening of a year, is in good agreement with the coral fossil data and the data calculated from the solar eclipse observations.

1.3.5 Timekeeping and near-space distances

SI Second and meter

In the Dynamic Universe, all processes in space are linked to space as a whole. All gravitationally bound systems expand in direct proportion to the expansion of space. All velocities in space and rates of physical processes are related to the velocity of the expansion, and consequently, to the velocity of light.

In present timekeeping, the unit of time, the **SI Second**, is based on the frequency of radiation from the transition between two hyperfine levels of the ground state of the cesium-133 atom on the Earth's geoid. The meter is defined as the length of the path traveled by light in a vacuum in 1/299 792 458 seconds (Section 5.7).

The velocity of light is a function of the gravitational potential – on the Earth's geoid at the equator, the velocity of light is higher than it is on the geoid at the poles. The frequency of the Ce clock, however, is the same at the poles and the equator due to the compensating effect of the rotational velocity. As a consequence, in absolute measures, the SI meter is longer at the equator than it is at the poles. Due to the effect of velocity on the Bohr radius, also, atoms and material objects are larger at the equator. In SI meters, the historical platinum rod standard of a meter, as well as all material objects, conserve their dimensions on the Earth's geoid at all latitudes.

Due to the eccentricity of the Earth's orbit, the length of the SI second, like the rates of all physical processes on the Earth, is subject to annual variation due to the variation of the orbital velocity and the gravitational state of the Earth in the Solar gravitational frame. At the perihelion, a clock runs slow, at aphelion fast, Figure 1.3.5-1. The effect is not observable on the Earth because the frequencies of all clocks change in parallel. Earth SI second works well as a practical standard on Earth and in near space.

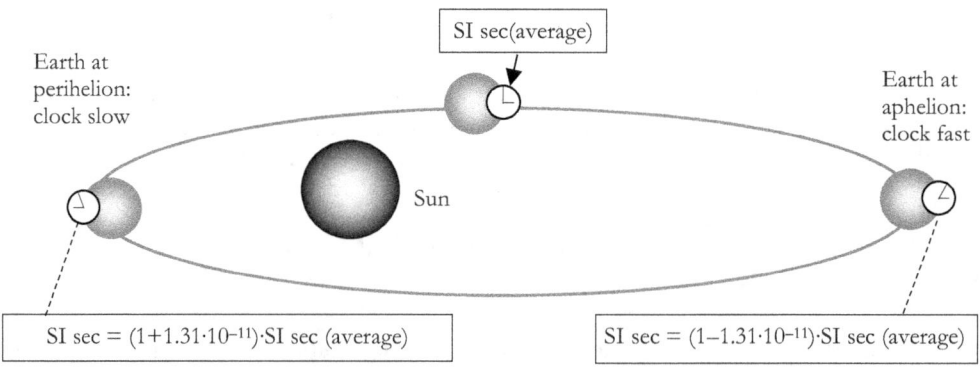

Figure 1.3.5-1. The effect of the annual changes in the orbital velocity and gravitational state of the Earth in the Solar gravitational frame on clocks and the SI second on the Earth.

Annual variation of the Earth to Moon distance

The Earth to Moon distance has been measured with high accuracy in the Laser Ranging Program [72]. The measurement is based on the two-way transmission time of a light pulse from the Earth to a reflector on the Moon and back to the Earth. Due to the eccentricity of the Earth's orbit around the Sun, the ticking frequency of the clock used in the measurement of the light propagation time, as well as the velocity of light, changes during the year. At the perihelion, the clock frequency is decreased by both the increased orbital velocity and the decreased solar gravitational potential, which also decreases the velocity of light. At the perihelion, the distance to the moon is increased. At the aphelion, the changes are opposite. Putting all the changes together, we enter into a null result; there is no observable variation in the Earth-to-moon distance due to the eccentricity of the Earth's orbit. The total variation in the clock frequency, the velocity of light, and the Earth to Moon distance between perihelion and aphelion can be listed as follows (Section 5.6.1):

Change of the Earth clock frequency	$\Delta F = \Delta f/f = 2 g_{Sun} \Delta r_{Sun}/c^2$
Change of the velocity light	$\Delta C = \Delta c/c = g_{Sun} \Delta r_{Sun}/c^2$
Change of the Earth-Moon distance	$\Delta R = \Delta r_{E\text{-}M}/r_{E\text{-}M} = -g_{Sun} \Delta r_{Sun}/c^2$
Actual change of the signal propagation time	$\Delta T = \Delta R - \Delta C = \Delta t/t = -2 g_{Sun} \Delta r_{Sun}/c$
Change in the observed signal propagation time	$\Delta N/N = \Delta F + \Delta T = 0$

In the list, g_{Sun} is the gravitational acceleration of the Sun at the Earth-Moon system. $\Delta r_{E\text{-}M}$ and $\Delta r_{S\text{-}E}$ are the differences in the Earth to Moon and the Sun to Earth distance, respectively. The actual decrease ΔT of the propagation time of the Earth-Moon-Earth signal from perihelion to aphelion is fully compensated by the corresponding decrease of the clock frequency ΔF, which means that no change is observed – which is confirmed in the Laser Ranging Program. The actual, undetected variation of the Earth's distance to the Moon due to the eccentricity of the Earth's orbit is 12.6 cm.

1.4 Summary

1.4.1 Hierarchy of physical quantities and theory structures

The postulates

Due to the empirically driven evolution in its different areas and the lack of a holistic, metaphysical basis, the development of contemporary physics has led to diversification, with specific postulates in different areas. The postulates behind relativity theory and quantum mechanics are listed in the corresponding boxes in Figure 1.4.1-1.

The main postulates in the Dynamic Universe are the spherically closed space, the zero-energy balance of motion and gravitation, and the use of time and distance as universal coordinate quantities. The DU postulates are defined at the base level, and they apply as such in all areas of physics and cosmology.

The force-based versus energy-based perspective

Figure 1.4.1-1 compares the hierarchy of some key quantities and theory structures in contemporary physics and the Dynamic Universe.

Contemporary physics, as it is today, can be seen as the result of the experimentally driven evolutionary development of our understanding of the observable physical reality. The turn from metaphysical conception to systematic scientific progress can be attributed to Isaac Newton, who, in the late 1600's, defined the concepts of mass and force and established the mathematical expressions for the primary interactions of gravitation and motion. Implicitly, Newton's equations define time and distance as coordinate quantities common to all events in space. Newton's second law can be seen as hiding an assumption of infinite Euclidean space; according to the second law, the velocity of an object increases linearly, without limits, as long as there is a constant force acting on an object. Newtonian physics is local by its nature; phenomena are studied in a local frame of reference where Newton's laws of motion and gravitation apply.

Over time, mismatches began to develop between theory and observations. The theory of relativity was needed to add effects of finiteness to the unlimited Newtonian space and to match the contradictions seen in electromagnetism between local frames in relative motion. Finiteness was introduced via modified metrics, which replaced the Newtonian universal coordinate quantities with the concept of space-time. Like Newtonian physics, relativistic physics is local by its nature. Newtonian empty space is replaced by a continuous field. Energy differences are calculated by integrating the force field.

The ultimate goal of the field concept is a unified field theory combining the four fundamental forces – strong interaction, electromagnetic interaction, weak interaction, and gravitational interaction – identified in contemporary physics.

In the DU, the hierarchy of force and energy is opposite to that in contemporary physics. Energy is a primary quantity. Force in the DU is defined as the gradient of energy, which shows a tendency toward minimum energy in an energy system.

Introduction

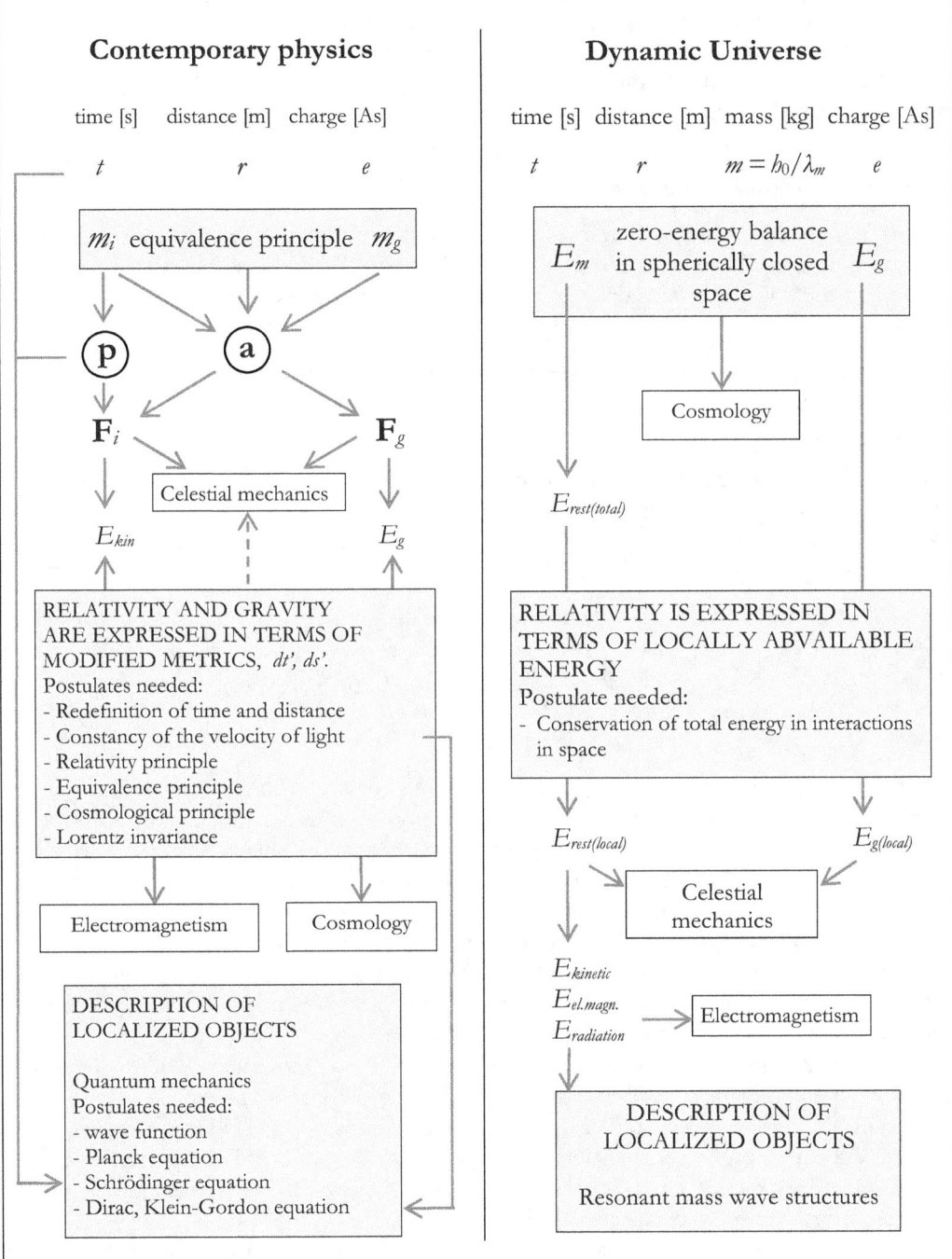

Figure 1.4.1-1. Hierarchy of some central physical quantities and theory structures in contemporary physics and Dynamic Universe.

In the GR, gravitational force is conveyed at the velocity of light by hypothetical gravitons. In the DU, gravitational force is local and immediate; it results from the tendency towards minimum potential, the actualization of the potentiality. Gravitational force is proportional to the gradient of the local gravitational potential. In the DU, the gravitational field is a scalar potential field.

The balance of gravitation and motion

In the Dynamic Universe framework, gravitational energy is understood as the potential energy energizing the contraction process, building up the rest energy as the energy of motion in the fourth dimension. The rest energy of matter is equal to the gravitational energy released.

DU space is characterized as the zero-energy continuum with the energies of motion and gravitation in balance.

The buildup of local structures within space is studied by conserving the overall zero-energy balance in space. Such an approach leads to a system of nested energy frames. Relativity appears as a consequence of the conservation of total energy in the system. The "lower" we are in the chain of energy frames, the smaller the energy available for local transactions.

Relativity in the DU is expressed in terms of locally available energy. Relativity does not need additional postulates; it is a direct consequence of the conservation of total energy in space, and an indivisible part of the overall energy balance in space. Relativity in the DU means relativity between the local and the whole. Any local state is related, via the system of nested energy frames, to the state of rest in a hypothetical homogeneous space, which serves as the universal frame of reference.

Starting from energy, instead of force, is essential for the holistic approach in the Dynamic Universe. The rest energy obtained in the contraction-expansion process serves as the source of energy in all local structures and expressions of energy. The buildup of elementary particles and mass centers in space means that a certain part of the momentum in the fourth dimension is turned toward space directions. As a consequence, the rest energy available in local structures becomes a function of the local gravitational environment and the local motion in space. The reduced rest energy reduces the rate of physical processes, e.g., the characteristic emission and absorption frequencies of atomic objects become functions of the state of gravitation and motion of the object.

All local expressions of energy, like kinetic energy, Coulomb energy, and the energy of electromagnetic radiation, are derivatives of the local rest energy. The energy of a quantum of radiation is derived from Maxwell's equations as the energy injected into a cycle of electromagnetic radiation by a single electron transition in the emitter. Localized mass objects in space can be described as resonant mass wave structures. Mass is characterized as the wavelike substance for the expression of energy.

Time and distance are basic quantities for human comprehension. Human orientation relies on definite time and distance – in the Dynamic Universe framework, time and distance are referred to as coordinate quantities and are the same for all observers, at any location at any moment. The rates of physical events and processes, as well as the dimensions of physical structures, however, are dependent on the local energy balance in space.

Introduction

1.4.2 Some fundamental equations

The velocity of light

In the Dynamic Universe, the velocity of light is not constant. The velocity of light is equal to the velocity of space in the fourth dimension, determined by the balance of the energies of motion and gravitation in space. In hypothetical homogeneous space, the velocity of light, c_0, is

$$M_\Sigma c_0^2 = GM_\Sigma M''/R_4 \qquad \Rightarrow \qquad c_0 = \sqrt{GM''/R_4}, \qquad (1.4.2{:}1)$$

where M_Σ is the total mass in space, M'' the mass equivalence of the total mass, R_4 the 4-radius of space, and G the gravitational constant. The velocity of light decreases with the expansion of space, the increasing 4-radius R_4 (see Section 3.3).

The rest energy of matter

Perhaps the most famous equation in physics is the rest energy of matter

$$E = mc^2. \qquad (1.4.2{:}2)$$

In the DU, the rest energy takes the form

$$E = c_0 mc, \qquad (1.4.2{:}3)$$

where c_0 is the velocity of light in a hypothetical homogeneous space determined by the current expansion velocity of space, and c is the local velocity of light (in the Earth's gravitational frame, c is estimated as $c \approx 0.999999 \cdot c_0$). Mass m and velocity c are functions of the local state of motion and gravitation

$$m = m_0 \prod_{i=0}^{n} \sqrt{1-\beta_i^2} \qquad ; \qquad c = c_0 \prod_{i=0}^{n}(1-\delta_i), \qquad (1.4.2{:}4)$$

where m_0 is the mass of the object at rest in hypothetical homogeneous space, $\beta_i = v_i/c_i$ is the velocity v_i relative to the local velocity of light in the i:th frame, $\delta_i = GM_i/r_i c_{0i}^2$ the gravitational factor in the i:th frame at distance r_i from the local mass center M_i (see Section 4.1.4).

The DU form of the rest energy conveys central relativistic effects in absolute time and distance that are used as universal coordinate quantities natural for human comprehension.

The total energy of motion

Motion and momentum in space are associated with the velocity and momentum due to the motion of space, the expansion of the 4D sphere in the fourth dimension. The rest energy is the energy related to the motion of space in the fourth dimension; it is expressed as the imaginary component of the complex total energy of motion

$$E_m^\square = c_0 \mathbf{p}^\square = c_0(\mathbf{p} + \mathbf{i}mc) = c_0\mathbf{p} + \mathbf{i}c_0 mc. \qquad (1.4.2{:}5)$$

In the complex quantity presentation of (1.4.2:5), momentum $\mathbf{p}^\square$ is directly proportional to the energy of motion. For mass m at rest in a local frame in space, momentum $\mathbf{p}$ (the real component) is zero, and the total energy of motion is the rest energy

$$E_m^\square = \mathbf{i}c_0 mc \quad ; \quad \left|E_m^\square\right| = E_{rest} = c_0 mc . \tag{1.4.2:6}$$

For mass m with momentum $\mathbf{p}$ in a local frame in space, the total energy of motion is

$$E_m^\square = c_0 (\mathbf{p} + \mathbf{i}mc) \quad ; \quad \left|E_m^\square\right| = c_0 \sqrt{p^2 + (mc)^2} . \tag{1.4.2:7}$$

For electromagnetic radiation, rest energy is zero, and the energy is

$$\left|E_m^\square\right| = c_0 p . \tag{1.4.2:8}$$

Substituting c_0 by c, equations (1.4.2:6–8) convey the complex function presentation of the energy of motion to the scalar energy of contemporary physics.

Kinetic energy

The kinetic energy of mass m is expressed as

$$E_{kin} = c_0 (\Delta m \cdot c + \Delta c \cdot m), \tag{1.4.2:9}$$

replacing the SR equation $E_{kin} = \Delta m \cdot c$. The last term in (1.4.2:9) is the kinetic energy obtained in a gravitational field, where the kinetic energy is obtained against the reduction of the local rest energy via the tilting of local space and the reduction of the local velocity of light, Δc. Kinetic energy obtained at constant gravitational potential requires the supply of mass Δm from the accelerating system.

The laws of motion

Newton's laws of motion marked the start of mathematical physics. The first law states the concept of an inertial frame of reference; an object preserves its state of rest or motion if there is no force acting on it. The second law states that a change in motion is proportional to the force impressed, and the third law states the balance of opposite actions.

The second law is expressed as

$$\mathbf{F}_{Newton} = \frac{d\mathbf{p}}{dt} = \frac{dm\mathbf{v}}{dt} = m\frac{d\mathbf{v}}{dt} = m\mathbf{a} , \tag{1.4.2:10}$$

where the first expression interprets Newton's *motion* as momentum. The last expression assumes classical constant mass, which reduces the change in momentum to acceleration, the change in velocity.

The theory of special relativity introduced the relativistic mass increase associated with velocity

$$\mathbf{F}_{SR} = \frac{d\mathbf{p}}{dt} = m(1-\beta^2)^{-3/2} \cdot \mathbf{a} . \tag{1.4.2:11}$$

In the DU framework, the equation of motion is derived from the change in the total energy of motion

$$E_m^\square = c_0 \mathbf{p}^\square = c_0 (\mathbf{p} + \mathbf{i}mc), \tag{1.4.2:12}$$

resulting in an equation

Introduction

$$\mathbf{F}_{DU} = \frac{c_0}{c}\frac{d\mathbf{p}_\delta}{dt} = m_0 \frac{\prod_{i=1}^{n-1}\sqrt{1-\beta_i^2}}{\prod_{i=1}^{n}(1-\delta_i)} \cdot (1-\beta_n^2)^{-3/2} \cdot \mathbf{a}_n, \qquad (1.4.2{:}13)$$

see Section 1.2.3, Energy structures in space.

Equation (1.4.2:13) conveys a mass increase as the energy contribution needed to build up motion and the effects of motion and gravitation in the parent frames, affecting the local rest mass and the velocity of light. Once we omit the effects of the parent frames and the local gravitational state, we enter the SR equation (1.4.2:11) and, by further omitting the effect of the relativistic mass of SR or the mass contribution of DU, the Newtonian equation (1.4.2:10).

An essential difference between SR and DU comes from the kinematic versus dynamic approaches. In the kinematic approach of the SR, the relativistic mass increase is a property of motion; in the dynamic approach of the DU, the increased mass expresses the energy contribution needed to obtain the motion. As a local approach based on the relativity principle, special relativity does not recognize the effects of the parent frames, like the orbital motion and gravitational state of the Earth in the solar frame and the effects of the motion and gravitation of the solar system in the Milky Way frame. In many local observations, the effects of the parent frames are canceled but become meaningful in frame-to-frame observations.

The Planck equation

Traditionally, Planck's equation, the energy (of a cycle) of electromagnetic radiation or a photon, is written in the form

$$E = hf \qquad (= h/dt), \qquad (1.4.2{:}14)$$

where h is Planck's constant, f is the frequency, and $dt = 1/f$ is the cycle time. Observing that the velocity of light is an internal factor in Planck's constant h, Planck's equation can be rewritten

$$E = \frac{h_0}{\lambda} c_0 c \approx \frac{h_0}{\lambda} c^2, \qquad (1.4.2{:}15)$$

where $h_0 = h/c$ and λ is the wavelength of radiation. The quantity h_0/λ has the dimension of mass [kg].

Rewriting of Planck's equation does not need any assumption tied to the Dynamic Universe. The interpretation of Planck's equation as the energy injected into a cycle of electromagnetic radiation by a single electron transition in the emitter is confirmed by the standard solution of Maxwell's equations for an electric dipole (see Section 5.1.1).

Physical and optical distance in space (cosmology)

In DU space, the physical distance to an object (corresponding to the *comoving distance* in standard cosmology) is expressed in terms of the separation angle θ seen from the 4-center of space, or in terms of the redshift z

$$D_{phys} = \theta \cdot R_4 = R_4\left(1-e^{-\theta}\right) = R_4 \ln(1+z), \tag{1.4.2:16}$$

where $R_4 = R_H$ is the 4-radius of space.

1.4.3 Dynamic Universe and Contemporary Physics

In spite of the very different theory structures and postulates, predictions for most local observables in the DU and contemporary physics are essentially the same. The cosmological appearance of space in the DU is quite different from that in standard Big Bang cosmology. In the DU, there is no instant start of physical existence or a "turn on" of the laws of nature. The laws of nature and the substance for the expression of energy are understood as eternal qualities. The buildup and release of the rest energy needed for the expression of physical existence and all material structures in space appear as a continuous process from infinity in the past to infinity in the future – or a cyclic process repeating the contraction-expansion cycles. Space is characterized as a zero-energy continuum with the energies of motion and gravitation in balance.

The system of nested energy frames is a characteristic feature of the Dynamic Universe. It allows the use of time and distance as absolute coordinate quantities and allows an analytical study of the linkage between the local and the whole, and thereby the linkage between local objects.

We may assume that the actual system of energy frames is more complicated than the simple hierarchical structure presented in this book. As in the case of Newtonian gravitational potential or the spacetime structure in general relativity, all mass objects or mass distribution in space contribute to a local condition. The hierarchical approach used in the system of nested energy frames, however, is illustrative and serves most practical needs.

In the DU, the picture of "quantum reality" is a derivative of the properties of mass as a wavelike substance, and the linkage of mass waves and electromagnetic radiation. As illustrated by the solution of the principal electron states in a hydrogen atom in Section 5.1.4, a quantum state can be understood as the energy minimum of a resonant mass wave state.

A quantum of radiation in the DU is the energy injected into a cycle of electromagnetic radiation by a single unit charge transition. A point emitter, such as an atom, can be approximated as a one-wavelength dipole in the fourth dimension. An isotropic emitter generates radiation spreading uniformly in all space directions. A directing emitter generates photon-like localized radiation, like a laser, as a macroscopic directing emitter.

Table 1.4.3-I summarizes some basic properties of contemporary physics (special relativity, general relativity, FLRW cosmology, quantum mechanics) – and the Dynamic Universe.

	Contemporary physics: SR, GR, FLRW cosmology, QM	**The Dynamic Universe**
Birth of the universe	The Big Bang turned on time and the laws of nature, producing the energy for physical existence.	Buildup of the rest energy of matter occurred in a contraction phase preceding the ongoing expansion phase.
Equality of the total gravitational energy and total rest energy in space	Coincidence.	Expression of the overall zero-energy balance of motion and gravitation in space = The zero-energy principle.
The velocity of light	Postulated to be the same (constant) for any observer.	Determined by the velocity of space in the fourth dimension.
Rest energy of matter	Property of mass.	The energy of motion mass is possessed due to the velocity of space in the fourth dimension.
Geometry of space	Undefined as a whole. Defined locally by spacetime metrics as an attribute of mass distribution in space.	Space is described as the 3-surface of a 4-sphere. Mass centers in space result in local dents in the fourth dimension.
Relativity	Consequence of spacetime metrics.	Consequence of the conservation of total energy in space.
Effect of motion and local gravitation on clock readings	The effect of motion and gravitation on clocks is due to dilated time.	The effect of motion and gravitation on clocks is a consequence of the conservation of total energy in space.
Planck's equation	Postulated as $E=hf$, where h is the Planck constant [Js].	Derived from Maxwell's equations into the form $E_\lambda = h_0/\lambda\, c_0 c$, where h_0 [kgm] is the intrinsic Planck constant, and the quantity h_0/λ [kg] is the elementary mass equivalence of a cycle of radiation.
Quantum objects	Structures described in terms of wave functions.	Resonant mass wave structures.
Approach to unified theory	Field theory for unifying primary interactions.	Unified expression of energies.

Table 1.4.3-I. Comparison of some fundamental features in contemporary physics and the Dynamic Universe.

Linkage of local and global

Any local object is linked to the rest of the space via the global gravitational energy. Any local gravitational system expands with the expansion of space. Any velocity *in* space is related to the velocity *of* space. When an object is accelerated in space, inertia appears as the work done against the global gravitational energy by the imaginary component of the kinetic energy (Section 4.1.3). This is the quantitative explanation of Mach's principle.

The buildup of local structures

The primary energy buildup in the Dynamic Universe is described in terms of the dynamics of a hypothetical homogeneous space, with motion only in the fourth dimension. There is no answer to what broke the ideal symmetry of homogeneous space to enable the buildup of radiation and material structures in space.

We may think that the turn of the contraction phase to expansion phase did not occur through an ideal single point, but by passing the 4-center at a finite radius, which meant conversion of, at least, part of energized mass into electromagnetic radiation in space — turning on the light in space — and triggering elementary particle buildup and the process of nucleosynthesis.

The destiny of the universe

The Dynamic Universe theory, as presented in this book, does not solve or define the ultimate beginning or end of physical existence. Mathematically, the cycle of physical existence and the zero-energy balance extends from infinity in the past to infinity in the future. It is natural to think about the possibility of closing also the fourth dimension, which would turn the expansion of space back to a new contraction and expansion cycle.

2. Basic concepts, definitions, and notations

2.1 Closed spherical space and the universal coordinate system

2.1.1 Space as a spherically closed entity

For a holistic view of space as an energy system, a basic assumption needed is that three-dimensional space is closed. Closing of a three-dimensional space requires the fourth dimension. With the three-dimensional space closed symmetrically through a fourth dimension, we obtain a three-dimensional "surface" of a four-dimensional sphere. On a cosmological scale, the curvature of spherically closed space can be expressed in terms of the radius of the structure in the fourth dimension, the 4-radius of space. Visualization of a four-dimensional sphere is difficult; we can approach the visualization by first thinking of an ordinary three-dimensional ball. In a three-dimensional ball, the surface is two-dimensional, but curvature in a third dimension makes it closed. Closing of three-dimensional space spherically through a fourth dimension makes it the "surface" of a 4-sphere with the radius perpendicular to all three space directions. In principle, space as the three-dimensional "surface" of a 4-sphere has no extension or "thickness" in the fourth dimension (see Figure 2.1.1-1).

A useful way of visualizing spherically closed space in the four-dimensional universe is to look at a plane passing through the center of the 4-sphere. On such a plane, the origin of the universal coordinate system is set to the center of the 4-sphere, and any space direction is seen as a circumference of a sphere with radius R_4 around the origin, Figure 2.1.1-2 (a).

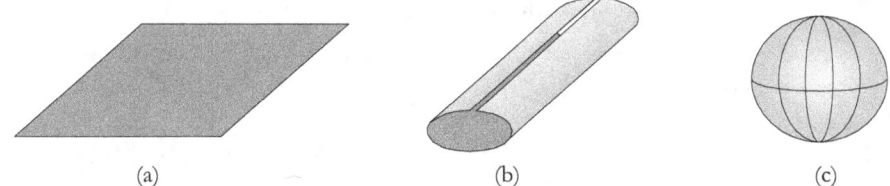

(a) (b) (c)

Figure 2.1.1-1. If we wish to eliminate the edges of a piece of paper and make its two-dimensional surface continuous, we need to wrap the paper around in some way. By forming it into a tube, we can eliminate two of the four edges, but then a third dimension is added as the radius of the tube. And the ends of the tube still have edges. The simplest structure that will also eliminate the edges of the tube is a sphere. Now the surface is symmetric and continuous in all directions. The third dimension we have added is perpendicular to the surface dimensions. The added dimension can be measured as distance, but it is not accessible without leaving the surface.

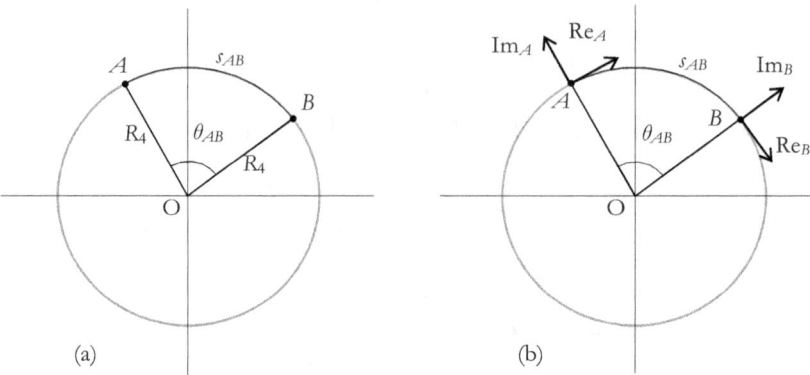

Figure 2.1.1-2. (a) Universal coordinate plane crossing points A and B in space, and the center of the 4-sphere inhabiting space. Any point in space is at a distance R_4 from the origin at point O. When, analogously, we eliminate the edges of a three-dimensional space by making it spherically continuous, we add a radius in the fourth dimension, perpendicular to the three space directions, which now appear as tangential directions in the structure. The shortest distance between points A and B in space is s_{AB} = arc[AB] along the circumference. (b) It is useful to apply complex coordinates in the study of local phenomena in space. In the local coordinates at points A and B, the real axes have the direction of arc[AB] connecting points A and B in space. Due to the curvature of space in the fourth dimension, the local real and imaginary axes at A and B have different directions. The path of light from A to B follows the curvature of space in the fourth dimension; for the viewer, it looks like light is coming along a straight line.

The distance between points A and B in Figure 2.1.1-2 can be expressed with the aid of the angle distance θ_{AB} as

$$\mathrm{arc}\left[AB\right] = s_{AB} = \theta_{AB} R_0 . \qquad (2.1.1{:}1)$$

If A and B stay at rest *in* space, the angle θ_{AB} remains constant, as it does also when space is expanding through an increase in R_4.

The present value of the 4-radius is about $R_4 \approx 14 \cdot 10^9$ [l.y.] = $1.3 \cdot 10^{26}$ [m]. The value of angle θ corresponding to the distance from the Earth to the Sun is about $\theta_{r(Sun)} = \Delta s/R_4 \approx 1.5 \cdot 10^{11}/1.3 \cdot 10^{26} \approx 10^{-15}$ radians. For the diameter of the Milky Way, the corresponding value of θ is $\theta_{r(MW)} = \Delta s/R_4 \approx 10^{-5}$ radians.

2.1.2 Time and distance

Time and distance are fundamental properties of the physical universe, and they serve as basic quantities for human conception. Time and distance are used as universal coordinate quantities applicable to all phenomena, independent of the observer and the local environment.

The frequency of a time standard like a Ce-clock is a function of the state of motion and gravitation in the frame in which the clock is running. The effects of the local gravitational state on the clock frequency and the local velocity of light are equal, which means

Basic concepts, definitions and notations

that the velocity of light is observed unchanged at any gravitational state. A distance standard based on the wavelength of a defined characteristic radiation of an atom is subject to the state of motion but not to the state of gravitation in the frame where it is used.

2.1.3 Absolute reference at rest, the initial condition

The center of a spherically closed space is the zero-momentum point of the system. It serves as the reference at rest for the contraction and expansion of space in the fourth dimension. Although not within three-dimensional space, the center at rest satisfies the intuitive view of Isaac Newton, "center of space at rest", expressed in the Principia [63].

In the initial condition of space, all mass is assumed to be at rest and homogeneously distributed in space with an essentially infinite 4-radius. Infinite distances in space mean zero gravitational energy, and the state of rest means zero energy of motion.

2.1.4 Notation of complex quantities

Local phenomena in space are described in locally defined complex coordinates where space dimensions appear as the real part of a complex function and the fourth dimension as the imaginary part. So long as space is assumed to be a fully homogeneous spherical structure with constant 4-radius R_4, the imaginary axis is aligned with the local 4-radius R_4 everywhere in space.

We will generally use superscript ($\square$) to denote a complex function. A single apostrophe ($'$) will denote the real part of the complex function in the selected space direction, and double apostrophes ($''$) the imaginary part. For example, the complex momentum of an object with momentum p' in space and momentum p'' in the fourth dimension is expressed (Fig. 2.1.4-2) as

$$p^{\square} = p' + ip'' = \mathbf{p} + i p'' \qquad (2.1.4:1)$$

For compatibility with the established use of symbols, however, the real part of momentum will usually be denoted as vector $\mathbf{p}$ or its scalar value p, instead of p', as shown in equation (2.1.4:2). In the two-dimensional complex plane presentation, the real axis is chosen in the direction of the phenomenon studied, which makes it possible to replace a vector quantity (in space) with its scalar value like $\mathbf{p} \Rightarrow p$. In the same way, velocity in space will be denoted as v instead of $\mathbf{v}$ or v', and the velocity of light propagating in space as c instead of $\mathbf{c}'$ or c'.

The local velocity of light in space is equal to the velocity of space in the local fourth dimension. The rest momentum of mass occurs in the local fourth dimension, the imaginary direction

$$\mathbf{p}_{rest} = \mathbf{i}\, mc \,. \qquad (2.1.4:2)$$

The rest energy of mass is expressed

$$E_{rest} = c_{4(0)} |\mathbf{p}_{rest}| = c_0 |\mathbf{p}_{rest}| = i c_0 mc \,, \qquad (2.1.4:3)$$

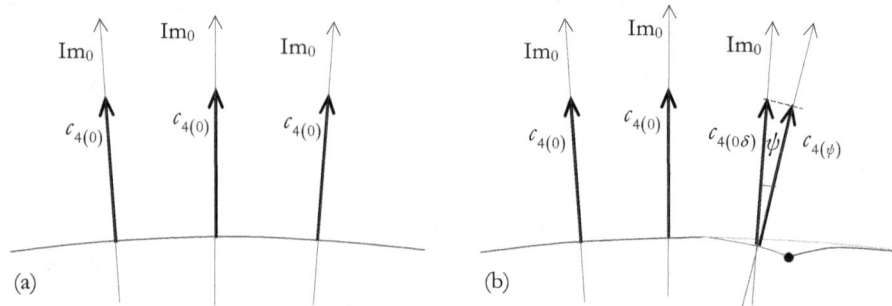

Figure 2.1.4-1. (a) The local imaginary axis in homogeneous space Im$_0$ follows the spherical shape. It always has the direction of the local 4-radius of space. (b) In locally tilted space in the vicinity of mass centers, the direction of the local imaginary axis Im$_\psi$ is tilted by an angle ψ relative to the imaginary axis Im$_0$. Local tilting of space means that the velocity of space in the local fourth dimension is reduced as $c_{4(\psi)} = c_{0(0\delta)} \cos\psi$. In real space, mass center buildup and the associated tilting of space occurs in several steps leading to a system of nested energy frames (Section 4.1.4).

where $c_{4(0)}$ means the imaginary velocity of homogeneous space, which is just the expansion velocity of the "surface" of a perfect 4-sphere. In locally tilted space (see Section 4.1.1), the velocity of space $c_{4(\psi)}$ in the local fourth dimension, and the related local velocity of light c are smaller than the velocity of light in non-tilted space $c_{4(0\delta)}$

$$c = c_{4(\psi)} = c_{4(0\delta)} \cos\psi, \qquad (2.1.4{:}4)$$

where ψ is the tilting angle of local space, Figure 2.1.4-1.

The notation c_0 is used for the velocity of light in hypothetical homogeneous space, $c_0 = c_{4(0)}$. Because we are used to using the velocity of light as the reference for velocities in space, c_0 is generally used as the notation for a velocity equal to the expansion velocity of space in the direction of the 4-radius R_4.

The rest energy in (2.1.4:3) is directly proportional to the rest momentum. The complex presentation of momentum and the energy of motion also reveals the linear linkage between momentum and the energy of motion when there is a real component of momentum, i.e., a momentum in a space direction. Generally, the complex presentation of the energy of motion is [see equation (2.2.2:4)]

$$E_m^\square = E'_m + iE''_m = c_0 |\mathbf{p}| = c_0 p^\square = c_0 (p' + ip''). \qquad (2.1.4{:}5)$$

The total energy of motion in equation (2.1.4:5) is the DU replacement of the concept of total energy in the special relativity framework. Figures 2.1.4-2 and 2.1.4-3 illustrate the use of a complex presentation of the momentum and the energy of motion.

The absolute value of the total energy becomes

$$\left|E_m^\square\right| = E_{m(tot)} = \sqrt{E'^2_m + E''^2_m} = c_0\sqrt{p'^2 + p''^2} = c_0\sqrt{p^2 + (mc)^2}, \qquad (2.1.4{:}6)$$

and the energy-momentum four-vector

$$E_m^2 = c_0^2 (mc)^2 + c_0^2 p^2. \qquad (2.1.4{:}7)$$

Basic concepts, definitions and notations

The increase in the total energy of motion due to motion in space is the kinetic energy

$$\left|\Delta E_M^{\square}\right| = E_{kin} = \left|E_{m(\varphi)}^{\square}\right| - \left|E_{m(0)}^{\square}\right| = c_0 \Delta \left|\mathbf{p}^{\square}\right| = E_{m(tot)} - E_{rest} \,. \tag{2.1.4:8}$$

A detailed derivation of kinetic energy is presented in Section 4.1.2.

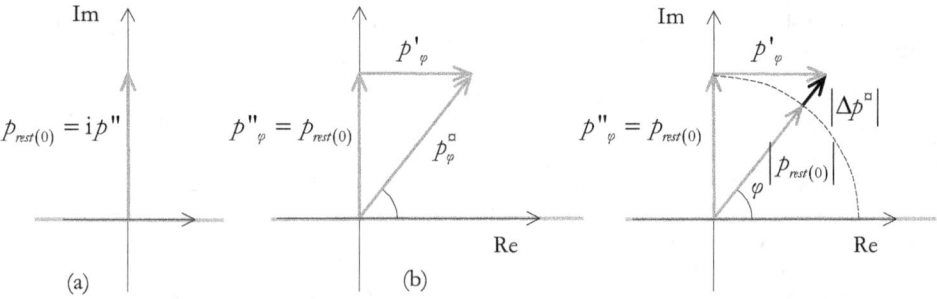

Figure 2.1.4-2. Complex presentation of momentum. (a) Momentum of a mass object at rest in space appears in the imaginary direction only $p^{\square} = p_{rest(0)} = i\,p''$. (b) The total momentum $p^{\square}_{\varphi}$ of an object moving in space is the sum of rest momentum ip''_{φ} and the momentum p'_{φ} in a space direction. (c) The increase of the absolute value of momentum due to momentum in space is $p^{\square}_{\varphi} = p_{rest(0)} + \Delta p^{\square}$, where $\Delta p^{\square}$ is the change in the absolute value of total momentum due to momentum in space (in the direction of the total momentum in the figure).

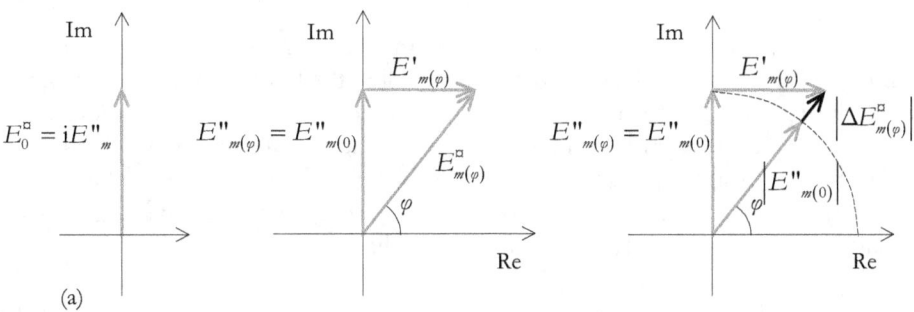

Figure 2.1.4-3. Complex presentation of the energy of motion. (a) The rest energy of a mass object at rest in space appears as imaginary energy of motion $E^{\square}_{m(0)} = i\,E''_{m(0)}$. (b) The total energy of motion $E^{\square}_{m(\varphi)}$ of an object moving in space is the sum of rest energy $iE''_{m(\varphi)}$ and the real part of the energy of motion (the energy equivalence of momentum in space) $E'_{m(\varphi)}$. (c) The increase of the absolute value of the energy of motion due to momentum in space is the kinetic energy $E_{kin} = \Delta E^{\square}_{m(\varphi)}$ (in the direction of the total energy of motion in the figure).

2.2 Base quantities

2.2.1 Mass

In the DU framework, mass has the meaning of the substance for the expression of energy.

The mass equivalence of a cycle of electromagnetic radiation, as derived from Maxwell's equations, is

$$m_\lambda = N^2 \frac{h_0}{\lambda} = N^2 \hbar_0 k, \qquad (2.2.1:1)$$

where N^2 is an intensity factor, and h_0 is the intrinsic Planck constant,

$$h_0 \equiv \frac{h}{c_0} \quad [\text{kg·m}]. \qquad (2.2.1:2)$$

The intrinsic Planck constant is derived from Maxwell's equations in Section 5.1.2. The derivation can be carried out from a general basis without assumptions tied to the Dynamic Universe model.

The other way round, the wavelength equivalence λ_m and the wave number equivalence k_m of mass m are

$$\lambda_m = \frac{h_0}{m} \quad \text{and} \quad k_m = \frac{2\pi}{\lambda} = \frac{m}{\hbar_0}, \qquad (2.2.1:3)$$

where $\hbar \equiv h/2\pi$.

The intensity factor N^2 in (2.2.1:1) comes from the solution of Maxwell's equation as the number of electrons oscillating in a one-wavelength dipole emitting electromagnetic radiation (Section 5.1.2). In the DU framework, with space moving at velocity c in the fourth dimension, a point emitter in space can be understood as a one-wavelength dipole in the fourth dimension with any space direction on a normal plane of the dipole. For $N = 1$ in equation (2.2.1:1), i.e., a single electron transition in a one-wavelength dipole, we get the minimum mass equivalence of a cycle of radiation

$$m_{\lambda(0)} = \frac{h_0}{\lambda} = \hbar_0 k, \qquad (2.2.1:4)$$

and the energy emitted into one cycle of electromagnetic radiation by a single electron transition in a point source

$$E_{\lambda(0)} = c_0 |\mathbf{p}| = c_0 |m_{\lambda(0)} \cdot \mathbf{c}| = c_0 \hbar_0 kc = \frac{h_0}{\lambda} c_0 c = h_0 c_0 \cdot f = h \cdot f, \qquad (2.2.1:5)$$

which is known as the *Planck equation*.

The mass presentation of wave and the wave presentation of mass allow unified expressions of the energy of a cycle in the forms

Basic concepts, definitions and notations

$$E = c_0 |\mathbf{p}| = c_0 mc = c_0 \hbar_0 kc = c_0 \frac{h_0}{\lambda} c , \qquad (2.2.1:6)$$

which applies equally for a cycle of mass wave and for the mass injected in a cycle of electromagnetic radiation at emission, when the wavelength is the emission wavelength $\lambda = \lambda_e$.

Electromagnetic radiation propagating in expanding space is subject to redshift, an increase of the wavelength; the mass equivalence of a cycle of radiation, $m\lambda = h_0/\lambda_e$, bound to the emission wavelength is conserved in the lengthened cycle, but the energy density of a cycle of radiation is diluted.

As a major difference to the prevailing concept of quantum as *a quantum of action*, the concept of quantum in the DU framework serves as the measure of the mass content of a cycle of electromagnetic radiation or a mass wave.

Mass is associated with gravitational potential extending, as a scalar gravitational field, throughout spherically closed space. The gravitational potential dilutes in inverse proportion to the distance from its source mass. The gravitational potential follows the motions of its source mass without delay. Mass senses the gravitational potential of all other mass as gravitational energy.

2.2.2 Energy and the conservation laws

Gravitational energy in homogeneous space

In homogeneous 3D space, gravitational energy is expressed in the form of the Newtonian gravitational energy

$$E_{g(0)} \equiv -mG \int_V \frac{\varrho\, dV(r)}{r} , \qquad (2.2.2:1)$$

where G is the gravitational constant, ϱ is the mass density, and r is the distance from m to dV. The total mass in homogeneous space is

$$M_\Sigma = -\varrho \int_V dV = \varrho V . \qquad (2.2.2:2)$$

In a spherically closed homogeneous 3D space, the total mass is $M_\Sigma = 2\pi^2 \varrho R_4^3$, where R_4 is the radius of space in the fourth dimension. The total gravitational energy in a spherically closed space becomes

$$E_{g(tot)} = -\frac{GM'' M_\Sigma}{R_4} , \qquad (2.2.2:3)$$

where M'' is the mass equivalence of the total mass at the barycenter of the structure (Section 3.2.2).

The energy of motion in homogeneous space

In homogeneous space, there is no motion in spatial directions. The only motion applicable to homogeneous space is the contraction or expansion of the 4D sphere. Denoting the velocity of homogeneous space in the fourth dimension by c_0, the energy of motion of mass m in space is expressed as the product of velocity c_0 and the momentum $\mathbf{p}=m c_0$

$$E_{m(0)} \equiv c_0 |\mathbf{p}| = c_0 |m \cdot \mathbf{c}_0| = mc_0^2 , \qquad (2.2.2{:}4)$$

substituting the total mass, $M_\Sigma = \Sigma m$, into (2.2.2:4), the total energy of motion is

$$E_{m(0),tot} \equiv c_0 |\mathbf{P}_{tot}| = c_0 |M_\Sigma \cdot \mathbf{c}_0| = M_\Sigma c_0^2 . \qquad (2.2.2{:}5)$$

In the non-homogeneous space with local mass centers, the direction of the local 4-velocity, c, may deviate from the 4-velocity, c_0, of homogeneous space. The general expression for the total energy of motion is expressed as a complex function with the momentum due to the motion of space as the imaginary part $\mathbf{i}mc$, and the momentum $\mathbf{p}$ in space as the real part

$$E_m^\square \equiv c_0 |\mathbf{p}^\square| = c_0 |\mathbf{p} + \mathbf{i}mc| = c_0 \sqrt{p^2 + (mc)^2} . \qquad (2.2.2{:}6)$$

The last form of equation (2.2.2:6) is formally equal to the expression of the total energy in special relativity. The last form of the energy of motion in (2.2.2:4) has the form of the first formulation of kinetic energy, *vis viva*, "*the living force*" suggested by Gottfried Leibniz in the late 1600's [8].

In the solar system, the local 4-velocity of space and the local velocity of light, c, can be estimated to be of the order of ppm (10^{-6}) smaller than the velocity of light, c_0, in hypothetical homogeneous space.

Conservation of total energy

The zero-energy balance of the energies of motion and gravitation created by the process of contraction and expansion of space is conserved in all energy interactions in space.

2.2.3 Force, inertia, and gravitational potential

Force is defined as the gradient of potential energy and, as the inertial force, a change in momentum. Force is local by its nature. Gravitational force means sensing the gradient of the local gravitational potential by local mass. Gravitational potential is an intrinsic property of its source mass. The gravitational potential is an intrinsic property of a mass object that extends throughout the spherically closed space; a motion of the source mass is conveyed to the gravitational potential without delay.

Inertia is the work done by an accelerated mass object against the global gravitational energy. Inertia in a local frame is equal to the imaginary component of the complex kinetic energy of an object (see Sections 4.1.3 and 4.1.7).

3. Energy buildup in spherical space

3.1 Volume of spherical space

The volume of spherically closed three-dimensional space is calculated as the surface "area" of a four-dimensional sphere. To do this, we start by calculating the surface area of an ordinary three-dimensional sphere.

With reference to Figure 3.1-1, observing that $r = R_3 \sin\theta$, we can calculate the surface area S_3 of a sphere in three dimensions as the integral

$$S_3 = \int_0^\pi 2\pi r\, R_3\, d\theta = 2\pi R_3^2 \int_0^\pi \sin\theta\, d\theta = 4\pi R_3^2, \quad (3.1:1)$$

where the circular differential surface unit is the circumference, $2\pi r$, times the differential width $R_3\, d\theta$. By following a similar procedure but replacing the circumference $(2\pi r)$ of a circle with radius $r = R_4 \sin\theta$ by the area of a sphere with radius r ($S_3 = 4\pi r^2$), as given by equation (3.1:1), we get

$$S_4 = \int_0^\pi 4\pi r^2 R_4\, d\theta = 4\pi R_4^3 \int_0^\pi \sin^2\theta\, d\theta = 2\pi^2 R_4^3 = V. \quad (3.1:2)$$

The "surface" $S_4 = 2\pi^2 R_4^3$ is equivalent to the volume of the closed three-dimensional surface of a four-dimensional sphere defined by the 4-radius R_4.

All objects in three-dimensional space are located at the surface of the four-dimensional sphere. At least on a macroscopic scale, the "thickness" of the surface in the direction of the 4-radius is zero. As a consequence, the fourth direction is not accessible to us. Any motion and energy interaction in three-dimensional space is described as a phenomenon on the surface of the sphere.

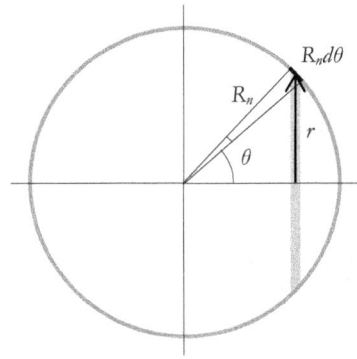

Figure 3.1-1. Calculation of the surface area of three- and four-dimensional spheres.

3.2 Gravitation in spherical space

3.2.1 Mass in spherical space

Gravitational interactions are assumed to take place in three-dimensional space. The gravitational field does not penetrate inside or extend outside space but follows the shape of space.

Referring to Figure 3.2.1-1, we can calculate the gravitational energy of the whole distributed mass at the surface of the sphere on a unit mass m at a selected location x_0, y_0, z_0.

On the cosmological scale, the total mass M_Σ is considered to be uniformly distributed in space, i.e., uniformly distributed on the three-dimensional surface of the sphere defined by the 4-radius R_4. Referring to equation (3.1:2), the derivation of the surface area S_4 of the sphere, we can express the mass dM in volume $dV = 4\pi r^2 R_4\, d\theta$ with the aid of mass density ϱ as

$$dM = \varrho dV = 4\pi\varrho r^2 R_4\, d\theta, \qquad (3.2.1:1)$$

and by replacing r by $R_4 \sin\theta$ as

$$dM = 4\pi\varrho R_4^3 \sin^2\theta\, d\theta, \qquad (3.2.1:2)$$

and by further applying expression (3.1:2) for the total volume of the three- dimensional surface, as

$$dM = \frac{2}{\pi}\varrho V \sin^2\theta\, d\theta. \qquad (3.2.1:3)$$

The factor ϱV in equation (3.2.1:3) is equal to the total mass M_Σ. Accordingly, equation (3.2.1:3) can be expressed as

$$dM = \frac{2M_\Sigma}{\pi} \sin^2\theta\, d\theta. \qquad (3.2.1:4)$$

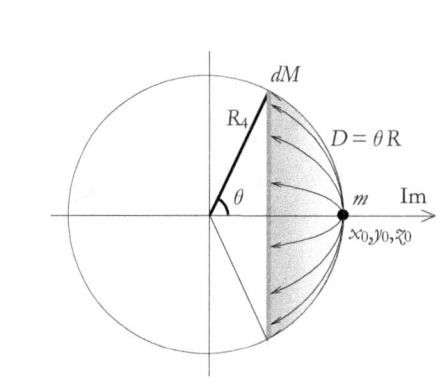

Figure 3.2.1-1. Calculation of the gravitational energy of an object with mass m, due to the effect of the total mass M_Σ in space. The total mass is considered to be uniformly distributed on the three-dimensional surface of a four-dimensional sphere with radius R_4.

3.2.2 Gravitational energy in spherical space

Based on the spherical symmetry, the gravitational energy of mass dM at distance $D = \theta R_4$ (see Figure 3.2.1-1) from mass m is expressed as inherent gravitational energy defined in equation (2.2.2:1)

$$dE_g = -\frac{Gm}{\theta R_4} dM, \qquad (3.2.2:1)$$

where distance θR_4 is the distance of dM from mass m along the spherical space. By applying equation (3.2.1:4) for mass dM, equation (3.2.2:1) can be expressed as

$$dE_g = -\frac{2}{\pi} \frac{GmM_\Sigma}{R_4} \frac{\sin^2 \theta}{\theta} d\theta. \qquad (3.2.2:2)$$

The gravitational energy due to the total mass in space is determined by integrating equation (3.2.2:2) for $\theta = 0$ to π, corresponding to Newtonian gravitation

$$E_g = -\frac{2}{\pi} \frac{GmM_\Sigma}{R_4} \int_0^\pi \frac{\sin^2 \theta}{\theta} d\theta = -\frac{GmM_\Sigma}{R_4} I_{g(\pi)}. \qquad (3.2.2:3)$$

The integral in equation (3.2.2:3) cannot be solved in closed mathematical form. Numerical integration of (3.2.2:3) gives

$$I_{g(\pi)} = \frac{2}{\pi} \int_0^\pi \frac{\sin^2 \theta}{\theta} d\theta \approx 0.776 *. \qquad (3.2.2:4)$$

Due to the spherical symmetry, equations (3.2.2:3) and (3.2.2:4) apply for mass m anywhere in homogeneous space. A direct interpretation of the equations is that the gravitational energy of mass m due to all other mass in space can be expressed as the gravitational energy due to mass $M" = I_g M_\Sigma$ at the center of the 4D sphere, inside the "hollow" space

$$E_g = -\frac{GmM"}{R_4}. \qquad (3.2.2:5)$$

The mass $M" = I_g \cdot M_\Sigma$ is referred to as the mass equivalence of space, Figure 3.2.2-4.

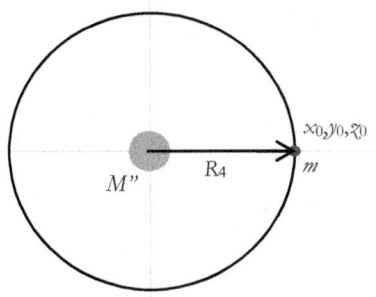

Figure 3.2.2-4. The gravitational energy due to the total mass M_Σ on mass m at any location x_0, y_0, z_0 in space, can be described as the gravitational energy due to the mass equivalence $M"$ at the center of the 4D sphere defining space.

---

* If the integration is made from 0 to 2π, the value of the integral is $I_{g(2\pi)} = 0.991$.

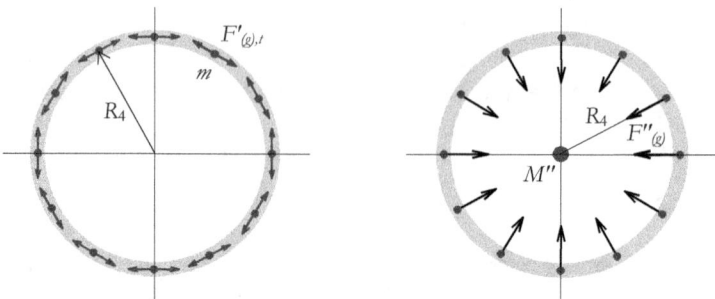

Figure 3.2.2-2. The tangential shrinking force, $F'_{g,t}$, due to the gravitation of uniformly distributed mass in spherical space, is equivalent to the gravitational effect, F''_g, of mass equivalence M'' at distance R_4 at the center of the structure.

The total mass M_Σ in space can be expressed as the integral of all masses dm' as

$$M_\Sigma = \int_0^M dm' . \qquad (3.2.2:6)$$

Substitution of $M_\Sigma = M''/\,0.776$ for m in equation (3.2.2:6) gives the total gravitational energy in space

$$E_{g(tot)} = -\frac{GM_\Sigma M''}{R_4} = -\frac{GI_g M_\Sigma^2}{R_4} . \qquad (3.2.2:7)$$

Gravitational force is defined as the gradient of gravitational energy. The gravitational force on mass m towards mass equivalence M'' is obtained as the derivative of the gravitational energy in equation (3.2.2:5)

$$\mathbf{F}_g = \frac{dE_g}{dR_4}\hat{\mathbf{r}}_4 = -\frac{GmM''}{R_4^2}\hat{\mathbf{r}}_4 , \qquad (3.2.2:8)$$

where the direction of the unit vector $\hat{\mathbf{r}}_4$ is in the direction of radius R_4.

3.3 Primary energy buildup of space

3.3.1 Contraction and expansion of space

The initial condition for the development of the energies of motion and gravitation in space is considered as a state of rest at infinite distances in space. In such a condition, both the gravitational energy and the energy of motion are zero. This situation occurs when the 4-sphere has an infinite radius R_4 at infinity in the past.

The primary energy buildup is described as the free fall of spherical space from the state where the 4-radius is infinite to the state where it is zero. In spherical geometry, the process means a homogeneous contraction of space, culminating in a singularity where space is reduced to a single point or a minimum radius. At the turning point, the mass in space has essentially infinite momentum, turning the process into expansion. In the expansion phase, the 4-radius increases back to infinity, while the energy of motion gained in the contraction is returned to gravitational energy. Free fall in the contraction phase and free escape in the expansion phase maintain zero total energy in the system.

In the contraction phase, mass in space gains energy of motion from its own gravitation. Space loses volume and gains motion. In the following expansion phase, space gains volume by losing motion. Space with infinite 4-radius continues to host all mass that has lost its energy: the energy of gravitation is zero because of the infinite distances, and the energy of motion is zero because all motion has ceased, Figure 3.3.1-1.

Because the sum of the energies of gravitation and motion remains zero throughout the process of energy buildup and release, the energy of motion in the imaginary direction is

$$iE''_m = -iE''_g, \tag{3.3.1:1}$$

or

$$E''_m + E''_g = 0. \tag{3.3.1:2}$$

In the primary energy buildup, mass within space is assumed to stay at rest. The only velocity of mass in the primary energy buildup is the contraction and expansion of spherical space in the imaginary direction. Accordingly, we can apply the inherent energy of motion to describe the energy of motion that mass has in the imaginary direction (the direction of R_4). With reference to equation (2.2.2:4), the energy of motion of mass m at rest in space has due to the motion of space at velocity c_0 is

$$E''_m = c_0 m c_0 = m c_0^2. \tag{3.3.1:3}$$

Substitution of equation (3.3.1:3) for E''_m and equation (3.2.2:5) for E''_g in equation (3.3.1:2) gives

$$m c_0^2 - \frac{GmM''}{R_4} = 0. \tag{3.3.1:4}$$

Observing that the total mass in space is the sum of all masses m

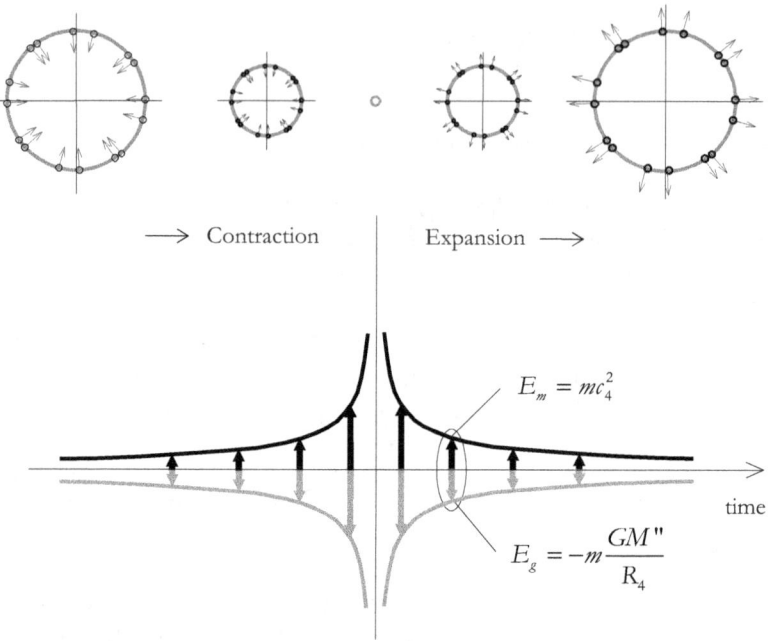

Figure 3.3.1-1. Energy buildup and release in spherical space. In the contraction phase, the velocity of the imaginary motion increases due to the energy gained from the loss of gravity. In the expansion phase, the velocity of the imaginary motion gradually decreases, while the energy of motion gained in contraction is returned to gravity.

$$M_\Sigma = \sum_V m, \qquad (3.3.1:5)$$

the total energies of motion and gravitation can be expressed as

$$M_\Sigma c_0^2 - \frac{GM_\Sigma M''}{R_4} = 0, \qquad (3.3.1:6)$$

where M'' is the mass equivalence of space defined in equation (3.2.2:6).

Velocity c_0 can be solved from equation (3.3.1:6) in terms of G, M_Σ, and R_4

$$c_0 = \pm\sqrt{\frac{GM''}{R_4}}. \qquad (3.3.1:7)$$

The negative value of c_0 in equation (3.3.1:7) refers to the velocity of contraction and the positive value to the velocity of expansion. The processes of contraction and expansion are symmetrical. All energy of motion gained in the contraction is returned to gravitational energy in the expansion.

Energy E''_m in equation (3.3.1:3) is the energy mass at rest in homogeneous space due to the velocity of contraction or expansion of space in the fourth dimension, i.e., it can be characterized as the rest energy of mass at rest in hypothetical homogeneous space. As will be shown in Section 4.1.2, the maximum velocity, c, obtainable in space is equal to the velocity of space in the local fourth dimension, which may deviate from the direction of the 4-radius (see Section 4.1.1). The general form of the rest energy, in accordance with (2.2.2:8) is

$$E_{rest} = E''_m = c_0 |\mathbf{p}| = c_0 mc .\tag{3.3.1:8}$$

> The rest energy of mass in space is the energy of motion due to the contraction or expansion of space.

3.3.2 Mass and energy in space

The 2006 CODATA recommended values of the gravitational constant and the present velocity of light at the surface of the Earth are:

$$G = 6.67428 \cdot 10^{-11} \quad [\text{Nm}^2/\text{kg}^2],\tag{3.3.2:1}$$

with a relative uncertainty, $|\Delta G|/G = 1.5 \cdot 10^{-3}$, and

$$c = 2.99792458 \cdot 10^8 \quad [\text{m/s}] \quad (\text{exact value defined}).\tag{3.3.2:2}$$

As shown in Section 4.1.1, the velocity of light is dependent on local gravitational conditions. As a consequence, the velocity of light on the Earth is slightly (presumably of the order of ppm) smaller than the velocity of light in hypothetical homogeneous space. The local velocity of light is denoted as c, in accordance with conventional notation.

Equation (3.3.1:7) shows the relationship between the velocity of light and the 4-radius of space. In the standard cosmology model, the constant velocity of light is related to the curvature of space through the Hubble constant and Hubble radius. In spherical space, the meaning of the Hubble radius is essentially the 4-radius, R_4.

A recent estimate of the Hubble constant derived from the Wilkinson Microwave Anisotropy Probe (WMAP) data combined with the distance measurements from the Type Ia supernovae (SN) and the Baryon Acoustic Oscillations (BAO) in the distribution of galaxies is $H_0 = 70.5 \pm 1.3$ [(km/s)/Mpc] [74].

Applying the Hubble constant $H_0 = 70$ [(km/s)/Mpc] and the local velocity of light, $c \approx c_0$, given in equation (3.3.2:2), the present length of the 4-radius R_4 is

$$R_4 = R_H = c/H_0 = 14.0 \cdot 10^9 \text{ light years } (=1.32 \cdot 10^{26} \text{ m}).\tag{3.3.2:3}$$

By substituting in equation (3.3.1:6) the values of G, c, and R_4 given in equations (3.3.2:1), (3.3.2:2), and (3.3.2:3), we obtain the total mass in space as

$$M_\Sigma = \frac{c^2 R_4}{GI_g} \approx 2.3 \cdot 10^{53} \ [\text{kg}],\tag{3.3.2:4}$$

and by applying equation (3.1:2) for the volume of space, we can express the density of mass in space as,

$$\varrho_{DU} = \frac{M_\Sigma}{V} \approx \frac{M_\Sigma}{2\pi^2 R_4^3}. \qquad (3.3.2{:}5)$$

Alternatively, by substituting equations (3.3.2:3) and (3.3.2:4) into equation (3.3.2:5), we can express the mass density in terms of the Hubble constant as

$$\varrho_{DU} = \frac{c^2}{2\pi^2 GI_g R_4^2} \approx 5.0 \cdot 10^{-27} \; \left[\frac{\mathrm{kg}}{\mathrm{m}^3}\right]. \qquad (3.3.2{:}6)$$

Applying the 4-radius R_4 given in equation (3.3.2:3) as the Hubble radius, R_H, of space in the expression of the Friedman critical mass density (consistent with Hubble constant $H_0 = 71$ [(km/s)/Mpc]), we get

$$\varrho_c = \frac{3c^2}{8\pi G R_H^2} \approx 9.5 \cdot 10^{-27} \; \left[\frac{\mathrm{kg}}{\mathrm{m}^3}\right]. \qquad (3.3.2{:}7)$$

The calculations of mass densities ϱ_c and ϱ_{DU} are illustrated in Figure 3.3.2-1(a) and 3.3.2-1(b), respectively.

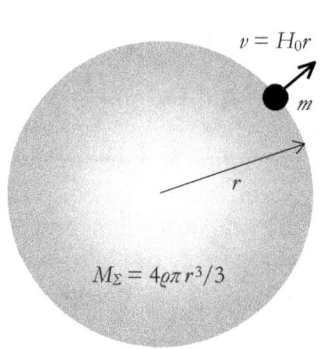

Figure 3.3.2-1 (a). The Friedman critical mass density, ϱ_c, can be calculated by determining the escape velocity $v = c$ of mass m from the surface of a three-dimensional sphere with radius r and the total mass

$$M = \varrho_c\, 4/3 \cdot \pi r^3$$

The Friedmann's the critical mass density is

$$\varrho_c = \frac{3H_0^2}{8\pi G} \approx 9.5 \cdot 10^{-27} \; \left[\frac{\mathrm{kg}}{\mathrm{m}^3}\right]$$

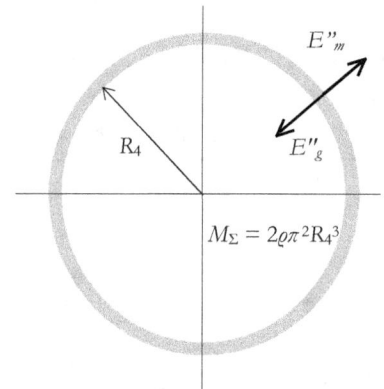

Figure 3.3.2-1 (b). In the DU, the density of matter is calculated from the total mass determined by the balance between the motion and the gravitation in the fourth dimension

$$M_\Sigma = M''/I_g = R_4\, c^2 / I_g G$$

resulting in mass density

$$\varrho = \frac{c_4^2}{2\pi^2 \cdot GI_g R_4^2} \approx 5.0 \cdot 10^{-27} \; \left[\frac{\mathrm{kg}}{\mathrm{m}^3}\right]$$

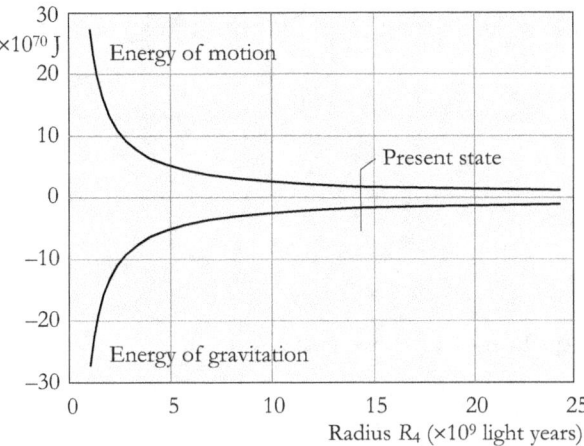

Figure 3.3.2-2. The energies of motion and gravitation of matter in space as functions of the 4-radius of space.

The DU prediction of the mass density corresponds to the "flat space" situation in Friedman-Lemaître-Robertson-Walker (FLRW) cosmology. Flat space in FLRW cosmology means that the sum of baryonic matter, dark matter, and dark energy is equal to the Friedman critical mass density. There is no place or need for dark energy in the DU framework. The predictions for magnitude versus redshift of Ia supernova standard candles in the DU are in good agreement with observations without dark energy (see Section 6.3).

Dark matter in the DU framework has the meaning of unstructured matter, which is considered the initial form of matter.

With reference to equation (3.3.1:5), the sum of the energies of gravitation and motion is zero all along the expansion of the 4-radius R_4 as presented in Figure 3.3.2-2 based on equations

$$E''_g = -\frac{GI_g M_\Sigma^2}{R_4} = -2.1 \cdot 10^{70} \; [J], \tag{3.3.2:8}$$

and

$$E''_m = M_\Sigma c_0^2 = 2.1 \cdot 10^{70} \; [J], \tag{3.3.2:9}$$

where the gravitational constant $G = 6.67 \cdot 10^{-11}$ [Nm²/kg²], the total mass in space $M_\Sigma = 1.8 \cdot 10^{53}$ [kg], and the velocity of space along the 4-radius $c_0 = c = 3 \cdot 10^8$ [m/s] at the present value of the 4-radius $R_4 = 13.8 \cdot 10^9$ [l.y.].

3.3.3 Development of space with time

The velocity of the expansion of space in the direction of the 4-radius can be expressed as

$$c_0 = \frac{dR_4}{dt} \; . \tag{3.3.3:1}$$

The time required for the 4-radius, R_4, to increase from the singularity ($R_4 = 0$, $t = 0$) to the present value of R_4 can be obtained by integration of dt solved from equation (3.3.3:1) as

$$dt = \frac{1}{c_0} dR_4 \qquad (3.3.3:2)$$

and

$$t = \int_0^{R_4} \frac{1}{c_0} dR_4. \qquad (3.3.3:3)$$

By applying equation (3.3.1:7) in equation (3.3.3:3) we get

$$t = \frac{1}{\sqrt{GM''}} \int_0^{R_4} \sqrt{R_4}\, dR_4, \qquad (3.3.3:4)$$

and

$$t = \frac{2}{3} \sqrt{\frac{R_4^3}{GM''}}, \qquad (3.3.3:5)$$

and by further applying equation (3.3.1:7)

$$t = \frac{2}{3} \frac{R_4}{c_0}. \qquad (3.3.3:6)$$

As a result of the higher expansion velocity close to the singularity, the age of the expanding space is two-thirds of the age estimate based on a constant value of c_0 as in the assumed inflation era in FLRW cosmology [64].

By applying the estimated value of the present 4-radius, $R_4 = 13.8 \cdot 10^9$ light years, we obtain the time since the singularity as $t = 9.2 \cdot 10^9$ current years. Solving equation (3.3.3:5) for R_4 gives

$$R_4 = (3/2)^{2/3} (GM'')^{1/3} t^{2/3}. \qquad (3.3.3:7)$$

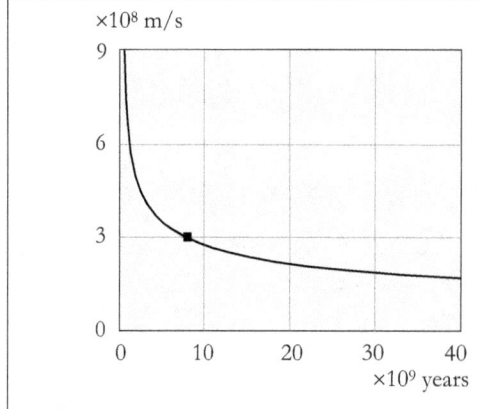

Figure 3.3.3-1. The decreasing expansion velocity of space in the R_4 direction. The present deceleration of the expansion velocity, and with it the velocity of light, is about 3.6 % per billion years. The velocity of light will drop to half of its present value in about 65 billion years and to 1 m/s in about $2 \cdot 10^{26}$ billion years.

Energy buildup in spherical space

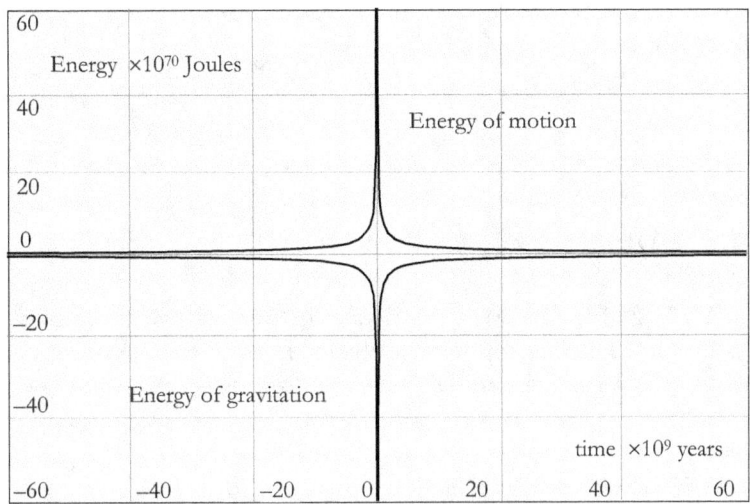

Figure 3.3.3-2. Development of the energy of the Universe as a zero-energy process.

The expansion velocity along the 4-radius can now be expressed as a function of the time from the singularity by differentiating equation (3.3.3:7) as

$$c_0 = \frac{dR_4}{dt} = \left(\frac{2}{3}GM''\right)^{1/3} t^{-1/3} = \frac{2}{3}\frac{R_4}{t}. \tag{3.3.3:8}$$

The development of the R_4 expansion velocity according to equation (3.3.3:8) is presented in Figure 3.3.3-1.

The change of the expansion velocity of space in the R_4 direction can be obtained from equation (3.3.3:8) as

$$\frac{dc_0}{dt} = -\frac{1}{3}\left(\frac{2}{3}GM''\right)^{1/3}\frac{t^{-1/3}}{t} = -\frac{1}{3}\frac{c_0}{t}, \tag{3.3.3:9}$$

or in terms of the relative change of the expansion velocity as

$$\frac{dc_0}{c_0} = -\frac{1}{3}\frac{dt}{t} \quad ; \quad \frac{dc_0}{c_0}\bigg/\Delta t = -\frac{1}{3t}. \tag{3.3.3:10}$$

From equation (3.3.3:7), we obtain the relative change in the R_4 radius of space:

$$\frac{dR_4}{R_4} = \frac{2}{3}\frac{dt}{t} \quad ; \quad \frac{dR_4}{R_4}\bigg/\Delta t = \frac{2}{3t}. \tag{3.3.3:11}$$

According to equations (3.3.3:11) and (3.3.3:10), the present ($t = 9.3 \cdot 10^9$ years) annual increase of the R_4 radius of space is $dR_4/R_4 \approx 7.2 \cdot 10^{-11}$/year and the deceleration rate of the expansion $dc_4/c_4 \approx -3.6 \cdot 10^{-11}$/year, which also means that the velocity of light slows down as $dc/c \approx -3.6 \cdot 10^{-11}$/year. In principle, the change is large enough to be detected.

However, the ticking frequency of an atomic clock used in such detection slows down at the same rate as the velocity of light, thus disabling the detection (see Section 5.1.4).

The energies of gravitation and motion can be expressed as functions of time by applying equations (3.3.3:7) and (3.3.3:8) in equations (3.3.2:8) and (3.3.2:9), respectively:

$$E''_g \approx -\left(\frac{2G}{3t}\right)^{2/3} M''^{5/3}, \qquad (3.3.3:12)$$

$$E''_m \approx \left(\frac{2G}{3t}\right)^{2/3} M''^{5/3}. \qquad (3.3.3:13)$$

Equations (3.3.3:12) and (3.3.3:13) can be applied for time symmetrically, from minus infinity to plus infinity. The development of the energies of motion and gravitation of the Universe as functions of time according to these equations is shown in Figure 3.3.3-2.

3.3.4 The state of rest and the recession of distant objects

As a consequence of the expansion of spherical space, objects at rest in space are subject to the hidden motion, the motion of space in the direction of the 4-radius. The increase of the 4-radius also means a stretching of space, so that objects in space have significant recession velocities with respect to one another.

As suggested by the pioneering work of Edwin Hubble in the 1920s, distant galaxies have a high recession velocity due to the expansion of space. Nevertheless, each of them may be at rest in space in its own space location in the universal coordinate system, Figure 3.3.4-1.

An observer in space observes the expansion of spherical space as the recession of all other objects at a velocity proportional to the expansion of the 4-radius and the distance of the objects from the observer along spherical space.

Objects A_1, A_2, and A_3 in Figure 3.3.4-1 are at rest in space. In other words, angles θ_1, θ_2, and θ_3 stay unchanged. The physical distances BA_n ($n = 1,2,3$) can be expressed in terms of angle θ_n and radius R_4 as

$$\text{arc}[BA_n] = D_n = \theta_n R_4 \qquad (3.3.4:1)$$

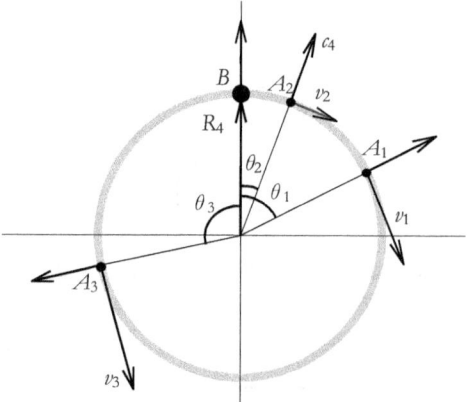

Figure 3.3.4-1. The expansion of the 4-radius R_4 causes an increase of all distances in space. The recession velocities v_1, v_2, and v_3 relative to point B are proportional to the distances BA_1, BA_2, and BA_3, along the curved space, respectively.

Distances arc$[BA_n]$ increase with the increase of R_4. The physical recession velocity can be expressed as

$$v_n = \frac{d(\theta_n R_4)}{dt} = \theta_n \frac{dR_4}{dt} = \theta_n c_0. \qquad (3.3.4:2)$$

When $\theta_n > 1$ radian ($\theta_n > 57.3°$), the physical recession velocity of the object exceeds the velocity of light.

Equation (3.3.4:2) shows the physical recession velocity at the time of the observation. Observations of distant objects are based on light propagation from the object. Since the 4-radius of space increases at the same velocity as light propagates in the tangential (space) direction, the actual path of light is a spiral in four dimensions, Figure 3.3.4-2.

The observed optical distance is the tangential length of the light path, i.e., the distance light travels in space. All the time during the propagation, the velocity of light in space, the tangential velocity component, c_{Re}, is equal to the velocity c_{Im} in the imaginary direction, which in homogeneous space is equal to the expansion velocity of space, c_4, along the R_4 radius. This means that the optical distance in space is equal to the increase in the 4-radius of space during the signal transmission time. Electromagnetic radiation carries momentum only in the direction of propagation. As shown in Section 4.1.8, light propagating in space has zero momentum in the imaginary direction.

The difference between the physical and optical distances is small as long as the distance is small compared to the length of the 4-radius, but becomes meaningful for objects at high distances. As a consequence, the linear Hubble law applies for objects at small distances but must be modified in the case of cosmologically distant objects (see Section 6.1.2).

The optical distance of stellar objects does not exceed the current length of the 4-radius of space but approaches it for observations of events close in time to the singularity of space.

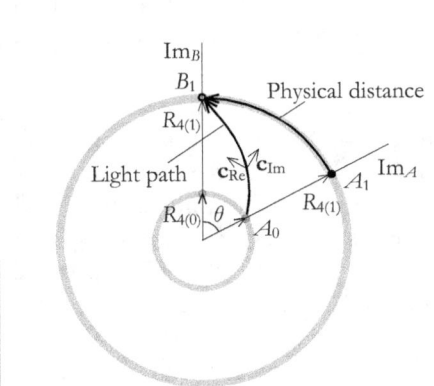

Figure 3.3.4-2. The physical distance from object A to observer B at the time T_1 when the 4-radius $R_4 = R_{4(1)}$ is equal to the arc $A_1B_{1phys} = s_{A1B1} = \theta_{AB}R_{4(1)}$. The optical distance is equal to the tangential component of the spiral light path from A_0 to B_1. The tangential component is the distance in space, in the direction of the real axis in the complex coordinate system. Because, throughout the traveling path, the velocity of light in space is equal to the velocity of space in the imaginary direction the optical distance light travels in space is equal to the increase of R_4 radius from $R_{4(0)}$.

3.3.5 From mass to matter

The process of the contraction and expansion of space in the four-dimensional Universe is referred to as the process of primary energy buildup and release. The process of primary energy buildup energizes mass by putting it into motion and into closer gravitational interaction with other mass in the contraction of space. The release of energy occurs in the expansion phase, restoring the pre-contraction state.

Matter is energized mass. In its initial form, matter is considered as unstructured *dark matter*.

Equation (3.3.1:4) shows the twofold nature of the energy of matter

$$E''_{tot(0)} = E''_{m(0)} + E''_{g(0)} = mc_0^2 - \frac{GmM''}{R''_0} = 0, \qquad (3.3.5:1)$$

where the distance to the mass equivalence of hypothetical homogeneous space is denoted as $R''_0 = R_4$.

Arithmetically, the total energy of matter is zero — the sum of the positive energy of motion and the negative energy of gravitation. The absolute values of the imaginary energies of motion and gravitation are thus a measure of the energy excitation of matter.

Matter with localized expression takes the form of elementary particles and material. The primary energy buildup is described as a process for hypothetical homogeneous space. Accordingly, the primary energy buildup may not create localized structures needed for the expression of baryonic matter or material forms.

Equation (3.3.5:1) describes the twofold nature of matter manifesting itself through the energy of motion and the energy of gravitation. The balance of the energies of motion and gravitation can be understood as the excited state of two complementary forms of energy. As shown by equations (3.3.3:12) and (3.3.3:13), the excitation amplitude of the energies of motion and gravitation decreases as the Universe expands, Figure 3.3.5-1.

Throughout the process, the rest energy is balanced by the energy of gravitation. In the course of the expansion, the rest energy of matter is fading away to zero at infinity when $R_4 \rightarrow \infty$.

At infinity in the future, all motion gained from gravity in the contraction will have been returned to the gravitational energy of the structure. Mass will no longer be observable because the energy excitation of matter will have vanished along with the cessation of motion. The energy of gravitation will also become zero owing to the infinite distances. The cycle of observable physical existence begins in emptiness and ends in emptiness, where the mass does not express itself as observable matter. Mass, as the substance of the expression of energy, however, is conserved throughout the cycle. The DU model does not exclude the possibility of a new cycle of physical existence.

The rest energy, the energy of motion due to the motion of space in the fourth dimension, can be considered as a localized manifestation of the energy of matter, which is in counterbalance with the non-localized manifestation of the energy of matter, the energy of gravitation.

We do not need to assume the existence of antimatter to balance the rest energy of matter.

At infinity in the past, as at infinity in the future, the 4-radius of space is infinite. Mass exists, but as it is not energized, it is not detectable. The energy of motion built up in the primary energy buildup is gained from the structural energy, the energy of gravitation. In contraction, space loses size and gains motion. In expansion, space loses motion and gains size.

> The buildup and disappearance of the physically observable Universe occurs as an inherently driven zero-energy process from emptiness at infinity in the past through singularity to emptiness at infinity in the future – or "essentially infinity" allowing a cycling universe.

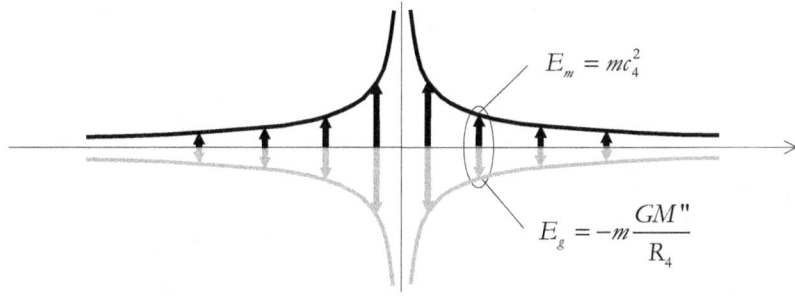

Figure 3.3.5-1. The twofold nature of matter at rest in space is manifested by the energies of motion and gravitation. The intensity of the energies of motion and gravitation declines as space expands along the 4-radius.

4. Energy structures in space

The primary energy buildup is described in terms of the dynamics of the whole space in the direction of the 4-radius. The primary energy buildup creates the energy of motion against the reduction of the global gravitational energy.

In the primary energy buildup, the total mass of the Universe is considered to be uniformly distributed throughout space. Mass in hypothetical homogeneous space is considered as unstructured wavelike dark matter energized by the motion of space in the fourth dimension. Conversion of dark matter into electromagnetic radiation and primordial nucleons may occur at the turnover of the contraction phase into the ongoing expansion phase, and further to atomic structures in nucleosynthesis as assumed in Big Bang cosmology. Such processes may also occur in local mass center buildup, in secondary energy buildup processes in space. The secondary energy buildup processes are assumed to conserve the total energy and the overall zero-energy balance in space.

The Dynamic Universe model does not give an unambiguous answer to the conversion of unstructured matter into electromagnetic radiation or the environment for nucleosynthesis. Such processes may occur as a consequence of a certain asymmetry in passing the singularity when the contraction of space is turned into expansion. Such a process could have much the same properties as assumed to have taken place in the first seconds of the Big Bang. It turns out that the conditions in the vicinity of local singularities in space, like in the centers of galaxies, may also be favorable for conversions of dark matter to radiation and baryonic matter conversions.

The energy structures of DU-space are described in terms of energy frames from galactic structures to atomic objects and elementary particles. Conservation of the energy excitation created in the contraction and expansion of space creates an unbroken chain of frames linked from the smallest elementary particle to the whole of spherical space. The Earth, along with the objects bound to its gravitational frame, can be regarded as an energy object in the solar gravitational frame, and an electron in an atom as an energy object in the electromagnetic frame of the nucleus.

While the dynamics of space as a homogeneous spherical structure produce the basis for predictions at a cosmological scale, the analysis of energy structures in space produces the basis for predictions for local phenomena and celestial mechanics.

DU space is characterized by a system of nested energy frames. Relativity in DU space appears as relativity of local to the whole – any local energy state is related, through the system of nested energy frames, to the state of rest in hypothetical homogeneous space, which serves as a universal frame of reference.

Relativity in DU space is a consequence of the conservation of total energy in space. Relativity is expressed in terms of locally available energy, not in terms of locally distorted metrics, as it is expressed in the theory of relativity.

4.1 The zero-energy balance

The initial condition produced by the primary energy buildup is regarded as a homogeneous spherical entity with all mass at rest, i.e., with momentum only in the direction of the 4-radius of the structure. Accordingly, the buildup of inhomogeneity requires motion of mass in space. The buildup of a local mass center in space is described in terms of free fall of mass, conserving the primary energies of motion and gravitation created in the primary energy buildup of space.

4.1.1 Conservation of energy in mass center buildup

Mass center buildup in homogeneous space

The primary energy buildup is based on spherical symmetry, which results in motion in the direction of the 4-radius of spherical space (the direction of the imaginary axis in hypothetical homogeneous space). The energy of the imaginary motion is balanced by gravitational energy from all mass in space, uniformly in all space directions relative to any space location. As a result of spherical symmetry, the gravitational energy of all mass is equivalent to inherent gravitational energy due to the mass equivalence M'' located in the direction of the imaginary axis at distance R'' in the imaginary direction, which in homogeneous space is equal to the direction of the 4-radius of spherical space, $R''_0 = R_4$. The zero-energy balance of motion and gravitation for a mass m at rest in hypothetical homogeneous space is expressed in equation (3.3.1:4) [see Figure 4.1.1-1(a)] as

$$E''_{tot(0)} = E''_{m(0)} + E''_{g(0)} = c_0 |\mathbf{p}''_0| - \frac{GmM''}{R''_0} = c_0 mc_0 - \frac{GmM''}{R''_0} = 0 . \tag{4.1.1:1}$$

In hypothetical homogeneous space, the energy excitation of motion and gravitation expressed in equation (4.1.1:1) appears in the direction of the Im_0 axis in the direction of the 4-radius of space.

In Section 3.2.2, the mass equivalence M'' and the gravitational energy $E''_{g(0)}$ were calculated by integrating the effects of masses dM in volume differentials in spherical shells surrounding a mass m at the center. Let's assume that a mass $M = dM(r_{0\delta})$ at distance r_0 ($m \ll M \ll M''$ and $r_0 \ll R_4$) is gathered up and condensed into a mass center at distance r_0 in a space direction denoted by the Re_0 axis. Due to the removal of mass M from the symmetry, the global gravitational energy, the gravitational energy due to the remaining, uniformly distributed mass, is reduced by (see equation (3.2.2:1))

$$dE''_g = -\frac{Gm}{D} dM = -\frac{Gm}{r_0} M = \Delta E''_g , \tag{4.1.1:2}$$

where $D = r_0$ is the distance (radius) of a spherical volume differential with mass $dM = \varrho \cdot 4\pi D^2 dr_0$ around the test mass m.

Energy structures in space 105

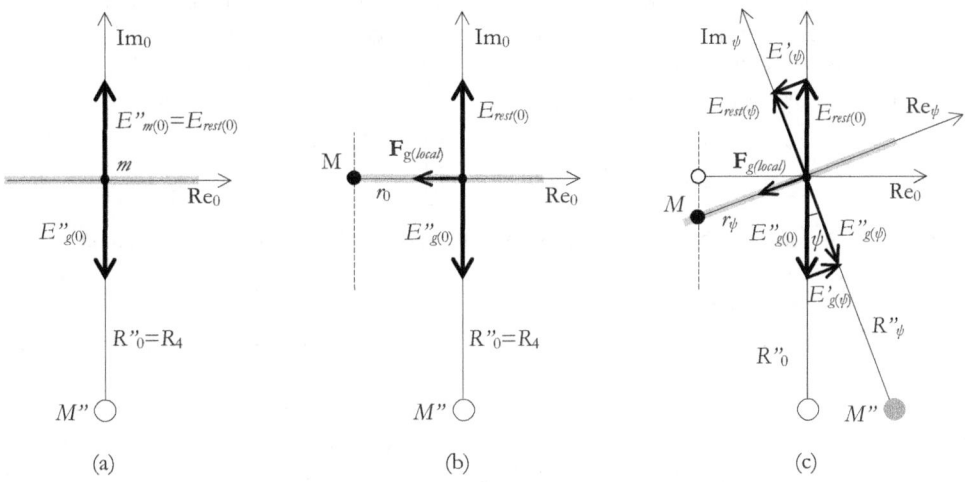

Figure 4.1.1-1. The balance of motion and gravitation. (a) The initial condition for energy interactions in space is the state of rest in hypothetical homogeneous space. In homogeneous space, mass is uniformly distributed throughout space, and the imaginary axis is in the direction of the 4-radius of space. An object at rest in homogeneous space has the energies of motion and gravitation in the imaginary direction only. (b) The uniformity of mass is disturbed, and the initial symmetry of motion and gravitation is broken when a mass center M is formed at a distance r_0 from a mass m in space in the direction of the Re_0 axis. Gravitational force $\mathbf{F}_{g(local)}$ towards the mass M is created. (c) The balance between the imaginary energies of motion and gravitation is re-established when local space is tilted by angle ψ. The rest energy $E_{rest(0)}$ is reduced through the buildup of the real part $E'_{(\psi)}$ as the energy equivalence of the momentum of free fall. An equivalent reduction $\Delta E''_g = \Delta E_{rest} = E''_{g(\psi)} - E''_{g(0)}$ occurs in the global gravitational energy. In the direction of the Re_ψ axis, the apparent distance from m to mass center M is denoted as r_ψ.

The formation of mass center M at distance r_0 from mass m in the direction of the real axis Re_0 in space creates a net gravitational force resulting in free fall of mass m towards mass M, Figure 4.1.1-1(b).

Creation of the momentum of free fall in space, orthogonal to the momentum in the fourth dimension, while simultaneously conserving the total primary momentum, requires that the direction of the fourth dimension becomes tilted. Tilting of space near a mass center creates the momentum of free fall by dividing the primary momentum, $p_0 = mc_0$, in the direction of the R_0-axis, into orthogonal components with the real part, the momentum of free fall, p_{ff}, in the direction of the tilted space, the Re_ψ-axis, and the imaginary part, p''_{ff}, in the direction of the Im_ψ axis perpendicular to the tilted space, Figure 4.1.1-1(c).

The total energy of motion is now expressed as the energy related to the vector sum of the local imaginary momentum (rest momentum) and the escape momentum back to homogeneous space (far enough from mass M)

$$\mathbf{i}_0 E''_{m(0)} = E_{m,tot(\psi)} = c_0 |\mathbf{p}''_0| = c_0 |\mathbf{p}'_{esc(\psi)} + \mathbf{i}_\psi p''_\psi|$$
$$= c_0 \sqrt{p^2_{esc(\psi)} + (mc_\psi)^2}, \qquad (4.1.1:3)$$

where $\mathbf{p'}_{esc(\psi)}$ is the escape momentum (opposite to the momentum of free fall, $\mathbf{p'}_{esc(\psi)}=-\mathbf{p'}_{ff(\psi)}$) from space tilted by angle ψ back to homogeneous space. This simply means that compared to the rest energy in homogeneous space, the rest energy of mass m at distance r_0 from mass M in space is reduced by the kinetic energy of free fall

$$E_{kin(ff)} = \Delta E''_{m(\psi)} = c_0 \left[\sqrt{p_{ff(\psi)}^2 + (mc_\psi)^2} - mc_\psi \right], \qquad (4.1.1{:}4)$$

where $|p_{ff(\psi)}| = |p_{esc(\psi)}|$, and the rest energy of mass m in space tilted by angle ψ is

$$E''_{m(\psi)} = E_{rest(\psi)} = c_0 m c_\psi = c_0 m c, \qquad (4.1.1{:}5)$$

where the local velocity of light c is equal to c_ψ, the local imaginary velocity of space in the direction of the Im$_\psi$ axis.

In complete symmetry with equation (4.1.1:3), the global energy of gravitation in space tilted by angle ψ can be expressed in complex form with the locally observed global gravitational energy opposite to the local imaginary energy of motion $E''_g = -E''_m = -c_0|\mathbf{p''}|$ and the real part opposite to the energy equivalence of the escape momentum $E'_g = -E'_m = -c_0|\mathbf{p}_{esc}|$, Figure 4.1.1:1(c)

$$i_0 E''_{g(0)} = E_{g,tot(\psi)} = E'_{g(\psi)} + i_\psi E''_{g(\psi)} = \sqrt{E'^2_{g(\psi)} + E''^2_{g(\psi)}}, \qquad (4.1.1{:}6)$$

where, with reference to equations (3.2.2:7) and (4.1.1:2), the locally observed global energy of gravitation $E''_{g(\psi)} = E''_{g(0)} - \Delta E_{g(\psi)}$ is

$$E''_{g(\psi)} = -\left(\frac{GM''m}{R''_0} - \frac{GMm}{r_0} \right) = -\frac{GM''m}{R''_0}\left(1 - \frac{MR''}{M''r_0}\right). \qquad (4.1.1{:}7)$$

The term $MR''/M''r_0$ in (4.1.1:7) is referred to as the gravitational factor δ. Applying the zero-energy balance in equation (3.3.5:1), the gravitational factor defining a gravitational state in tilted space at distance r_0 from mass center M formed in hypothetical homogeneous space can be expressed in the forms

$$\delta = \frac{MR''}{M''r_0} = \frac{GM}{r_0 c_0^2}. \qquad (4.1.1{:}8)$$

Substitution of (4.1.1:8) into (4.1.1:7) gives

$$E''_{g(\psi)} = -\frac{GM''M_\Sigma}{R''_0}\left(1 - \frac{GM}{r_0 c_0^2}\right) = E_{g(0)}(1 - \delta). \qquad (4.1.1{:}9)$$

In terms of the tilting angle ψ, the global gravitational energy $E''_{g(\psi)}$ in tilted space is

$$E''_{g(\psi)} = E''_{g(0)} \cos\psi, \qquad (4.1.1{:}10)$$

Combining equations (4.1.1:9) and (4.1.1:10), the cosine of the tilting angle can be expressed in terms of the gravitational factor δ

$$\cos\psi = 1 - \delta. \qquad (4.1.1{:}11)$$

Equations (4.1.1:3) and (4.1.1:6) express the conservation of the primary energies of motion and gravitation as a consequence of the tilting of local space near mass center M.

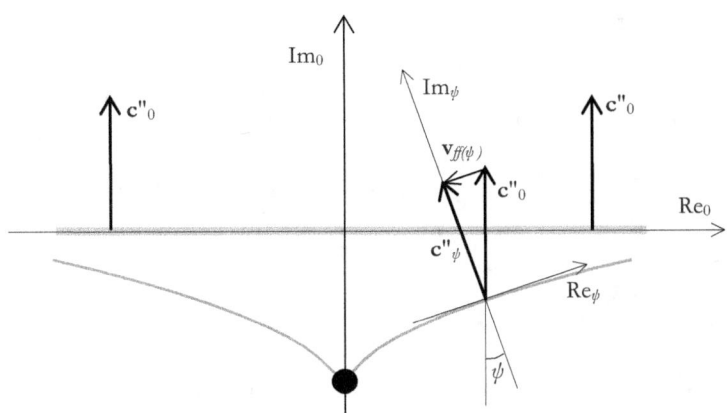

Figure 4.1.1-2. As a consequence of the conservation of the primary energies of motion and gravitation, the buildup of a mass center in space bends the spherical space locally, causing a tilting of space near the mass center. The local imaginary axis is always perpendicular to local space. As a consequence, the local imaginary velocity of space is reduced in tilted space.

Conservation of mass and the primary energy in free fall in space through tilting of space near mass centers means that the velocity of free fall is obtained from the expansion velocity of space

$$v_{f\!f(\psi)} = c_0 \sin\psi. \qquad (4.1.1{:}12)$$

The local velocity of light equal to the imaginary velocity in tilted space can be expressed (Figure 4.1.1-2)

$$c = c_\psi = c_0 \cos\psi = c_0(1-\delta), \qquad (4.1.1{:}13)$$

where the last form is obtained by substitution of (4.1.1:11) for $\cos\psi$. For consistency with common praxis, the local velocity of light is denoted as c ($c = c_\psi$).

Substitution of (4.1.1:13) for c in (4.1.1:5) gives the locally available rest energy of an object at rest at gravitational state δ in tilted space

$$E_{rest(\psi)} = c_0 m c_0 \cos\psi = E_{rest(0)} \cos\psi = E_{rest(0)}(1-\delta). \qquad (4.1.1{:}14)$$

For an object at rest in space tilted by angle ψ, the zero-energy balance of the local rest energy and global gravitational energy is expressed as the equality of equations (4.1.1:9) and (4.1.1:14) as

$$E_{rest(\psi)} = E''_{g(\psi)} \quad \Rightarrow \quad E''_{rest(0)}(1-\delta) = E''_{g(0)}(1-\delta). \qquad (4.1.1{:}15)$$

When related to the local velocity of light in tilted space (4.1.1:13), the velocity of free fall (4.1.1:12) becomes

$$v_{f\!f(\psi)} = c_0 \sin\psi = \frac{c\sin\psi}{\cos\psi} = c\tan\psi. \qquad (4.1.1{:}16)$$

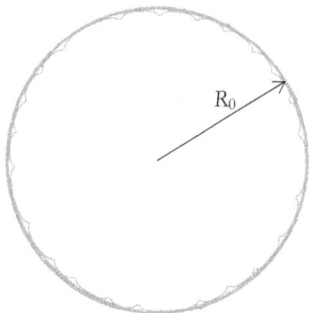

Figure 4.1.1-3. Real space is not a smooth 4-sphere but textured by dents around mass centers in space. The radius R_0 of homogeneous space is interpreted as average 4-radius of "free space" between mass centers.

Mass center buildup in real space

As a result of the conservation of total gravitational energy in the buildup of mass centers, real space "the smooth 4-sphere" becomes textured by dents formed around mass centers, Figure 4.1.1-3.

Mass center buildup occurs in many steps. Gathering of mass into a mass center in tilted space can be described in full analogy to the buildup of a "first-order" mass center in hypothetical homogeneous space. The imaginary energies of motion and gravitation at a distance r_A from mass a center M_A, where space is tilted by angle ψ_B relative to homogeneous space, are

$$E''_{m(B)} = E''_{m(0)} \cos\psi_B = c_0 m c_0 \cos\psi_B, \tag{4.1.1:17}$$

and

$$E''_{g(B)} = E''_{g(0)} \cos\psi_B = -\frac{GM''m}{R''_0}\cos\psi_B. \tag{4.1.1:18}$$

When a mass center M_B is created at a distance r_A from M_A via accumulation of nearby mass, a local sub-dent is formed in tilted space in the gravitational frame M_A. The tilted space at distance r_A from M_A serves as the apparent homogeneous space for the sub-dent formed around mass center M_B. The buildup of M_B occurs in full analogy to the buildup of mass center M_A in hypothetical homogeneous space, Figure 4.1.1-4.

The imaginary energies to be conserved in the accumulation of mass into a local mass center M_B at distance r_B from location B in tilted space are the imaginary energies of motion and gravitation $E''_{m(B)}$ and $E''_{g(B)}$ in equations (4.1.1:17) and (4.1.1:18), respectively. For mass m in the sub-dent around M_B at distance r_B from mass center M_B, the imaginary energies of motion and gravitation, the local rest energy, and global gravitational energy are

$$E''_m = E''_{m(B)} \cos\psi = E''_{m(0)} \cos\psi_B \cos\psi, \tag{4.1.1:19}$$

and

$$E''_g = E''_{g(B)} \cos\psi = E''_{g(0)} \cos\psi_B \cos\psi, \tag{4.1.1:20}$$

where ψ is the tilting angle of the local space at distance r_B from mass center M_B.

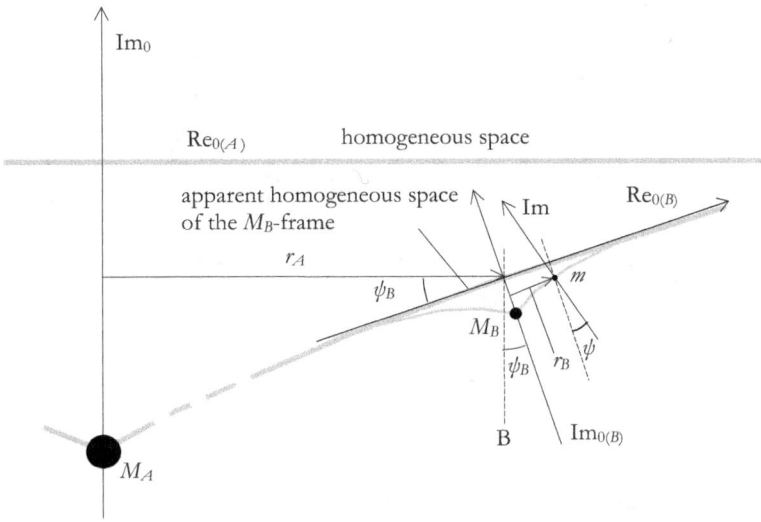

Figure 4.1.1-4. The profile of space in the vicinity of local mass centers. Each mass center causes local tilting of space in its neighborhood relative to the surrounding space, referred to as apparent homogeneous space and, finally, to hypothetical homogeneous space. In the figure, the M_A-frame has been formed in hypothetical homogeneous space where all mass is uniformly distributed and where the imaginary axis has the direction of the 4-radius of space, $\mathrm{Im}_{0(A)} \| \mathrm{Im}_0 \| R_0$. The local imaginary axis at test mass m is denoted as Im, and the distance from m to the local mass center M_B is denoted as r_B.

Generally, the imaginary energies of motion and gravitation of mass m in the n:th subframe can be related to imaginary energies in the $(n-1)$:th frame, which serves as the parent frame and the apparent homogeneous space to the local frame. The local imaginary energy of motion at a location where space in the local frame has tilted by an angle ψ_n is

$$E''_{m(n)} = E''_{m(n-1)} \cos\psi_n = c_0 m c_{n-1}(1-\delta_n) = c_0 m c, \qquad (4.1.1:21)$$

where the local velocity of light c is determined by the velocity of space in the local fourth dimension. The local velocity of light is related to the velocity of light in the parent frame as

$$c_n = c = c_{n-1}\cos\psi_n = c_{n-1}(1-\delta_n), \qquad (4.1.1:22)$$

or

$$c = c_\delta = c_{0\delta}(1-\delta), \qquad (4.1.1:23)$$

where the gravitational factor δ means the gravitational factor of the object in the local frame. The velocity c_δ, which is generally denoted as c, means the local velocity of light at a gravitational state defined by δ, and the velocity $c_{0\delta}$ means the velocity of light in an apparent homogeneous space of the local frame, Figure 4.1.1-5.

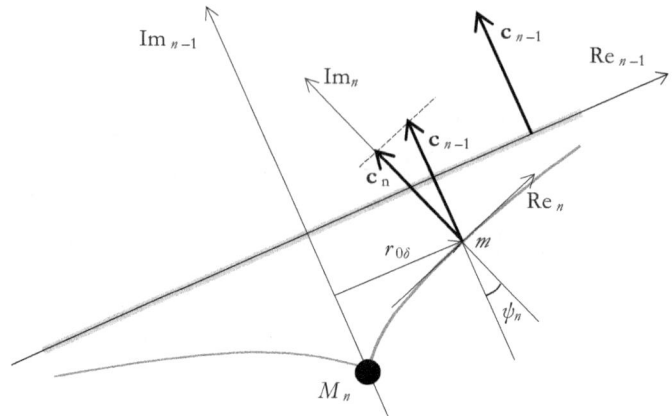

Figure 4.1.1-5. The velocity of light is determined by the velocity of space in the local fourth dimension. Following the conservation of the total energy in local mass center buildup, the local velocity of light is related to the velocity of light in apparent homogeneous space of the local frame. Using notations based on the local gravitational factor δ, the local velocity of light is $c = c_\delta = c_n$, and the velocity of light in apparent homogeneous space of the local frame $c_{0\delta} = c_{n-1}$.

The imaginary energy of gravitation in the n:th frame is

$$E''_{g(n)} = E''_{g(n-1)} \cos\psi_n = -\frac{GM''m}{R''_{n-1}}(1-\delta) = -\frac{GM''m}{R''_n}, \qquad (4.1.1{:}24)$$

where the local apparent 4-radius of space R'', which is the local apparent distance to mass equivalence M'', is

$$R'' = R''_n = \frac{R''_{n-1}}{1-\delta} \quad \text{or} \quad R'' = R''_\delta = \frac{R''_{0\delta}}{1-\delta}. \qquad (4.1.1{:}25)$$

Following the same procedure for the imaginary energies in the parent frame and "the grandparent frames", the local imaginary energies of motion and gravitation are finally related to the imaginary energies of motion and gravitation in hypothetical homogeneous space as

$$E''_{m(n)} = E''_{m(0)} \prod_{i=1}^{n} \cos\psi_i = c_0 m c_0 \prod_{i=1}^{n}(1-\delta_i) = c_0 mc \qquad (4.1.1{:}26)$$

and

$$E''_{g(n)} = E''_{g(0)} \prod_{i=1}^{n} \cos\psi_i = -\frac{GM''m}{R''_0}\prod_{i=1}^{n}\cos\psi_i$$
$$= -\frac{GM''m}{R''_0}\prod_{i=1}^{n}(1-\delta_i) = -\frac{GM''m}{R''}, \qquad (4.1.1{:}27)$$

respectively.

As implicitly stated in equations (4.1.1:26) and (4.1.1.27), the local velocity of light c, and the local apparent distance $R"$ to mass equivalence $M"$ are

$$c = c_n = c_0 \prod_{i=1}^{n}(1-\delta_i), \qquad (4.1.1:28)$$

and

$$R" = R"_n = R"_0 \Big/ \prod_{i=1}^{n}(1-\delta_i), \qquad (4.1.1:29)$$

where the gravitational factor δ_i is

$$\delta_i = 1 - \cos\psi_i = \frac{M_i R"_{i-1}}{M" r_{i-1}} = \frac{GM_i}{r_{i-1} \cdot c_0 c_{i-1}} = \frac{GM_i}{r_{0\delta} \cdot c_0 c_{0\delta_i}} \approx \frac{GM_i}{r_i c^2}. \qquad (4.1.1:30)$$

The notation c means generally the velocity of light in the local frame, and $c_{0\delta}$ the velocity of light in the apparent homogeneous space of the local frame. The notation $R"$ means the apparent local distance to mass equivalence of space $M"$, and r means the flat space distance to the mass center in the local gravitational frame. In space directions, the distance $r_{0\delta}$ means the flat space distance to the mass center of the local gravitational frame, i.e., the distance in the direction of the apparent homogeneous space of the local gravitational frame, and the distance r_δ means the distance in the direction of local space, Figures 4.1.1-6 and 4.1.1-7.

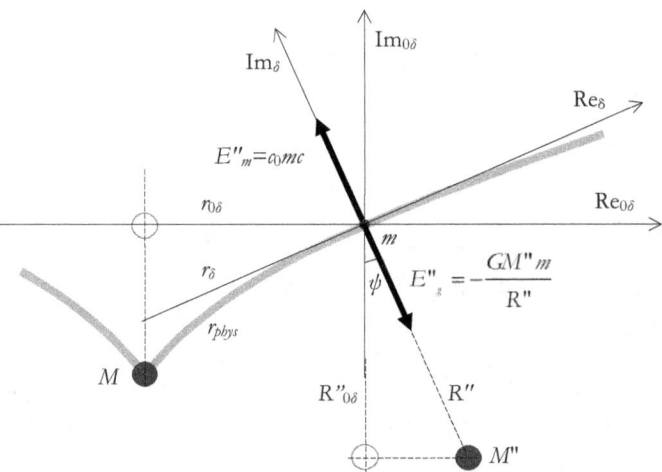

Figure 4.1.1-6. The imaginary energies of motion and gravitation in a δ state have the direction of the local imaginary axis $\mathrm{Im} = \mathrm{Im}_\delta$ tilted by an angle ψ from the direction of the imaginary axis $\mathrm{Im}_{0\delta}$ in apparent homogeneous space. The local rest energy $E_{rest} = E"_m$ is balanced by the locally observed global gravitational energy $E"_g$. The distance $r_{0\delta}$ is the flat space distance from m to the local mass center M measured in the direction of the apparent homogeneous space of the local gravitational frame, the $\mathrm{Re}_{0\delta}$ axis. The distance r_δ is the apparent distance to M in the direction of the local Re_δ axis. The physical distance following the curved shape of space is r_{phys}.

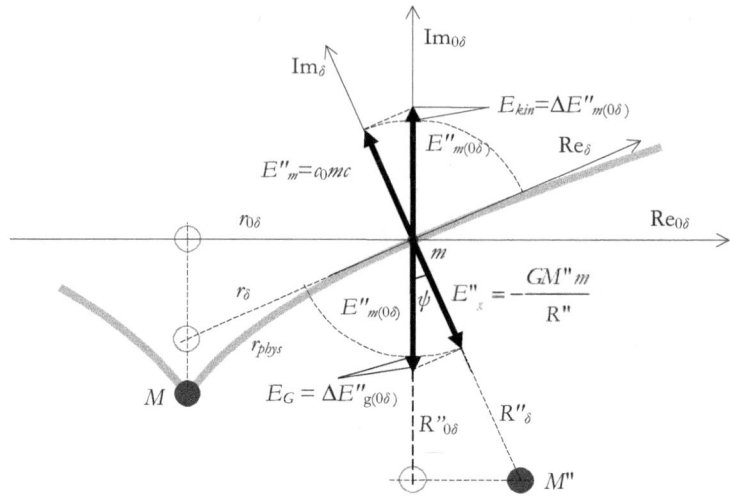

Figure 4.1.1-7. As demanded by the conservation of the total momentum and the energies of motion and gravitation, space is tilted in the direction of the fourth dimension near mass centers. The imaginary axis of local space makes an angle ψ with the imaginary axis of apparent homogeneous space. The total momentum $\mathbf{p}''_0$ of mass m in homogeneous space is conserved as the vector sum of the local imaginary momentum $\mathbf{p}''_\delta$ and the escape momentum $\mathbf{p}'_{esc(\delta)}$ in the direction of the local real axis.

The distance definitions, the apparent distance r_δ in the direction of the local Re_δ axis, the flat space distance $r_{0\delta}$ in the direction of the $\mathrm{Re}_{0\delta}$ axis, and the physical distance measured along the curved space, are illustrated in Figure 4.1.1-6.

The local gravitational energy, the energy of gravitation converted into kinetic energy in free fall from an infinite distance to a distance $r_{0\delta}$ in the local gravitational frame, is

$$E_G = \Delta E_{g(\delta)} = -\frac{GMm}{r_{0\delta}}. \qquad (4.1.1{:}31)$$

E_G describes the release of the global gravitational energy due to the tilting of space as a consequence of the buildup of mass center M, Figure 4.1.1-7.

As illustrated in equation (4.1.1:31), the local gravitational energy E_G has the Newtonian form for distance $r_{0\delta}$ measured in the flat space direction. Newtonian gravitation is expressed in terms of the distance r_δ measured in the direction of the local Re axis

$$E_{Newton} = -\frac{GMm}{r_\delta} = \frac{GMm}{r_{0\delta}}(1-\delta). \qquad (4.1.1{:}32)$$

4.1.2 Kinetic energy

The buildup of kinetic energy in free fall and at constant gravitational potential is compared, and a general expression for kinetic energy is introduced. In free fall, the velocity in space is obtained against the reduction of the velocity of space in the local fourth

Energy structures in space

dimension, and the kinetic energy against the reduction in the locally available rest energy. Kinetic energy at constant gravitational potential requires an insertion of energy from the local accelerating system, such as Coulomb energy, which is described as an insertion of mass equivalence, increasing the mass of the object in motion.

The connection between kinetic energy and momentum is analyzed. It is shown that the imaginary part of kinetic energy is the work done against the gravitational energy of the total mass in spherical space, thus giving a quantitative expression to Mach's principle.

Kinetic energy obtained in free fall

The *kinetic energy* of an object moving in a local frame is defined as the total energy of motion minus the energy of motion the object has at rest in the local frame (2.1.4:8). The total energy of motion of an object in free fall from the state of rest far from the local mass center is, Figure 4.1.2-1

$$E_{m(total)} = c_0 |\mathbf{P}_{total}| = c_0 |\mathbf{P}_{\delta(Re)} + \mathbf{P}_{\delta(Im)}| = c_0 |\mathbf{P}_{0(Im)}| = c_0 m c_{0\delta}. \tag{4.1.2:1}$$

The energy of motion of an object at rest in a gravitational state δ is the imaginary energy of motion in the local fourth dimension

$$E_{rest(\delta)} = c_0 |\mathbf{P}_{\delta(Im)}| = c_0 m c_{\delta} = c_0 m c, \tag{4.1.2:2}$$

and the kinetic energy obtained in free fall from the state of rest far from the local mass center is

$$\begin{aligned} E_{kin(ff)} &= E_{m(total)} - E_{rest} = c_0 \Delta |\mathbf{p}^\square| = c_0 m (c_{0\delta} - c) \\ &= c_0 m \Delta c = c_0 m c_{0\delta} [1 - (1 - \delta)] = c_0 m c_{0\delta} \delta. \end{aligned} \tag{4.1.2:3}$$

Equation (4.1.2:3) means that kinetic energy in free fall is obtained against the reduction of the local rest energy via tilting of space and the associated reduction in the local velocity of light. The total energy of motion, as the sum of local rest energy and the kinetic energy of free fall, is conserved.

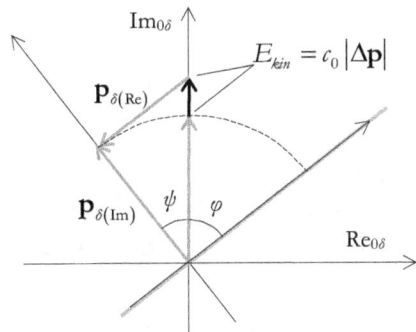

Figure 4.1.2-1. Kinetic energy in free fall by change in the local rest momentum via tilting of space by $\psi = \pi/2 - \varphi$. The total energy of motion is conserved. The local rest energy is reduced.

Kinetic energy obtained via the insertion of mass

In free fall, kinetic energy is obtained against the reduction of the local rest energy via the reduction of the velocity of light in tilted space. In free fall, mass is conserved. Buildup of kinetic energy at constant gravitational potential, when the velocity of light is constant, requires the insertion of local energy in the form of mass or mass equivalence to create momentum in a spatial direction. Insertion of a mass Δm via acceleration of a charged mass object initially at rest in a Coulomb energy frame (see Section 5.1.2) adds to the total energy of motion

$$E_{m(tot)} = E_{rest} + \Delta E_{EM} = E''_0 + \frac{q_1 q_2 \mu_0}{4\pi}\left(\frac{1}{r_2} - \frac{1}{r_1}\right) c_0 c \qquad (4.1.2:4)$$
$$= E_{rest} + \Delta m_{EM} \cdot c_0 c,$$

where Δm_{EM} is the mass equivalence released by Coulomb energy. The kinetic energy gained is equal to the Coulomb energy released. As given in the last term of (4.1.2:4) the kinetic energy can be expressed in terms of Coulomb mass equivalence Δm_{EM}

$$E_{kin} = \Delta E_{EM} = c_0 c \Delta m_{EM} = c_0 c \Delta m, \qquad (4.1.2:5)$$

and the total energy of motion can be expressed in the form

$$E_{m(tot)} = E_{rest} + E_{kin} = c_0 mc + c_0 \Delta m \cdot c = c_0 c (m + \Delta m). \qquad (4.1.2:6)$$

A complex presentation of the total energy of motion illustrates the buildup of the momentum and the total energy of motion as the orthogonal sum of the momentum at rest in the imaginary direction and the momentum created in space

$$E_{m(tot)} = c_0 |\mathbf{p}^\square| = c_0 |\mathbf{p}' + i\,\mathbf{p}_0''| = c_0 |(m + \Delta m)\mathbf{v} + i\,mc| \qquad (4.1.2:7)$$
$$= c_0 \sqrt{(mc)^2 + (m + \Delta m)^2 (\beta c)^2},$$

where the velocity in space is denoted as $\mathbf{v} = \beta c \hat{\mathbf{r}}$. The increased mass $(m+\Delta m)$ contributes to the real component of the momentum via acceleration in Coulomb field, Figure 4.1.2-2.

Combining of equations (4.1.2:6) and (4.1.2:7) gives

$$(m + \Delta m)^2 = m^2 + (m + \Delta m)^2 \beta^2, \qquad (4.1.2:8)$$

and further, by solving the total mass, m_β, of the moving object,

$$m_\beta = m + \Delta m = \frac{m}{\sqrt{1-\beta^2}} = m_{rel}. \qquad (4.1.2:9)$$

As shown by equation (4.1.2:10) the increased mass resulting from the additional mass Δm needed to obtain velocity $v = \beta c$ in space is equal to the relativistic mass or *relativistic mass* m_{rel} in the theory of relativity.

The increase of the mass of an object in motion in space is not a property of the velocity, but the contribution of mass or mass equivalence from the system releasing the energy converted into kinetic energy.

Energy structures in space

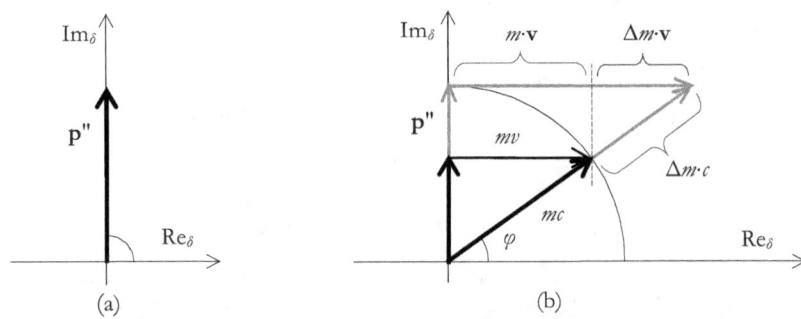

Figure 4.1.2-2. The momentum $\mathbf{p'} = (m+\Delta m)\cdot \mathbf{v}$ in a space direction results in velocity $v = c\cdot\cos\varphi$ in space. Velocity c is the local velocity of light equal to the local imaginary velocity of space.

Conversion of gravitational energy into kinetic energy in free fall is not associated with the exchange of mass, but the kinetic energy is obtained against reduction of the rest energy via reduction of the velocity of space in the local fourth dimension due to tilting of space.

Applying the increased mass $m_\beta = m+\Delta m = m_{rel}$, the total energy of motion can be expressed

$$E_{tot} = c_0 \left| p^{\square} \right| = c_0 \left| p' + \mathrm{i}\, p'' \right| = c_0 \left| m_\beta \beta c + \mathrm{i}\, mc \right|$$

$$= c_0 mc \sqrt{\frac{\beta^2}{1-\beta^2}+1} = \frac{c_0 mc}{\sqrt{1-\beta^2}}. \tag{4.1.2:10}$$

Substitution of (4.1.2:10) for $E_{m(tot)}$ in (4.1.2:6) gives the kinetic energy

$$E_{kin} = E_{m(tot)} - E_{rest} = c_0 mc \left(\frac{1}{\sqrt{1-\beta^2}} - 1 \right). \tag{4.1.2:11}$$

The expression for the total energy of motion in (4.1.2:10) and kinetic energy in (4.1.2:12) are equal to the total energy and kinetic energy derived based on the Lorentz transformation in the special theory of relativity (assuming $c_0 \approx c$). In the DU framework, there is no need or role for the Lorentz transformation. Following the conservation of the total energy, the mass increase Δm in the buildup of kinetic energy is just the mass or mass equivalence transferred from the system, releasing the energy for the buildup of kinetic energy.

In (4.1.2:10), the real component of the complex energy of motion is

$$E'_{m(tot)} = c_0 \left| \mathbf{p'} \right| = c_0 \left| m_\beta \beta c \right| = \frac{c_0 mc \cdot \beta}{\sqrt{1-\beta^2}}, \tag{4.1.2:12}$$

where the momentum in space is

$$\mathbf{p} = \mathbf{p}' = \frac{m}{\sqrt{1-\beta^2}} \mathbf{v} = m_\beta \mathbf{v} = m_\beta \beta c \hat{\mathbf{r}}, \qquad (4.1.2:13)$$

which corresponds to the momentum in the framework of special relativity, but without the Lorentz transformation, the relativity principle, or postulated invariance of the velocity of light.

Kinetic energy obtained in free fall and via the insertion of mass

Obtaining kinetic energy in free fall in gravitation and via the insertion of mass equivalence at constant gravitational potential, where the local velocity of light is constant, can be compared by studying equations (4.1.2:3) and (4.1.2:6)

$$E_{kin(ff)} = E_{kin(\Delta c)} = c_0 \Delta |\mathbf{p}| = c_0 \cdot m\Delta c, \qquad (4.1.2:14)$$

$$E_{kin(\Delta m)} = c_0 \Delta |\mathbf{p}| = c_0 \cdot c\Delta m = c_0 m c_0 \left(\frac{1}{\sqrt{1-\beta^2}} - 1 \right). \qquad (4.1.2:15)$$

Equation (4.1.2:14) describes the kinetic energy obtained from gravitational energy in free fall from apparent homogeneous space to gravitational state δ in the local gravitational frame, and equation (4.1.2:15) describes the kinetic energy obtained from local potential energy, such as Coulomb energy in the local energy frame. In the case of free fall, the kinetic energy is obtained against the reduction of the velocity of space in the local fourth dimension, which also means a reduction of the local velocity of light. In order to acquire velocity at a constant gravitational potential where the velocity of light is constant, there must be a source for mass exchange to supply the mass increase Δm, Figure 4.1.2-3.

The two mechanisms for the building up of kinetic energy can be expressed as

$$E_{kin} = c_0 \Delta |\mathbf{p}| = c_0 \left(|m\Delta c| + |c\Delta m| \right), \qquad (4.1.2:16)$$

where the first term refers to kinetic energy obtained in free fall in a local gravitational frame, and the second term refers to kinetic energy obtained via insertion of mass in a local energy frame.

Buildup of kinetic energy in free fall in a gravitational field conserves the total energy of the falling object. Buildup of kinetic energy via insertion of mass increases the total energy of the object put into motion.

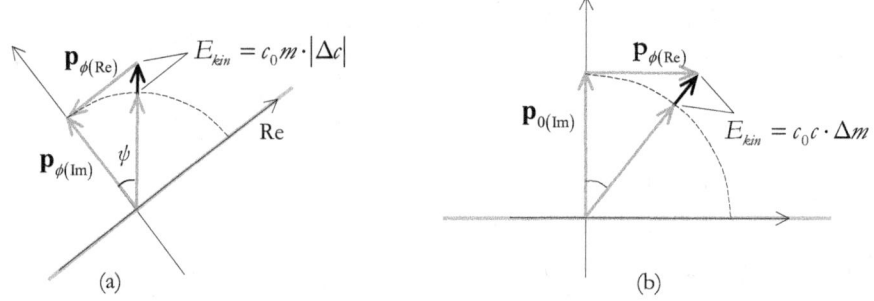

Figure 4.1.2-3. (a) Kinetic energy in free fall by change in the local rest momentum via tilting of space. (b) Kinetic energy by the insertion of excess mass.

4.1.3 Inertial work and a local state of rest

Energy as a complex function

In the DU framework, it is useful to study energy as a complex function. The complex presentation of the energy of motion gives an energy vector character that allows a direct linkage of the energy of motion to momentum. In the case of gravitational energy, the vector presentation shows the direction of the gradient of the energy. The absolute values of the complex energies restore the conventional concept of scalar energy.

The real and imaginary parts of the complex energy of motion can be referred to as energy equivalences of momenta in the direction of the real axis and the imaginary axis, respectively. *In the complex presentation, momentum is directly proportional to its energy equivalence.*

Momentum is presented in terms of the vector components in the direction of the imaginary and real axes. Choosing the real axis in the direction of the real component of the momentum, complex momentum can be expressed in terms of its scalar components in the imaginary and real directions

$$\mathbf{p}^\square = \mathbf{p}' + i\,\mathbf{p}'' \quad \text{or} \quad p^\square = p' + i\,p''. \tag{4.1.3:1}$$

The complex energy of motion is expressed

$$E_m^\square = c_0 p^\square = c_0\left(p' + i\,p''\right) = c_0 p' + i\,c_0 p'' = E' + i\,E'', \tag{4.1.3:2}$$

where the two last forms show the energy equivalences of momentum in the direction of the real axis and imaginary axis. The complex presentation allows the polar coordinate expression

$$E_m^\square = E_{m(tot)}\left(\cos\varphi + i\sin\varphi\right), \tag{4.1.3:3}$$

which relates the real and imaginary components of the complex energy to the total energy via the phase angle φ. The complex presentation of energy is essential for the study of the balance between the energy of motion and the global energy of gravitation in the imaginary direction and for a detailed analysis of the energy balances in space (in the direction of the real axis) and in the direction of the imaginary axis.

The concept of internal energy

The total energy of motion in (4.1.2:11) can be expressed in complex form

$$E_{m(tot)}^\square = c_0\left(p_\varphi^\square\right) = c_0\left(\frac{m\beta c}{\sqrt{1-\beta^2}} + imc\right) = c_0 mc\left(\frac{\beta}{\sqrt{1-\beta^2}} + i\right), \tag{4.1.3:4}$$

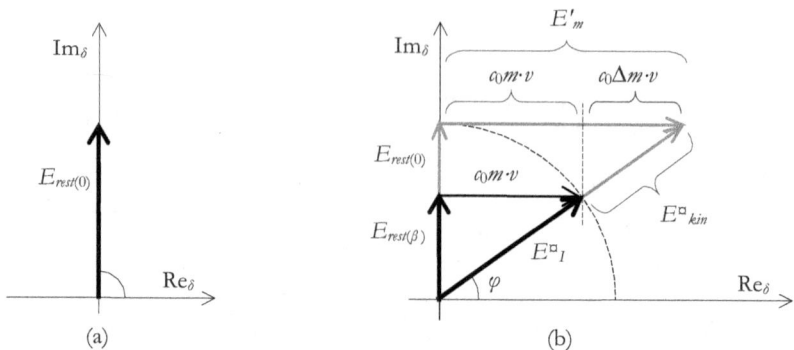

Figure 4.1.3-1. The turn of the total momentum due to momentum $\mathbf{p}'=(m+\Delta m)\cdot\mathbf{v}$ added in a space direction results in velocity $v = c\cos\varphi$ in space. Velocity c is the local velocity of light equal to the local imaginary velocity of space.

or

$$\begin{aligned} E^{\square}_{m(tot)} &= c_0(m+\Delta m)c\cdot(\cos\varphi + i\sin\varphi) \\ &= c_0 mc(\cos\varphi + i\sin\varphi) + c_0\Delta m\cdot c(\cos\varphi + i\sin\varphi) \\ &= E^{\square}_I + E^{\square}_{kin}, \end{aligned} \qquad (4.1.3{:}5)$$

where the first term on the last line of (4.1.3:5) is referred to *as the internal energy of motion*, E_I, with the absolute value equal to the energy of motion the object possesses at rest in the local frame (the rest energy at $\varphi = \pi/2$), Figure 4.1.3-1,

$$\begin{aligned} E^{\square}_I &= c_0 mc(\cos\varphi + i\sin\varphi) = c_0 m\cdot c\beta + ic_0 mc\sqrt{1-\beta^2} \\ &= c_0 m\cdot v + ic_0 mc\sqrt{1-\beta^2}. \end{aligned} \qquad (4.1.3{:}6)$$

The corresponding complex expression for the kinetic energy is

$$E^{\square}_{kin} = c_0\Delta mc(\cos\varphi + i\sin\varphi) = c_0\Delta m\cdot v + ic_0 c\Delta m\sqrt{1-\beta^2}. \qquad (4.1.3{:}7)$$

Figure 4.1.3-2 illustrates the structure of the total energy of motion as the sum of the complex internal energy and the kinetic energy as obtained by regrouping the real and imaginary parts in equation (4.1.3:5)

$$E^{\square}_{m(tot)} = (E'_I + E'_{kin}) + i(E''_I + E''_{kin}). \qquad (4.1.3{:}8)$$

The scalar value of the internal energy is equal to the rest energy $E_{rest\,(0)}$ of the object at rest. As a complex quantity, the internal energy is "turned" to angle φ relative to the real axis. The real part E'_I of the internal energy contributing to the momentum in space is created against a reduction in the imaginary part E''_I.

As the counterpart of the internal energy E_I, the *internal momentum* $p^{\square}_I$ is

$$p^{\square}_I = m\cdot v + imc\sqrt{1-\beta^2} = mc(\cos\varphi + i\sin\varphi), \qquad (4.1.3{:}9)$$

see Figure 4.1.5-3.

Energy structures in space

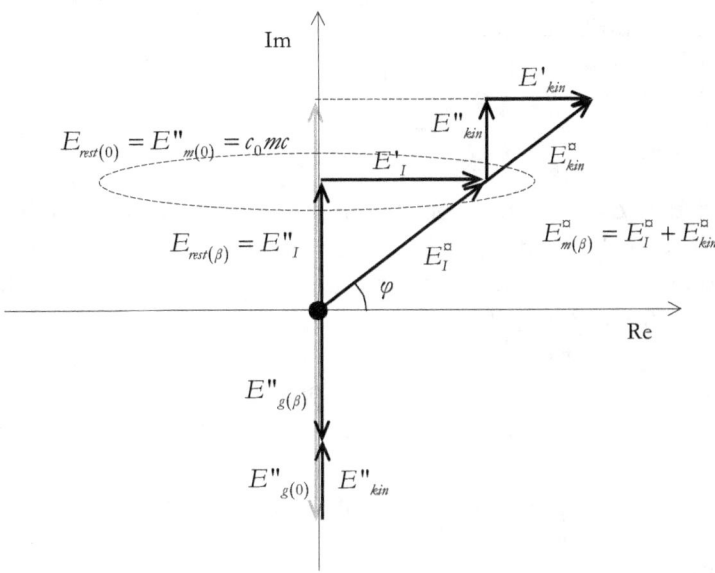

Figure 4.1.3-2. Illustration of the components of the internal energy and kinetic energy of an object moving at velocity βv in a local energy frame. The effect of the imaginary part of the kinetic energy E''_{kin} is a reduction of the global energy of gravitation of the moving object; it is the inertial work done against the global gravitation via central acceleration relative to the equivalence M'' at the center of the spherically closed space.

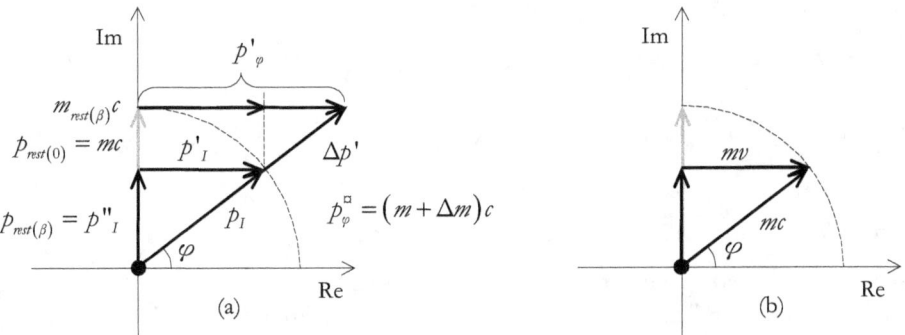

Figure 4.1.3-3. (a) The real part of the total momentum is the momentum observed in space. The internal momentum can be illustrated as the rest momentum $p_{rest(0)}$ of the object turned by angle φ with respect to the real axis. (b) The real part of the internal momentum contributes to the momentum in space by $p'_I = mv$. The imaginary part of the internal momentum serves as the rest momentum of the object $p''_I = m_{rest(\beta)} c$.

The absolute value of the internal momentum is equal to the absolute value of the momentum of the object at rest, the rest momentum $p_{rest(0)}$. The real part of the internal momentum, $p'_I = mv$, contributes to the real component of the momentum of the object in a space direction. The imaginary part of the internal momentum is identified as the rest momentum of the moving object

$$p''_I = p_{rest(\beta)} = mc\sqrt{1-\beta^2} = p_{rest(0)}\sqrt{1-\beta^2} \ . \tag{4.1.3:10}$$

The imaginary velocity of an object in space is determined by the velocity of space in the fourth dimension, which means that the reduction of the imaginary momentum due to the buildup of momentum in space means a reduction of the rest mass of the moving object.

The imaginary part of the internal momentum is the rest momentum of the object moving at velocity $v = \beta c$ in the local frame in space

$$m_{rest(\beta)} = m_{rest(0)}\sqrt{1-\beta^2} = m\sqrt{1-\beta^2} \ . \tag{4.1.3:11}$$

Applying rest mass $m_{rest(\beta)}$, the rest energy of an object moving at velocity βc in the local frame is

$$E_{rest(\beta)} = c_0 p_{rest(\beta)} = c_0 m_{rest(0)} c\sqrt{1-\beta^2} = c_0 m_{rest(\beta)} c \ . \tag{4.1.3:12}$$

The reduction of the imaginary part of the internal energy due to a reduction of the rest mass $m_{rest(\beta)}$ means that the reduction also affects the global gravitational energy $E''_{g(\beta)}$

$$E''_{g(\beta)} = -\frac{GM'' m_{rest(\beta)}}{R''} = -\frac{GM'' m}{R''}\sqrt{1-\beta^2} \ . \tag{4.1.3:13}$$

Reduction of the global gravitational energy due to motion in space does not require "an immediate interaction" with all other mass in space — it is just the consequence of the reduction of the local rest mass of the moving object.

For the moving object, the balance of the imaginary energies of motion and gravitation is obtained as the sum of (4.1.3:12) and (4.1.3:13) as

$$E''_{rest(\beta)} + E''_{g(\beta)} = 0 \ . \tag{4.1.3:14}$$

Reduction of rest mass as a dynamic effect

In spherically closed space, any motion in space is central motion relative to mass equivalence M'' at distance R'' in the fourth dimension. Accordingly, the reduction of the rest mass and the related rest momentum and rest energy of the moving object can be interpreted as consequences of the central force caused by motion in space.

In a simplified analysis, we can express the central force created by momentum **p** in homogeneous space due to the curvature of space by radius R''_0 perpendicular to momentum **p** (a more detailed analysis is given in Section 4.1.8)

Energy structures in space

$$\mathbf{F}_{4(\beta)} = \frac{d\mathbf{p}}{dt} = c_0^2 \frac{\beta^2 m_\beta}{R"_0} \hat{\mathbf{r}}_4 = \frac{c_0^2 \beta^2 \, m/\sqrt{1-\beta^2}}{R"_0} \hat{\mathbf{r}}_4 \, . \tag{4.1.3:15}$$

The global gravitational force as the gradient of the global gravitational energy of mass m_β is

$$\mathbf{F}_{4(g)} = -d\left(-\frac{GM"m_\beta}{R"_0}\right)\!\Big/dR"\hat{\mathbf{r}}_4 = -\frac{GM"m_\beta}{R"^2_0} \hat{\mathbf{r}}_4 = -\frac{c_0^2 m_\beta}{R"_0} \hat{\mathbf{r}}_4 \, , \tag{4.1.3:16}$$

where the last form is based on the zero-energy balance of motion and gravitation in DU space (3.3.1:4)

$$\frac{GM"m_\beta}{R"_0} = c_0^2 m_\beta \, . \tag{4.1.3:17}$$

The net force in the fourth dimension is obtained as the sum of the centrifugal force in (4.1.3:15) and the gravitational force in (4.1.3:16)

$$\begin{aligned}\mathbf{F}_{4,tot} &= \mathbf{F}_{4(g)} + \mathbf{F}_{4(\beta)} = -\frac{c_0^2 m_\beta}{R"_0} + \frac{m_\beta v^2}{R"_0} = -\frac{c_0^2}{R"_0} m_\beta \left(1-\beta^2\right) \\ &= -\frac{c_0^2}{R"_0} \frac{m}{\sqrt{1-\beta^2}} \left(1-\beta^2\right) = -\frac{c_0^2}{R"_0} m\sqrt{1-\beta^2} = \frac{c_0^2 m_{rest(\beta)}}{R"_0} \, , \end{aligned} \tag{4.1.3:18}$$

which means the balance of the imaginary energies of motion and gravitation

$$\frac{E_{rest(\beta)}}{R"_0} = \frac{E"_{g(\beta)}}{R"_0} \qquad \text{or} \qquad E_{rest(\beta)} = E_{g(\beta)} \, . \tag{4.1.3:19}$$

The zero-energy balance of motion and gravitation in the fourth dimension is obtained equally for mass $m_\beta = m_{rel} = m/\sqrt{1-\beta^2}$ moving at velocity βc in its parent frame, and for mass $m_{rest(\beta)} = m\sqrt{1-\beta^2}$ at rest in a local frame moving at velocity βc in the parent frame space, Figure 4.1.3-4.

The local state of rest in the DU is obtained against the reduction of the locally available rest energy in the moving frame. The local state of rest is characterized by the zero-energy balance between motion and gravitation in the fourth dimension.

The imaginary part of the kinetic energy is the work done in reducing the global gravitational energy, and equally, the rest energy of the object in motion

$$E"_{kin} = E"_{g(0)} - E"_{g(\beta)} = \Delta E_{g(global)} = \Delta E_{rest} \, . \tag{4.1.3:20}$$

Equation (4.1.3:20) means a quantitative expression of Mach's principle by identifying the inertial work as the imaginary part of kinetic energy. The real part of kinetic energy contributes to the momentum in space

$$\Delta p' = \Delta m \cdot v \, . \tag{4.1.3:21}$$

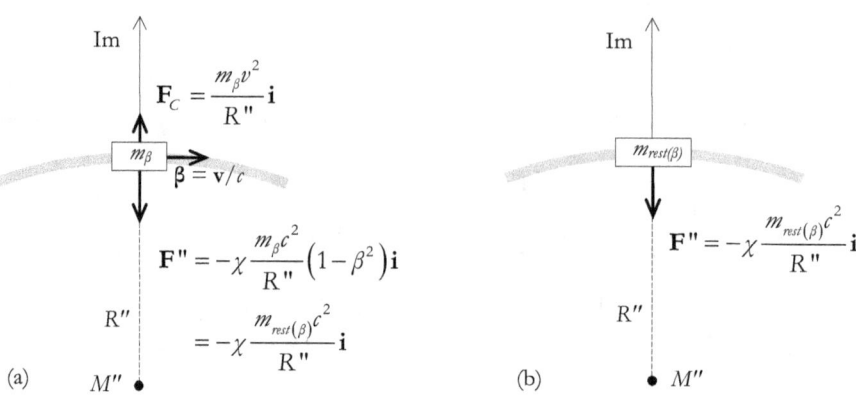

Figure 4.1.3-4. (a) The gravitational force of mass equivalence M'' on mass m_β moving at velocity **v** is reduced by the central force F_C, which makes it equal to the gravitational force of mass equivalence M'' on mass $m_{rest(\beta)}$ at rest in the local frame, as illustrated in figure (b).

4.1.4 The system of nested energy frames

With reference to equation (4.1.3:12), the rest energy of object m at rest in frame n moving at velocity β_n in its parent frame $n-1$ is

$$E_{rest(n)} = E_{rest(n-1)}\sqrt{1-\beta_n^2}, \qquad (4.1.4:1)$$

where $E_{rest(n-1)}$ is the rest mass of the object at rest in frame $n-1$. Frame $n-1$, carrying mass m in frame n, moves at velocity β_{n-1} in frame $n-2$. The rest energy of mass m can now be related to the rest energy mass m has at rest in frame $n-2$ as

$$E_{rest(n)} = E_{rest(n-2)}\sqrt{1-\beta_{n-1}^2} \cdot \sqrt{1-\beta_n^2}, \qquad (4.1.4:2)$$

Figure 4.1.4-1.

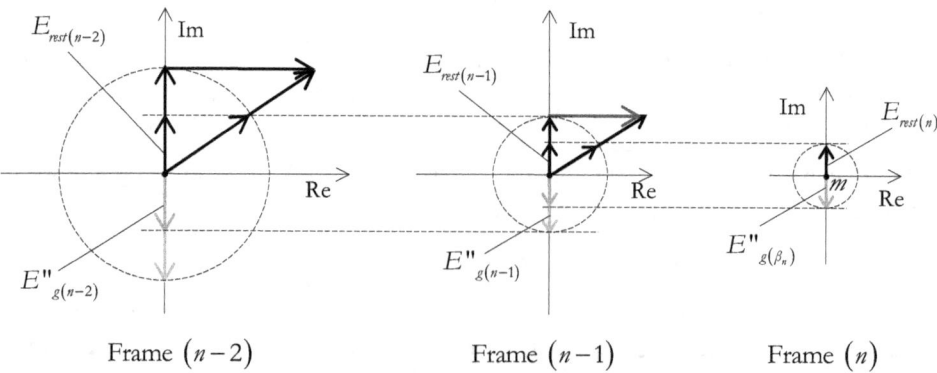

Figure 4.1.4-1 Motion of frame n with mass m at velocity β_n in frame $n-1$, which is moving at velocity β_{n-1} in its parent frame $(n-2)$.

Equation (4.1.4:2) can be expressed in terms of mass $m_{(n)}$ as

$$E_{rest(n)} = c_0 c \cdot m_{rest(n)} = c_0 c \cdot m_{rest(n-2)} \sqrt{1-\beta_{n-1}^2} \cdot \sqrt{1-\beta_n^2} \,, \qquad (4.1.4:3)$$

where, at constant gravitational potential, both c_0 and c are constants and mass $m_{(n)}$ is related to mass $m_{(n-2)}$ as

$$m_{rest(n)} = m_{rest(n-2)} \sqrt{1-\beta_{n-1}^2} \sqrt{1-\beta_n^2} \,. \qquad (4.1.4:4)$$

When frame $(n-2)$ is in motion at velocity $\beta_{(n-2)}$ in frame $(n-3)$ which is the parent frame to frame $(n-2)$, frame $(n-3)$ at velocity in frame $n-4$ ... etc., rest mass $m_{rest(n)}$ can be finally related to the rest mass m_0 of the object at rest in hypothetical homogeneous space

$$m_{rest(n)} = m_0 \prod_{i=1}^{n} \sqrt{1-\beta_i^2} \,. \qquad (4.1.4:5)$$

Applying the rest mass in (4.1.4:5) the rest energy $E_{rest(n)}$ becomes

$$E_{rest(n)} = c_0 m_{rest(0)} c = c_0 m_0 c \prod_{i=1}^{n} \sqrt{1-\beta_i^2} \,, \qquad (4.1.4:6)$$

where c is the local velocity of light determined by the local gravitational state and the gravitational states of each of the nested frames in their parent frames, as described by equation (4.1.1:28). Substitution of (4.1.1:28) for c in (4.1.4:6) gives a general expression for the rest energy of an object

$$E_{rest(n)} = c_0 m_{rest(n)} c = m_0 c_0^2 \prod_{i=0}^{n} \left[(1-\delta_i) \sqrt{1-\beta_i^2} \right], \qquad (4.1.4:7)$$

or simply as

$$E_{rest} = c_0 mc \,, \qquad (4.1.4:8)$$

where

$$m = m_{rest(n)} = m_0 \prod_{i=0}^{n} \sqrt{1-\beta_i^2} \,, \qquad (4.1.4:9)$$

and

$$c = c_n = c_0 \prod_{i=1}^{n} (1-\delta_i) \,. \qquad (4.1.4:10)$$

The complementary counterpart of the rest energy in equations (4.1.4:7) and (4.1.4:8) is the global gravitational energy [see equations (4.1.1:27–29)]

$$E_{g(global)} = E''_g = -\frac{GM'' m_0}{R''_0} \prod_{i=1}^{n} \left[(1-\delta_i) \sqrt{1-\beta_i^2} \right] = -\frac{GM'' m}{R''} \,, \qquad (4.1.4:11)$$

where m is the local rest mass given in (4.1.4:9) and R'' is the local apparent distance to M'' given in equation (4.1.1:29) as

$$R'' = R''_n = R''_0 \Big/ \prod_{i=1}^{n} (1-\delta_i) \,. \qquad (4.1.4:12)$$

By defining the frame factor χ

$$\chi \equiv \frac{c_0}{c} = 1 \bigg/ \prod_{i=1}^{n}(1-\delta_i), \qquad (4.1.4{:}13)$$

equation (4.1.4:8) for the rest energy can be written in the form

$$E_{rest} = c_0 mc = \chi \cdot cmc = \chi \cdot mc^2. \qquad (4.1.4{:}14)$$

The expression of local rest energy in equation (4.1.4:14) is formally close to the expression of the rest energy in the formalism of the theory of relativity, which postulates the velocity of light and the rest mass of an object as being invariants and independent of the gravitational environment and velocities that local mass is subject to in space. An estimate of the value of χ on the Earth is of the order of $\chi \approx 1+10^{-6}$ (= 1.000001), which summarizes the effects of our gravitational state in the Earth, the Sun, the Milky Way, and the local galaxy group gravitational frames. In practice, in measurements of the effect of χ becomes included in the value of the rest mass.

The system of nested energy frames is a central feature in the Dynamic Universe model. The nested energy frames create a link from any local energy frame to hypothetical homogeneous space, which serves as a universal reference to all energy states in space. The system of nested energy frames is a consequence of the zero-energy principle and the conservation of the energy excitation built up in the primary energy build-up process. The conservation of primary energy in energy interactions in space is illustrated by the chain of nested energy frames in Figure 4.1.4-2.

The state of rest in hypothetical homogeneous space serves as the universal reference for a state of rest in space. Each energy frame has its local state of rest characterized by the local rest mass, rest momentum, and rest energy. In the state of rest in a local energy frame, an *energy object* has its momentum in the local imaginary direction only.

In a kinematic sense, for observing velocity as the rate of change in the distance between two objects, any object or state of motion, independent of the energy frame it belongs to, can be chosen as the reference for relative velocities. Relative velocity, however, is not the basis for the energy of motion or kinetic energy related to the observed velocity. Kinetic energy in a local system is always related to velocity relative to the state of rest of the local frame.

The barycenter of hypothetical homogeneous space is in the center of the spherically closed space. It is the reference at rest for the contraction and expansion of space in the direction of the 4-radius.

4.1.5 Effect of location and local motion in a gravitational frame

Local rest energy of orbiting bodies

Let's assume that a solid body M_B rotates about a central mass M_A ($M_A \gg M_B$) at distance $r_{0\delta} = r_A$ at angular velocity ω_A. The rest energy of mass in the rotating body at distance r_A from the central mass M_A is

Energy structures in space

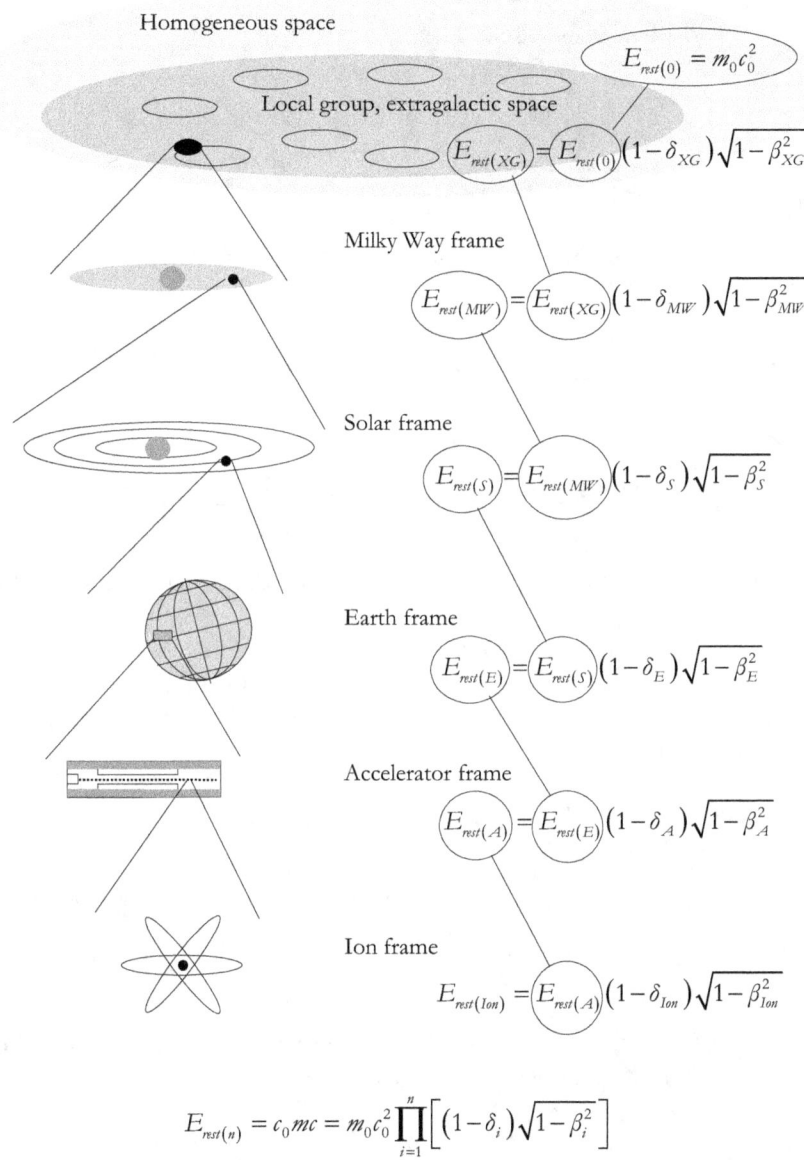

$$E_{rest(n)} = c_0 mc = m_0 c_0^2 \prod_{i=1}^{n}\left[(1-\delta_i)\sqrt{1-\beta_i^2}\right]$$

Figure 4.1.4-2. The rest energy of an object in a local frame is linked to the rest energy of the local frame in its parent frame. The system of nested energy frames relates the rest energy of an object in a local frame to the rest energy of the object in homogeneous space.

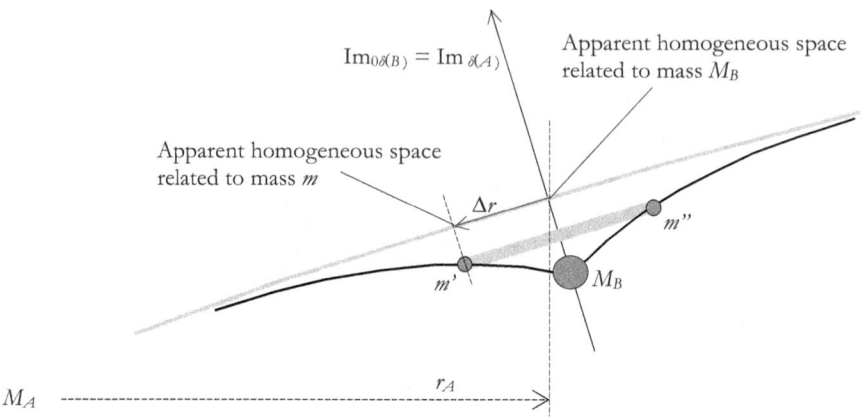

Figure 4.1.5-1. The local gravitational frame around mass M_B orbits central mass M_A. Mass m orbits mass M_B in the local gravitational frame at distance Δr from mass M_B.

$$E_{rest(B)} = E_{rest(A)}\left(1 - \frac{GM_A}{r_{0\delta}c_{0\delta}c_0}\right)\sqrt{1-\beta_A^2} \approx E_{rest(A)}\left(1 - \frac{GM_A}{r_A c^2} - \frac{\omega_A^2 r_A^2}{2c^2}\right), \quad (4.1.5{:}1)$$

where $E_{rest(A)}$ means the rest energy of mass at rest in apparent homogeneous space of the M_A gravitational frame, Figure 4.1.5-1.

From different locations in an orbiting body M_B with radius Δr ($\Delta r \ll r_A$) the distance to the central mass varies within $r_A \pm \Delta r$. The difference in the rest energy of mass m in the orbiting body can be related to the difference in the distance to the central mass by differentiating (4.1.5:1)

$$\Delta E_{rest(B)} \approx E_{rest(A)}\left(\frac{GM_A}{r_A c^2} - \frac{\omega^2 r_A^2}{c^2}\right)\frac{\Delta r}{r_A} = \left(\frac{GM_A}{r_A c^2} - \beta_A^2\right)\frac{\Delta r}{r_A}, \quad (4.1.5{:}2)$$

where β_A is the orbital velocity that, in the case of a circular Keplerian orbit, is

$$v_A^2 = \frac{GM_A}{r_A} \quad \Rightarrow \quad \beta_A^2 = \frac{\omega^2 r_A^2}{c^2} = \frac{GM_A}{r_A c^2}. \quad (4.1.5{:}3)$$

Substitution of (4.1.5:3) for β_A^2 in (4.1.5:2) suggests that the rest energy in the orbiting body is independent of its location within $r_A \pm \Delta r$, i.e.,

$$\Delta E_{rest(B)} \approx E_{rest(A)}\left(\frac{GM_A}{r_A c^2} - \frac{GM_A}{r_A c^2}\right)\frac{\Delta r}{r_A} = 0. \quad (4.1.5{:}4)$$

Instead of a solid body, an object orbiting the central mass M_A can be interpreted as a platform or local frame hosting a subsystem with central mass M_B and "satellites" orbiting M_B within distance $r_A \pm \Delta r$ from the central mass M_A in the parent frame. The rest energy of mass m rotating the local mass center M_B in the local frame becomes (Figure 4.1.5-1)

$$E_{rest(m)} \approx E_{rest(A)}\left(1-\frac{GM_A}{r_A c^2}-\frac{\omega^2 r_A^2}{2c^2}\right)\cdot\left(1-\frac{GM_B}{\Delta r c^2}\right)\sqrt{1-\beta_B^2}, \qquad (4.1.5:5)$$

where $\beta_B = \Delta r \omega / c$ is the local orbital velocity of the satellite orbiting mass M_B at radius Δr. As shown by equation (4.1.5:4) the first factor in parentheses in (4.1.5:5) is independent of Δr, thus allowing the substitution of (4.1.5:1) for the first factor in

$$E_{rest(m)} \approx E_{rest(B)} \cdot \left(1-\frac{GM_B}{\Delta r c^2}\right)\sqrt{1-\beta_B^2}, \qquad (4.1.5:6)$$

which suggests that the fluctuation of distance $r_A \pm \Delta r$ to M_A does not affect the rest energy observed in the satellite orbiting mass M_B in the frame rotating mass M_A. The velocity of light at the satellite's location, however, is a function of the momentary distance to masses M_B and M_A

$$c_\delta = c_{0\delta(A)}\left(1-\frac{GM_A}{(r_A+\Delta r)c^2}\right)\left(1-\frac{GM_B}{\Delta r \cdot c^2}\right), \qquad (4.1.5:7)$$

where distance $r_A + \Delta r$ is

$$r_A + \Delta r = r_A + \Delta r \cos\theta, \qquad (4.1.5:8)$$

where angle θ is the angle between $\Delta\mathbf{r}$ and $\mathbf{r}_A$. Substitution of (4.1.5:8) for distance $r_A+\Delta r$ in (4.1.5:7) gives

$$\begin{aligned}c_{\delta(\theta)} &\approx c_{(r_A)}\left[1-\frac{GM_B}{\Delta r \cdot c^2}-\frac{GM_A}{r_A c^2}\left(1-\frac{\Delta r}{r_A}\cos\theta\right)\right] \\ &= c_{(r_A,\Delta r)}\left(1+\frac{GM_A}{r_A^2 c^2}\Delta r \cos\theta\right) = c_{(r_A,\Delta r)}\left(1+\frac{g_A}{c^2}\Delta r \cos\theta\right),\end{aligned} \qquad (4.1.5:9)$$

where g_A is the gravitational acceleration at distance r_A from mass M_A.

Energy object

Gravitational frames around mass centers in space can be regarded as energy objects in their parent frame. Any local frame with internal interaction of potential energy and motion can be regarded as an *energy object* in its parent frame. A closed container with gas atoms inside is an example of an energy object. When the container is at rest in a local frame, the rest energy of the gas molecules with average thermal velocity β_G in the container is

$$E_{rest(G),0} = E_{rest(G0),0}\sqrt{1-\beta_G^2}, \qquad (4.1.5:10)$$

where $E_{rest(G0),0}$ is the rest energy of the gas molecules at rest in the container. When the container is put into motion at velocity β in the local frame, the rest energy of the gas molecules inside the container is reduced as, Figure 4.1.5-2

$$E_{rest(G),\beta} = E_{rest(G),0}\sqrt{1-\beta^2} = E_{rest(G0),0}\sqrt{1-\beta_G^2}\sqrt{1-\beta^2}. \qquad (4.1.5:11)$$

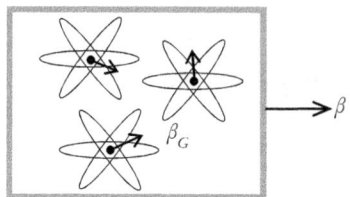

Figure 4.1.5-2. The rest energy of electrons and nuclei of atoms in a closed box is affected both by the motion of the atoms in the box, and the motion of the box in the local gravitational frame, as well as gravitation and motions of all parent frames of the local gravitational frame.

> In the DU framework, there are no independent objects in space.
> Every object is linked to the rest of space.

4.1.6 Free fall and escape in a gravitational frame

In free fall, the buildup of momentum in space occurs against the reduction of the imaginary velocity of space via a turn of the imaginary axis in tilted space (see Section 4.1.1). Escape of mass m from the state of rest in a δ state to the state of rest in apparent homogeneous space releases the kinetic energy of escape into the increase of the imaginary momentum and rest energy

$$E_{kin(esc)} = E_{rest(0\delta)} - E_{rest(\delta)}. \tag{4.1.6:1}$$

The kinetic energy needed by an object at a state characterized by gravitational factor δ in the local frame to escape to the apparent homogeneous space is equal to the kinetic energy of free fall from the apparent homogeneous space to state δ (see equation (4.1.2:3))

$$E_{kin(esc)} = c_0 m \left(c_{0\delta} - c_\delta \right) = c_0 m \Delta c , \tag{4.1.6:2}$$

which illustrates that in the case of escape, kinetic energy is needed to restore the higher velocity of light in the apparent homogeneous space. In other words, the kinetic energy in escape is used in "climbing" towards apparent homogeneous space, Figure 4.1.6-1.

Obviously, the kinetic energy needed to climb from gravitational state δ_1 to δ_2 ($\delta_1 > \delta_2$) can be expressed as the difference of the kinetic energies for escape from δ_1 to δ_2 as

$$E_{kin(\delta 1, \delta 2)} = c_0 m \left(c_{0\delta} - c_{\delta 1} \right) - c_0 m \left(c_{0\delta} - c_{\delta 2} \right) = c_0 m \left(c_{\delta 2} - c_{\delta 1} \right). \tag{4.1.6:3}$$

Substituting equation (4.1.1:23) for $c_{\delta 1}$ and $c_{\delta 2}$ equation (4.1.6:3) obtains the form

$$E_{kin(\delta 1, \delta 2)} = c_0 m c_{0\delta} \left[(1-\delta_2) - (1-\delta_1) \right], \tag{4.1.6:4}$$

and further by substituting equation (4.1.1:30) for δ_1 and δ_2 the form

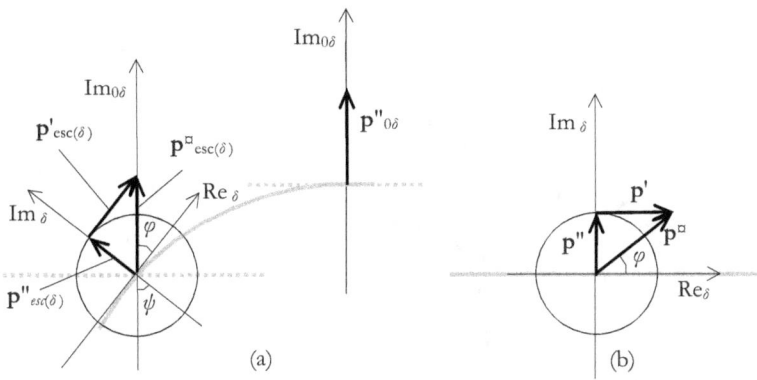

Figure 4.1.6-1. (a) Escape momentum in gravitational state δ in a local gravitational frame. The total momentum $\mathbf{p}^{\square}{}_{esc(\delta)}$ has the direction of the imaginary axis in apparent homogeneous space. Motion towards apparent homogeneous space reduces δ to zero, which gradually reduces $\mathbf{p}'{}_{esc(\delta)}$ to zero, making $\mathbf{p}^{\square}{}_{esc(\delta)}$ equal to the imaginary momentum in apparent homogeneous space $\mathbf{p}''{}_{0\delta}$. (b) Momentum $\mathbf{p}'$ in gravitational state δ "stores" the extra momentum needed for escape as increased relativistic mass.

$$E_{kin(\delta1,\delta2)} = -c_0 mc_{0\delta}\left(\frac{GM}{c_0 c_{0\delta} r_{0\delta(2)}} - \frac{GM}{c_0 c_{0\delta} r_{0\delta(1)}}\right) = -GMm\left(\frac{1}{r_{0\delta(2)}} - \frac{1}{r_{0\delta(1)}}\right), \quad (4.1.6:5)$$

which is equal to the gravitational energy restored through escape.

With reference to equation (4.1.1:3), escape momentum can be expressed as

$$\mathbf{p}^{\square}_{esc} = \mathbf{p}'_{esc(\delta)} + \hat{\mathbf{i}}_\delta p''_\delta = mv_{esc}\hat{\mathbf{r}} + mc_\delta \hat{\mathbf{i}}_\delta . \quad (4.1.6:6)$$

The escape velocity is given in terms of the tilting angle in equation (4.1.1:12) as

$$v_{esc(\delta)} = c_{0\delta}\sin\psi = c_{0\delta}\sqrt{1-\cos^2\psi} . \quad (4.1.6:7)$$

Substitution of equation (4.1.1:22) for cosψ in equation (4.1.6:7) gives the velocity of free fall in the form

$$v_{esc(\delta)} = c_{0\delta}\sqrt{1-(1-\delta)^2} . \quad (4.1.6:8)$$

To solve for the velocity and acceleration of an object in free fall, it is useful to define the *critical radius* r_c, which is the distance from the local mass center corresponding to gravitational factor δ = 1, when space has been tilted by 90°,

$$r_c \equiv \frac{GM}{c_0 c_{0\delta}} . \quad (4.1.6:9)$$

In terms of the critical radius, the gravitational factor δ can be expressed as

$$\delta = \frac{r_c}{r_{0\delta}} . \quad (4.1.6:10)$$

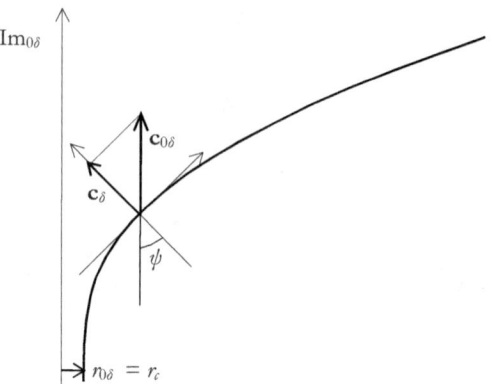

Figure 4.1.6-2. The shape of space close to a local singularity at $r_{0\delta} = r_c$ where space has tilted 90°. At the singularity, the local velocity of light c_δ goes to zero.

For $r_{0\delta} = r_c$ we have

$$\cos\psi = 1 - \delta = 0 \quad \Rightarrow \quad \psi = \frac{\pi}{2}, \qquad (4.1.6\!:\!11)$$

which, as illustrated in Figure 4.1.6-2, means a local singularity (a black hole) in space.

Substituting equation (4.1.6:10) for δ in equation (4.1.6:8), we get

$$v_{esc(\delta)} = c_{0\delta}\sqrt{1 - \left(1 - \frac{r_c}{r_{0\delta}}\right)^2}. \qquad (4.1.6\!:\!12)$$

Substitution of equation (4.1.6:12) for $v_{esc(\delta)}$ in equation (4.1.6:6) gives the real part of the total momentum in the form

$$\mathbf{p}'_{esc(\delta)} = mv_{esc(\delta)}\,\hat{\mathbf{r}} = mc_{0\delta}\sqrt{1 - \left(1 - \frac{r_c}{r_{0\delta}}\right)^2}\,\hat{\mathbf{r}}_\delta. \qquad (4.1.6\!:\!13)$$

The time derivative of momentum $\mathbf{p}'_{esc(\delta)}$ in the direction of the local Re_δ-axis can be written as

$$\frac{dp'_{esc(\delta)}}{dt}\,\hat{\mathbf{r}}_\delta = \frac{dp'_{esc(\delta)}}{dr_{0\delta}}\frac{dr_{0\delta}}{dt}\,\hat{\mathbf{r}}_\delta = \frac{dp'_{esc(\delta)}}{dr_{0\delta}}\,v_{esc(0\delta)}\,\hat{\mathbf{r}}_\delta, \qquad (4.1.6\!:\!14)$$

where $v_{esc(0\delta)}$ is the component of $v_{esc(\delta)}$ in the direction of radius $r_{0\delta}$ in the direction of the $\mathrm{Re}_{0\delta}$-axis (Figure 4.1.6-3)

$$v_{esc(0\delta)} = v_{esc(\delta)}\cos\psi = v_{esc(\delta)}(1-\delta) = v_{esc(\delta)}\left(1 - \frac{r_c}{r_{0\delta}}\right). \qquad (4.1.6\!:\!15)$$

Substituting equation (4.1.6:13) for $p'_{esc(\delta)}$ and equation (4.1.6:15) for $v_{esc(0\delta)}$, Figure 4.1.6-3, equation (4.1.6:14) obtains the form

Energy structures in space

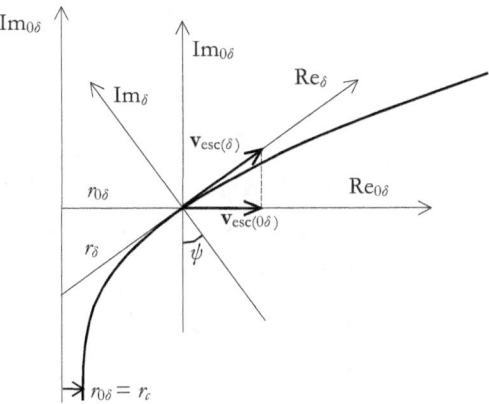

Figure 4.1.6-3. Velocity $\mathbf{v}_{esc(0\delta)}$ is the velocity component of velocity $\mathbf{v}_{esc(\delta)}$ in the direction of the $Re_{0\delta}$-axis.

$$\frac{dp'_{esc(\delta)}}{dt}\hat{\mathbf{r}}_\delta = mc_{0\delta}\frac{d\sqrt{1-(1-r_c/r_{0\delta})^2}}{dr_{0\delta}}v_{esc(0\delta)}\hat{\mathbf{r}}_\delta$$
$$= \frac{mc_{0\delta}}{2\sqrt{1-(1-r_c/r_{0\delta})^2}}\frac{2(1-r_c/r_{0\delta})}{r_{0\delta}^2}\frac{r_c}{r_{0\delta}^2}v_{esc(\delta)}(1-r_c/r_{0\delta})\hat{\mathbf{r}}_\delta.$$
(4.1.6:16)

Substitution of the momentum $p'_{esc(\delta)}$ in equation (4.1.6:13) back to equation (4.1.6:16) gives

$$\frac{dp'_{esc(\delta)}}{dt}\hat{\mathbf{r}}_\delta = \frac{mc_{0\delta}c_{0\delta}}{v_{esc(\delta)}}\frac{r_c}{r_{0\delta}^2}v_{esc(\delta)}(1-r_c/r_{0\delta})^2\hat{\mathbf{r}}_\delta = \frac{mc_{0\delta}c_{0\delta}}{r_{0\delta}}\frac{r_c}{r_{0\delta}}(1-r_c/r_{0\delta})^2\hat{\mathbf{r}}_\delta,$$
(4.1.6:17)

and further

$$\frac{dp'_{esc(\delta)}}{dt}\hat{\mathbf{r}}_\delta = \frac{mc_{0\delta}^2}{r_{0\delta}}\delta(1-\delta)^2\hat{\mathbf{r}}_\delta = \frac{c_{0\delta}}{c_0}\frac{GMm}{r_{0\delta}^2}(1-\delta)^2\hat{\mathbf{r}}_\delta.$$
(4.1.6:18)

With reference to equation (4.1.1:31), equation (4.1.6:18) can be written in terms of the gravitational force as the gradient of the local gravitational energy E_G,

$$\frac{dp_{esc(\delta)}}{dt}\hat{\mathbf{r}}_\delta = -\frac{dp_{ff(\delta)}}{dt}\hat{\mathbf{r}}_\delta = -\frac{c_{0\delta}}{c_0}\frac{dE_{G(\delta)}}{dr_{0\delta}}(1-\delta)^2\hat{\mathbf{r}}_\delta$$
$$= -\frac{c_\delta/(1-\delta)}{c_0}\frac{dE_{G(\delta)}}{dr_\delta}(1-\delta)\hat{\mathbf{r}}_\delta = -\frac{1}{\chi}\mathbf{F}_{G(\mathbf{r})},$$
(4.1.6:19)

where the direction of the gravitational force $\mathbf{F}_{G(\mathbf{r})}$ acts in the direction of local space towards mass M, i.e., $-\hat{\mathbf{r}}_\delta$.

Substitution of equation (4.1.6:13) for p'_{esc} in equation (4.1.6:18) gives the acceleration in free fall in the direction of local space

$$\frac{dv_{esc(\delta)}}{dt}\hat{\mathbf{r}}_\delta = \frac{1}{m}\frac{dp_{esc(\delta)}}{dt}\hat{\mathbf{r}}_\delta = -\frac{c_{0\delta}}{c_0}\frac{GM}{r_{0\delta}^2}(1-\delta)^2\,\hat{\mathbf{r}}_\delta = -\frac{c_{0\delta}}{c_0}\frac{GM}{r_\delta^2}\,\hat{\mathbf{r}}_\delta,\qquad(4.1.6{:}20)$$

or in terms of the local velocity of light $c = c_\delta = c_{0\delta}(1-\delta)$, as

$$\frac{dv_{esc(\delta)}}{dt}\hat{\mathbf{r}}_\delta = -\frac{c_\delta}{c_0}\frac{GM}{r_{0\delta}^2}(1-\delta)\,\hat{\mathbf{r}}_\delta = -\frac{1}{\chi}\frac{GM}{r_{0\delta}^2}(1-\delta)\,\hat{\mathbf{r}}_\delta = -\frac{1}{\chi}\frac{GM}{r_\delta^2(1-\delta)}\,\hat{\mathbf{r}}_\delta,\qquad(4.1.6{:}21)$$

where χ is the frame conversion factor $\chi = c_0/c$ defined in equation (4.1.4:13). The velocity of free fall in the direction of apparent homogeneous space is

$$v_{esc(0\delta)}\hat{\mathbf{r}}_{0\delta} = v_{esc(\delta)}(1-\delta)\hat{\mathbf{r}}_{0\delta} = -\frac{1}{\chi}\frac{GM}{r_\delta^2}\,\hat{\mathbf{r}}_{0\delta} = -\frac{1}{\chi}\frac{GM}{r_{0\delta}^2}(1-\delta)^2\,\hat{\mathbf{r}}_{0\delta}.\qquad(4.1.6{:}22)$$

4.1.7 Inertial force of motion in space

In the preceding Section, the time derivative of momentum was derived for the case of free fall in a local gravitational frame. By applying the expressions of the energy of motion derived in Section 4.1.2 and the effects of nested energy frames derived in Section 4.1.4, we can relate the time derivative of momentum to the gradient of energy in the direction of motion. As a result, we get a general expression for inertial force. We find that the inertial force given by the theory of special relativity is an approximation of the general expression in the case in which the effects of the whole space and the nested energy frames are ignored. By further ignoring the extra mass needed to obtain the motion, we end up with Newton's equation of motion.

In the DU, the energies of motion and gravitation are postulated "at rest in hypothetical homogeneous space" (see Section 2.2.2); energies and forces in real space are derived quantities.

With reference to equations (4.1.2:7), and (4.1.4:14), the total energy of motion can be written as

$$E_m^\square = c_0\left|\mathbf{p}^\square\right| = \chi c \cdot \left|\mathbf{p}^\square\right| = \chi c \cdot (m + \Delta m)c = \frac{\chi mc^2}{\sqrt{1-\beta^2}} = \chi m_\beta c^2,\qquad(4.1.7{:}1)$$

where χ is the frame conversion factor defined in equation (4.1.4:13) and the local velocity of light c is determined by the local gravitational state. For constant c, which means staying in a particular gravitational state and ignoring the deceleration of the expansion of space in the direction of the 4-radius (3.3.3:10), differentiation of equation (4.1.7:1) gives

$$dE_m^\square\,\hat{\mathbf{r}}^\square = d\left(\chi c \cdot p^\square\right)\hat{\mathbf{r}}^\square = \chi c \cdot dp^\square\,\hat{\mathbf{r}}^\square = \chi \frac{dx_{p^\square}}{dt}\cdot dp^\square\,\hat{\mathbf{r}}^\square,\qquad(4.1.7{:}2)$$

where $\chi \cdot c = c_0$ is constant (when ignoring the reduction of c_0 with the expansion of space) and dx_p is the distance differential in the direction of the complex total momentum $\mathbf{p}^\square$ shown by the unit vector $\hat{\mathbf{r}}^\square$. Equation (4.1.7:2) can be written in the form

$$\frac{d\mathbf{E}^\square_m}{dx^\square_{\mathbf{p}^\square}} = \chi\frac{d\mathbf{p}^\square}{dt} = \mathbf{F}^\square_m = -\mathbf{F}^\square_i, \tag{4.1.7:3}$$

which defines the force $\mathbf{F}^\square_m$ resulting in a change in the momentum. According to equation (4.1.7:3), force $\mathbf{F}^\square_m$ is the time derivative of momentum times the local frame conversion factor χ. Inertial force $\mathbf{F}^\square_i$ resisting a change in momentum is opposite to force $\mathbf{F}^\square_m$. Substitution of the complex form of momentum in equation (4.1.3:1) for $\mathbf{p}^\square$ in equation (4.1.7:3) gives

$$\begin{aligned}\mathbf{F}^\square_m &= \chi\frac{d\mathbf{p}^\square}{dt} = \chi\frac{d\mathbf{p}'}{dt} + \chi\frac{d\mathbf{p}''}{dt} = -\left(\mathbf{F}'_i + i\,\mathbf{F}''_i\right) \\ &= -\left(\chi m_{rest(0)}c\frac{d\left[\boldsymbol{\beta}(1-\beta^2)^{-1/2}\right]}{dt} + \chi c\frac{dm_{rest(0)}}{dt}\mathbf{i}_\delta\right),\end{aligned} \tag{4.1.7:4}$$

The real component of the inertial force is the force observed in the direction of acceleration in (4.1.7:4) (the rest mass $m_{rest(0)}$ is denoted as m)

$$\mathbf{F}'_m = -\mathbf{F}'_i = \chi mc\frac{d\left[\boldsymbol{\beta}(1-\beta^2)^{-1/2}\right]}{d\beta}\frac{d\beta}{dt}, \tag{4.1.7:5}$$

In terms of acceleration $a = c\cdot d\beta/dt$, equation (4.1.7:5) obtains the form

$$\begin{aligned}\mathbf{F}'_m &= \chi ma\left[\frac{d(1-\beta^2)^{-1/2}}{d\beta}\boldsymbol{\beta} + (1-\beta^2)^{-1/2}\frac{d\boldsymbol{\beta}}{d\beta}\right] \\ &= \chi ma\left[\frac{d(1-\beta^2)^{-1/2}}{d\beta}\beta\,\hat{\mathbf{r}}_\beta + (1-\beta^2)^{-1/2}\hat{\mathbf{r}}_a\right],\end{aligned} \tag{4.1.7:6}$$

where $\hat{\mathbf{r}}_\beta$ and $\hat{\mathbf{r}}_a$ are the unit vectors in the directions of the velocity $\boldsymbol{\beta}$ and acceleration $\mathbf{a}$, respectively. Derivation of equation (4.1.7:6) gives

$$\begin{aligned}\mathbf{F}'_m &= \chi m\,a\left[\beta^2(1-\beta^2)^{-3/2}\hat{\mathbf{r}}_v + (1-\beta^2)^{-1/2}\hat{\mathbf{r}}_a\right] \\ &= \chi m_\beta a\left[\frac{\beta^2}{1-\beta^2}\hat{\mathbf{r}}_v + \hat{\mathbf{r}}_a\right].\end{aligned} \tag{4.1.7:7}$$

For rectilinear motion in a local gravitational state, $\hat{\mathbf{r}}_v \| \hat{\mathbf{r}}_a$, the inertial force can be expressed as

$$\mathbf{F}'_{m(rectilinear)} = \chi m_\beta a\left[1 + \frac{\beta^2}{1-\beta^2}\right]\hat{\mathbf{r}}_v = c_0 m_\beta \frac{1}{1-\beta^2}\frac{d\beta}{dt}\hat{\mathbf{r}}_v. \tag{4.1.7:8}$$

The second term in equation (4.1.7:7) shows that, in the case of uniform circular motion when velocity is constant, and acceleration is perpendicular to the velocity, $\hat{\mathbf{r}}_v \perp \hat{\mathbf{r}}_a$, the inertial force is

$$\mathbf{F}'_{m(a)} = \frac{\chi m\, a}{\sqrt{1-\beta^2}} \mathbf{a} = \chi m_\beta \mathbf{a} = c_0 m_\beta \frac{d\beta}{dt} \hat{\mathbf{a}}. \tag{4.1.7:9}$$

The derivation of the inertial force in equations (4.1.7:1-9) assumed a fixed gravitational state with constant velocity of light. By writing the frame conversion factor χ in equation (4.1.7:4) into the form $\chi = c_0/c$, we get

$$\mathbf{F}'_m = \chi \frac{d\mathbf{p}_\delta}{dt} = \frac{c_0}{c} mc \frac{d\left[\boldsymbol{\beta}\left(1-\beta^2\right)^{-1/2}\right]}{dt} = c_0 m \frac{d\left[\boldsymbol{\beta}\left(1-\beta^2\right)^{-1/2}\right]}{dt}, \tag{4.1.7:10}$$

where the local velocity of light serves only as the reference for the local velocity in $\beta = \mathbf{v}/c$. Substitution of equation (4.1.4:5) for the local rest mass m in equation (4.1.7:10) gives

$$\mathbf{F}'_m = \chi \frac{d\mathbf{p}_\delta}{dt} = c_0 m_0 \prod_{i=1}^{n-1} \sqrt{1-\beta_i^2} \frac{d\left[\boldsymbol{\beta}\left(1-\beta^2\right)^{-1/2}\right]}{dt}. \tag{4.1.7:11}$$

Approximating $c \approx c_0$ and $m_0 \approx m$, equation (4.1.7:11) obtains the form of the law of motion in the special theory of relativity

$$\mathbf{F}_{(SR)} = \frac{d\mathbf{p}}{dt} = mc \frac{d\left[\boldsymbol{\beta}\left(1-\beta^2\right)^{-1/2}\right]}{dt} = m \frac{d\left[\mathbf{v}\left(1-\beta^2\right)^{-1/2}\right]}{dt}, \tag{4.1.7:12}$$

and by further ignoring the excess of mass needed in the buildup of the local motion equation (4.1.7:11) obtains the form of Newton's law of motion

$$\mathbf{F}_{(Newton)} = \frac{d\mathbf{p}}{dt} = m\frac{d\mathbf{v}}{dt} = m\mathbf{a}, \tag{4.1.7:13}$$

which can both be interpreted as local approximations of equation (4.1.7:11).

4.1.8 Inertial force in the imaginary direction

A special feature of the DU-model is the motion of space in the imaginary direction, which is the motion resulting in the rest energy of matter in space. The balance between motion and gravitation in the imaginary direction is affected by motion in space through the reduction of the rest mass. It can be shown that a similar reduction in the interaction in the imaginary direction can be derived by interpreting motion in space as central motion relative to the mass equivalence of spherical space. The latter approach shows the propagation at the velocity of light in space as propagation in a "satellite orbit" in spherical space, where the central acceleration of motion cancels the gravitational effect of the central mass.

The inertial force of motion in space was determined from the time derivative of total momentum, assuming velocity c in the imaginary direction to be constant. The price to be paid for the buildup of the real component of the internal momentum $p'_I = mv$ contributing to the momentum in space is a reduction in the rest momentum via a reduction of the rest mass $m_{rest(\beta)}$. Reduction of the rest mass and the imaginary momentum of an object in motion appears as a reduction in the inertial force in the imaginary direction, as

$$\mathbf{F}''_{i(n)} = -\chi \frac{d\mathbf{p}''}{dt} = -\chi m \sqrt{1-\beta^2} \frac{d\mathbf{c}''}{dt}. \qquad (4.1.8:1)$$

where β is the velocity of the object in the local frame.

The reduction of the rest mass can also be deduced by studying the motion of an object in a local frame as central motion relative to the mass equivalence of space in the imaginary direction. An object moving at velocity β at gravitational state δ has relativistic mass $m_\beta = m_{eff} = m/\sqrt{1-\beta^2}$. The imaginary gravitational force on the object due to the mass equivalence M'' is

$$\mathbf{F}''_{\delta(g)} = -\frac{dE''_{\delta(g),tot}}{dR''}\hat{\mathbf{i}}_\delta = -\frac{E''_{\delta(g),tot}}{R''}\hat{\mathbf{i}}_\delta = -Gm_{eff}\frac{M''}{R''^2}\hat{\mathbf{i}}_\delta, \qquad (4.1.8:2)$$

where R'' is the local imaginary radius of space (the distance to the mass equivalence of space). The ratio $GM''/R'' = \omega c$ in a local gravitational state can be solved from the zero-energy balance of the local rest energy and the global gravitational energy in (4.1.4:8) and (4.1.4:11). By further applying the frame conversion factor χ, equation (4.1.8:2) can be expressed in the form

$$\mathbf{F}''_{\delta(g)} = \frac{-\chi m_\beta c^2}{R''}\hat{\mathbf{i}}_\delta. \qquad (4.1.8:3)$$

Central acceleration due to motion in space is generated in the imaginary direction due to the turn of the $\mathbf{c}''$ vector (Figure 4.1.8-1)

$$\mathbf{a}''_{(v)} = -\frac{d\mathbf{c}''}{dt} = -\frac{v^2}{R''}\hat{\mathbf{i}}. \qquad (4.1.8:4)$$

The inertial force generated by mass m_β due to the acceleration $\mathbf{a}''_{(v)}$ in the direction of the Im_δ-axis can be expressed as

$$\mathbf{F}''_i = \chi m_\beta \frac{d\mathbf{c}''}{dt} = \chi m_\beta \frac{v^2}{R''}\hat{\mathbf{i}}_\delta. \qquad (4.1.8:5)$$

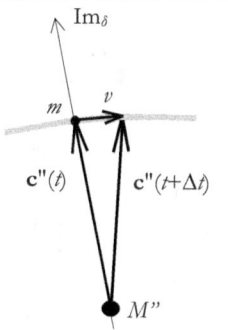

Figure 4.1.8-1. Velocity v in space results in the acceleration $a''_{(v)} = v^2/R''$ in the direction of the local imaginary axis, Im_δ. If the gravitational state is conserved, also c'' and the distance R''_δ to the mass equivalence M'' are conserved.

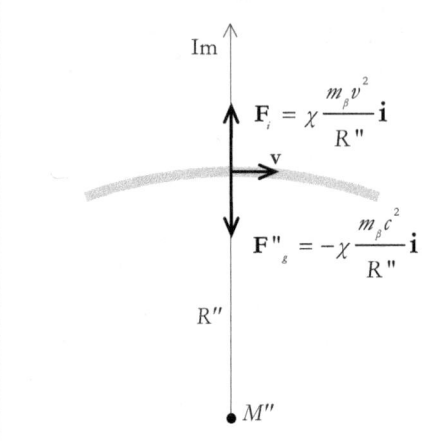
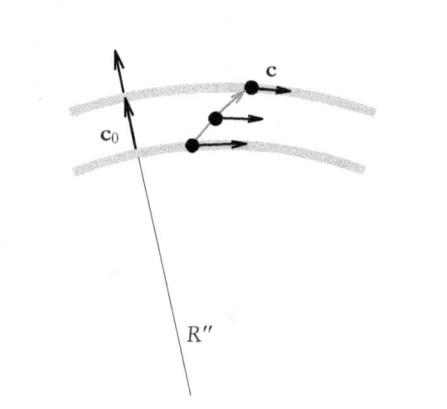

Figure 4.1.8-2. Motion in space reduces the gravitational force of mass equivalence M'' by the amount of the central force F_C created by the motion.
The apparent imaginary radius R'' is perpendicular to the space directions everywhere in space. In hypothetical homogeneous space $R'' = R_4$.

Figure 4.1.8-3. Propagation at the velocity of light "in a satellite orbit" in spherical space. The velocity of light decreases with the increase in R''.

The force $\mathbf{F}''_i$ is in the opposite direction to the imaginary gravitational force given in equation (4.1.8:3). With equations (4.1.8:3) and (4.1.8:5) combined, the total imaginary force on mass m_β moving at velocity $\mathbf{v}$ in space can now be expressed as

$$\mathbf{F}''_{\delta(g,a)} = -\chi m_\beta \left(\frac{c^2}{R''} - \frac{v^2}{R''} \right) \hat{\mathbf{i}}_\delta = -\frac{\chi mc^2}{R''\sqrt{1-\beta^2}} \left(1 - \beta^2 \right) \hat{\mathbf{i}}_\delta . \qquad (4.1.8{:}6)$$

As shown by equation (4.1.8:6), the effect of the imaginary acceleration due to motion in space is to reduce the total imaginary force of gravitation of the object by a factor $(1-\beta^2)$, which reduces (4.1.8:6) into form

$$\begin{aligned} \mathbf{F}''_{\delta(g,a)} &= -\frac{\chi mc^2 \sqrt{1-\beta^2}}{R''} \hat{\mathbf{i}}_\delta = -\frac{c_0 mc\sqrt{1-\beta^2}}{R''} \hat{\mathbf{i}}_\delta \\ &= -\frac{c_0 m_{rest(\beta)} c}{R''} = -\frac{E_{rest(\beta)}}{R''} . \end{aligned} \qquad (4.1.8{:}7)$$

The imaginary gravitational force on mass $m_{rest(\beta)}$ at rest is

$$\begin{aligned} \mathbf{F}''_{\delta(g,a)} &= -\frac{\chi mc^2 \sqrt{1-\beta^2}}{R''} \hat{\mathbf{i}}_\delta = -\frac{c_0 mc\sqrt{1-\beta^2}}{R''} \hat{\mathbf{i}}_\delta \\ &= -\frac{c_0 m_{rest(\beta)} c}{R''} = -\frac{E_{rest(\beta)}}{R''} = -\frac{E''_{g(\beta)}}{R''} , \end{aligned} \qquad (4.1.8{:}8)$$

Energy structures in space

where the last form describes the gravitational force as the gradient of global gravitational energy of mass $m_{rest(\beta)}$, Figure 4.1.8-2.

If an object is moving at the local velocity of light in space ($\beta \Rightarrow 1$), the effective imaginary gravitational force goes to zero as is obvious from equation (4.1.8:8). In such a case, the object moves similarly to a satellite in expanding spherical space, Figure 4.1.8-3. The rest energy and rest momentum of an energy object moving at the velocity of light in space is zero.

4.1.9 Topography of space in a local gravitational frame

The curvature of space near local mass centers is a consequence of the conservation of the energy balance created in the primary energy buildup of space. Because the fourth dimension is a geometrical dimension, the shape of space can be solved in distance units, including the topography of the fourth dimension.

As a local mass center in space is approached, the growing contribution of the local gravitational effect causes an increase in the tilting angle of space, ψ. The slope of the curvature of space can be expressed as

$$\frac{dR''_{0\delta}}{dr_{0\delta}} = \tan \psi, \qquad (4.1.9:1)$$

where ψ is the tilting angle of space at distance $r_{0\delta}$ from the local mass center, Figure 4.1.9-1.

The total curvature of space due to tilting close to a local gravitational center can be calculated as the integrated effect of $dR''_{0\delta}$

$$dR''_{0\delta} = \tan \psi \, dr_{0\delta} = \frac{\sin \psi}{\cos \psi} dr_{0\delta} = \frac{\sqrt{1-(1-\delta)^2}}{1-\delta} dr_{0\delta}. \qquad (4.1.9:2)$$

When $\delta \ll 1$, $dR''_{0\delta}$ can be approximated as

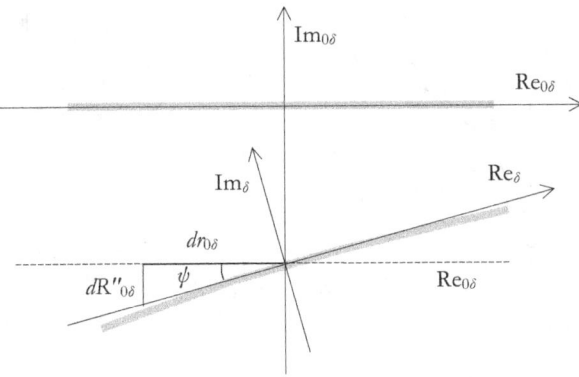

Figure 4.1.9-1. Coordinate system for calculating the topography of space.

$$dR''_{0\delta} \approx \frac{\sqrt{2\delta(1-\delta/2)}}{1-\delta} dr_{0\delta} \approx \sqrt{2\delta}\, dr_{0\delta} \approx \sqrt{\frac{2GM}{r_{0\delta}c_0 c_{0\delta}}}\, dr_{0\delta}, \qquad (4.1.9:3)$$

which gives the local curvature as a function of the distance from the gravitational center M as

$$\Delta R''_{0\delta} \approx \int_{r01}^{r02} \sqrt{\frac{2r_c}{r_{0\delta}}}\, dr_{0\delta} = 2\sqrt{2}\, r_c \left(\sqrt{\frac{r_{02}}{r_c}} - \sqrt{\frac{r_{01}}{r_c}} \right), \qquad (4.1.9:4)$$

where $r_c = GM/c_0 c_{0\delta}$ is the critical radius as defined in equation (4.1.6:9).

Equation (4.1.9:4) applies for $r_{0\delta} \gg r_c$, which is the case for "ordinary" mass centers in space. For example, the critical radius for the mass of the Earth, $M_e \approx 6 \cdot 10^{24}$ kg, is $r_{c(Earth)} \approx$ 4.5 mm. Figure 4.1.9-2 illustrates the actual dimensions of the local curvature of space in our planetary system. The calculation is based on equation (4.1.9:4). As can be seen, the Sun dips about 26,000 km further into the fourth dimension than does the Earth, which is about 150,000 km "deeper" than the planet Pluto.

Close to a local singularity in space, where $r_{0\delta} \approx r_c$, we can denote

$$r_{0\delta} = r_c + \Delta r_{0\delta}. \qquad (4.1.9:5)$$

Applying equations (4.1.9:5) and (4.1.6:10) allows us to express the gravitational factor δ as

$$\delta = \frac{r_c}{r_{0\delta}} = \frac{r_c}{r_c + \Delta r_{0\delta}}, \qquad (4.1.9:6)$$

and equation (4.1.9:2) as

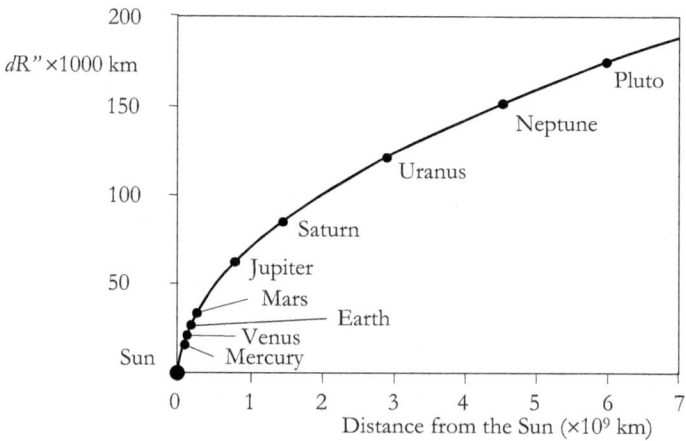

Figure 4.1.9-2. The topography of the Solar System in the fourth dimension. Observe the different scales in the vertical and horizontal axes.

Energy structures in space

$$dR''_{0\delta} = \frac{\sqrt{1-(1-\delta)^2}}{1-\delta} dr_{0\delta} = \sqrt{\frac{1}{(1-\delta)^2} - 1}\ dr_{0\delta}$$

$$= \sqrt{\left(\frac{r_c}{\Delta r_{0\delta}}+1\right)^2 - 1}\ dr_{0\delta} = \sqrt{\left(\frac{r_c}{\Delta r_{0\delta}}+1\right)^2 - 1}\ d(\Delta r_{0\delta}).$$

(4.1.9:7)

When $\Delta r_{0\delta} \ll r_c$, equation (4.1.9:7) can be approximated as

$$dR''_{0\delta} \approx \sqrt{\left(\frac{r_c}{\Delta r_{0\delta}}\right)^2 - 1}\ d(\Delta r_{0\delta}) \approx \frac{r_c}{\Delta r_{0\delta}} d(\Delta r_{0\delta}),$$

(4.1.9:8)

which can be integrated in closed form as

$$\Delta R''_{0\delta(r_{0\delta}=r_c)} \approx r_c \int_{\Delta r_{01}}^{\Delta r_{02}} \frac{d(\Delta r_{0\delta})}{\Delta r_{0\delta}} = r_c \ln\frac{\Delta r_{01}}{\Delta r_{02}}.$$

(4.1.9:9)

Inspection of equation (4.1.9:9) shows that the flat space radius, $r_{0\delta}$, never reaches the critical radius r_c and space has a tube-like form in the fourth dimension, Figure 4.1.9-3. The formation of infinite "worm holes" must be considered merely a hypothetical possibility (see Section 4.2.8 for orbital velocity near a local singularity).

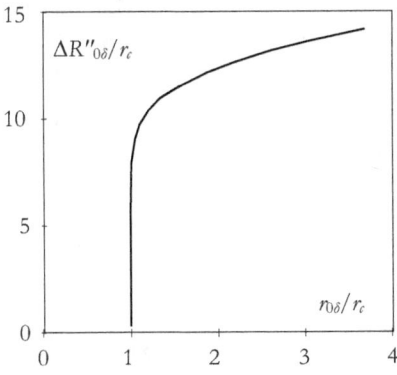

Figure 4.1.9-3. The geometry of a singularity in space in the fourth dimension. The curve is based on numerical integration of equation (4.1.9:2). At $r_{0\delta} \approx r_c$, $\Delta R''_{0\delta}$ can be approximated by equation (4.1.9:9) and at $r_{0\delta} \gg r_c$ by equation (4.1.9:4). The vertical scale corresponds to $\Delta r_{0\delta}(\min) = 10^{-6}\ r_c$. Here a perfect symmetry in the buildup of the singularity is assumed.

4.1.10 Local velocity of light

The local velocity of light is a function of the distance from mass centers in space. At the surface of the Earth, the velocity of light is reduced by about 20 cm/s compared to the velocity of light at the distance of the Moon from the Earth. The velocity of light at the Earth's distance from the Sun is about 3 m/s lower than the velocity of light far from the Sun.

The local velocity of light is determined by the gravitational state, as expressed in equation (4.1.4:10). The velocity of light is known best on the Earth, in the local gravitational frame of the Earth. The farther away we go, the less accurate our knowledge of the gravitational frames we are bound to.

The apparent homogeneous space around the Earth is the space at Earth's distance from the Sun, as it would be with the effect of the Earth's gravity removed. The velocity of light in the apparent homogeneous space of the Earth is affected by the gravitation of the Sun, the Milky Way, and the galaxy group the Milky Way belongs to.

If we initially consider only the effect of the gravitation of the Earth itself, the local velocity of light at distance $r_{0\delta}$ from the center of the Earth can be expressed in accordance with equations (4.1.1:23) and (4.1.1:30) as

$$c = c_{0\delta}(1-\delta_e) = c_{0\delta}\left(1 - \frac{GM_e}{r_{0\delta}c_0 c_{0\delta}}\right), \qquad (4.1.10:1)$$

where M_e is the mass of the Earth, $r_{0\delta}$ is the flat space distance from the center of the Earth, and $c_{0\delta}$ is the velocity of light in apparent homogeneous space. The effect of the gravitation of the Earth on the velocity of light can be calculated by subtracting $c_{0\delta}$ from c given in equation (4.1.10:1). Thus

$$\Delta c_r = c - c_{0\delta} = c_{0\delta}(1-\delta) - c_{0\delta} = -c_{0\delta}\frac{GM_e}{r_{0\delta}c_0 c_{0\delta}} = -\frac{GM_e}{r_{0\delta}c_0} \approx -\frac{GM_e}{rc}. \qquad (4.1.10:2)$$

Figure 4.1.10-1 illustrates the effect of the Earth and the Moon on the velocity of light in the solar gravitational frame. The "tilting" of the velocity of light in apparent homogeneous space around the Earth in Figure 4.1.10-1 is due to the gravitation of the Sun.

The gravitation of the Sun reduces the velocity of light in apparent homogeneous space around the Earth, $c_{0\delta(Earth)}$, by about 2.96 [m/s] relative to the velocity of light in apparent homogeneous space around the Sun in the Milky Way. Distance to the Sun from a fixed location on the rotating Earth is a function of the time of day and the latitude.

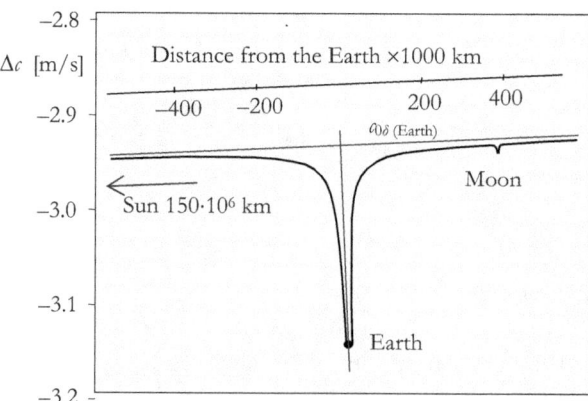

Figure 4.1.10-1. Effect of the gravitation of the Sun, Earth, and Moon on the velocity of light. The tilted baseline at the top shows the effect of the Sun on the velocity of light, which is the apparent homogeneous space velocity of light for the Earth, $c_{0\delta(Earth)}$. The Moon is shown in its "Full Moon" position, opposite to the Sun. The curves in the figure are based on equation (4.1.10:2) as separately applied to the Earth and the Sun. The effect of the mass of the Milky Way on the velocity of light in our planetary system is about $\Delta c \approx -300$ m/s.

Energy structures in space

There is also an annual variation due to the eccentricity of the orbit of the Earth and the inclination angle of the Earth's rotation axis. Generally, a difference in the distance to the barycenter of the gravitational frame studied results in a difference in the velocity of light

$$\frac{dc}{c_{0\delta}} = d\left(1 - \frac{GM}{r_{0\delta}c_0 c_{0\delta}}\right) = \frac{GM}{r_{0\delta}^2 c_0 c_{0\delta}} dr_{0\delta} \approx \delta \frac{dr}{r}, \qquad (4.1.10:3)$$

or

$$dc \approx g\frac{dr}{c}, \qquad (4.1.10:4)$$

where g is the gravitational acceleration at a distance r from the barycenter.

The orbital radius of the Earth in the solar frame is about $1.5 \cdot 10^{11} \pm 2.5 \cdot 10^9$ m with a daily perturbation of about $\pm 6.4 \cdot 10^6$ m at the equator. The average gravitational factor is $\delta \approx 9.85 \cdot 10^{-9}$.

The annual fluctuation in the velocity of light due to the eccentricity of the Earth's orbit is

$$\frac{dc}{c} \approx \delta\frac{dr}{r} \approx \pm 9.85 \cdot 10^{-9} \cdot \frac{2.5 \cdot 10^9}{1.5 \cdot 10^{11}} \approx \pm 1.6 \cdot 10^{-10}. \qquad (4.1.10:5)$$

The daily perturbation of the velocity of light at the equator is

$$\frac{dc}{c} \approx \delta\frac{dr}{r} \approx \pm 9.85 \cdot 10^{-9} \cdot \frac{6.4 \cdot 10^6}{1.5 \cdot 10^{11}} \approx \pm 4.2 \cdot 10^{-13}. \qquad (4.1.10:6)$$

The effects of the variation of the velocity of light on the ticking frequency and the synchronization of atomic clocks on the Earth and in Earth satellites are discussed in Section 5.7.3.

4.2 Celestial mechanics

Because of the dents around mass centers, the geometry of DU-space has features in common with the Schwarzschild metric based on four-dimensional spacetime. The precise geometry of space makes it possible to solve for the effect of the 4-D geometry on Kepler's laws and the orbital equation in closed mathematical form. A perihelion shift, equal to that predicted by the general theory of relativity, can be derived as the rotation of the orbit relative to a non-rotating reference coordinate system. In addition to the perihelion shift, the length of the radius of the orbit is subject to a perturbation with a maximum at the aphelion. The DU model does not predict gravitational radiation; gravitational energy is potential energy by its nature. All mass in space contributes to the local gravitational potential. Orbits of local gravitational systems are subject to expansion with the expansion of the whole spherical space.

In DU-space, orbits around mass centers are stable down to the critical radius, which is half of the critical radius in Schwarzschild space. This means a major difference to orbits around local singularities in Schwarzschild space, where orbits become unstable at radii below $3 \cdot r_{c(Schwd)}$ $3 \cdot r_{c(Schwd)}$. Slow orbits below the radius of the minimum period maintain the mass of the local singularity.

4.2.1 The cylinder coordinate system

In all observations in a local gravitational frame, the reference space moves in the local fourth dimension, the $Im_{0\delta}$-direction, at the same velocity as the objects studied. For the study of orbital equations, it is therefore convenient to choose a cylinder coordinate system with the base plane parallel to the apparent homogeneous space of the gravitational frame studied. The z-coordinate shows the distance in the direction of the $Im_{0\delta}$-axis drawn through the center of the central mass of the frame, Figure 4.2.1-1.

With reference to equation (4.1.9:2), the distance differential $dz_{0\delta}$ can be expressed as

$$ds_{Im(0\delta)} = dz_{0\delta} = -dR"_{0\delta} = -\tan\psi \, dr_{0\delta} = -\frac{\sqrt{1-(1-\delta)^2}}{1-\delta} dr_{0\delta}, \tag{4.2.1:1}$$

where ψ is the tilting angle of local space.

The cylinder coordinate system allows the orbital equations to be solved by first studying the flat space projection of the orbits in planar polar coordinates on the base plane parallel to the apparent homogeneous space. The real space orbit can then be constructed by adding the $z_{0\delta}$-coordinate given in equation (4.2.1:1).

4.2.2 The equation of motion

Equation (4.1.6:20) gives the gravitational acceleration in the direction of the local Re_δ-axis. In order to utilize the cylinder coordinate system defined in Section 4.2.2, we apply the "flat space" component of the gravitational acceleration to first solve the equation of

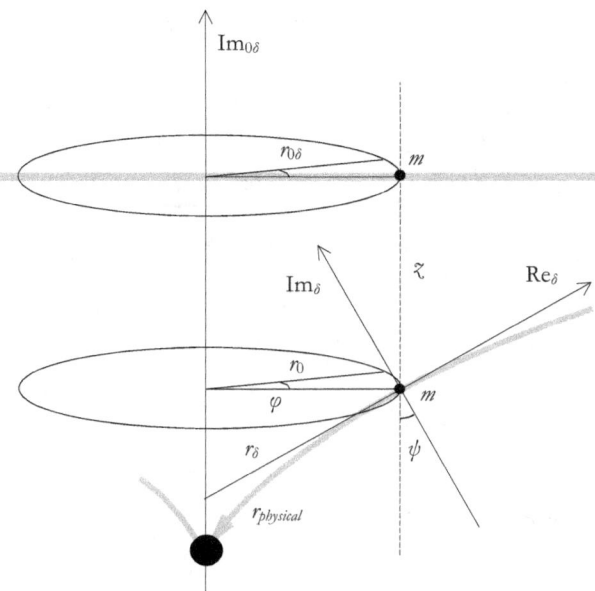

Figure 4.2.1-1. Apparent homogeneous space and tilted (actual) local space. The local complex coordinate system, Im_δ –Re_δ, at object m is illustrated. The imaginary velocity of apparent homogeneous space, appearing in the direction of the $\text{Im}_{0\delta}$-axis, is $c_{0\delta}$, and the imaginary velocity of local space, the component of $c_{0\delta}$ in the direction of the Im_δ-axis, is $c_\delta = c_{0\delta} \cdot \cos\psi$.

motion as a plane solution in the direction of apparent homogeneous space. Based on equation (4.1.6:20), the component of the acceleration of free fall in the direction of distance $r_{0\delta}$ along the $\text{Re}_{0\delta}$-axis is

$$\mathbf{a}_{ff(0\delta)} = -\mathbf{a}_{esc(0\delta)} = -\frac{dv_{esc(\delta)}}{dt}(1-\delta)\,\hat{\mathbf{r}}_{0\delta} = \frac{c_{0\delta}}{c_0}\frac{GM}{r_{0\delta}^2}(1-\delta)^3\,\hat{\mathbf{r}}_{0\delta}\,. \quad (4.2.2{:}1)$$

On the flat space plane, the centripetal acceleration in central motion on a plane in the direction of apparent homogeneous space can be expressed as

$$\mathbf{a}_{0\delta} = \frac{d\mathbf{v}_{\perp(0\delta)}}{dt} = \ddot{\mathbf{r}}_{0\delta}\,, \quad (4.2.2{:}2)$$

where velocity $\mathbf{v}_{\perp(0\delta)} = \mathbf{v}_{\perp(\delta)}$ is the velocity component perpendicular to radius $r_{0\delta}$ (and also to radius r_δ) in the local gravitational frame, Figure 4.2.2-1.

Combining equations (4.2.2:1) and (4.2.2:2) gives the balance of the gravitational and kinematic accelerations on the flat space plane

$$\mathbf{a}_{0\delta} = \ddot{\mathbf{r}}_{0\delta} = \frac{c_{0\delta}}{c_0}\frac{GM}{r_{0\delta}^2}(1-\delta)^3\,\hat{\mathbf{r}}_{0\delta}\,. \quad (4.2.2{:}3)$$

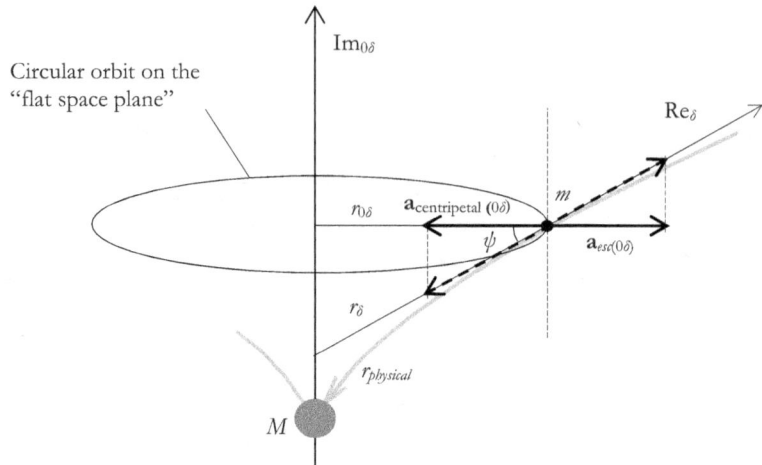

Figure 4.2.2-1. Acceleration $\mathbf{a}_{ff(0\delta)}$ is the flat space component of acceleration $\mathbf{a}_{ff}$. Acceleration $a_{ff(0\delta)}$ has the direction of $r_{0\delta}$.

Equation (4.2.2:3) has the form of the classical equation of motion in a gravitational frame but is corrected by the factor $(1-\delta)^3$ originating from the effect of the local curvature of space on the gravitational acceleration. By applying the system mass $M = M+m$ for mass combined with the frame conversion factor

$$\mu = \frac{c_{0\delta}}{c_0} G(M+m), \qquad (4.2.2:4)$$

and equation (4.1.6:10) for δ, equation (4.2.2:3) can be expressed in the form

$$\ddot{\mathbf{r}}_{0\delta} = \frac{\mu}{r_{0\delta}^2}\left(1 - \frac{r_c}{r_{0\delta}}\right)^3 \hat{\mathbf{r}}_{0\delta}. \qquad (4.2.2:5)$$

Equation (4.2.2:5) can be solved following the procedure used in deriving Kepler's equations.

4.2.3 Perihelion direction on the flat space plane

Equation (4.2.2:5) differs by factor $(1-r_c/r_{0\delta})^3$ from the classical equation of motion

$$\ddot{\mathbf{r}}_{0\delta} = \frac{\mu}{r_{0\delta}^2} \hat{\mathbf{r}}_{0\delta}, \qquad (4.2.3:1)$$

used in deriving Kepler's orbital equation. In order to find out the effect of the factor $(1-r_c/r_{0\delta})^3$, we follow the procedure used in deriving Kepler's orbital equation.

The angular momentum per unit mass (related to the orbital velocity in the direction of the flat space plane) can be expressed as

Energy structures in space

$$\mathbf{k}_{0\delta} = \mathbf{r}_{0\delta} \times \dot{\mathbf{r}}_{0\delta}. \quad (4.2.3{:}2)$$

The time derivative of $\mathbf{k}_{0\delta}$ is

$$\dot{\mathbf{k}}_{0\delta} = \mathbf{r}_{0\delta} \times \ddot{\mathbf{r}}_{0\delta} + \dot{\mathbf{r}}_{0\delta} \times \dot{\mathbf{r}}_{0\delta} = \mathbf{r}_{0\delta} \times \ddot{\mathbf{r}}_{0\delta}. \quad (4.2.3{:}3)$$

Substituting (4.2.2:5) for $\ddot{\mathbf{r}}_{0\delta}$ in (4.2.3:3) we get

$$\dot{\mathbf{k}}_{0\delta} = \mathbf{r}_{0\delta} \times \mathbf{r}_{0\delta} \frac{-\mu\left(1 - r_c/r_{0\delta}\right)^3}{r_{0\delta}^3} = 0. \quad (4.2.3{:}4)$$

To determine vector $\mathbf{e}_\delta$ we form the vector product $\mathbf{k}_{0\delta} \times \ddot{\mathbf{r}}_{0\delta}$,

$$\begin{aligned}\mathbf{k}_{0\delta} \times \ddot{\mathbf{r}}_{0\delta} &= \left(\mathbf{r}_{0\delta} \times \dot{\mathbf{r}}_{0\delta}\right) \times \mathbf{r}_{0\delta} \frac{-\mu\left(1 - r_c/r_{0\delta}\right)^3}{r_{0\delta}^3} \\ &= \frac{-\mu\left(1 - r_c/r_{0\delta}\right)^3}{r_{0\delta}^3}\left[\left(\mathbf{r}_{0\delta} \cdot \mathbf{r}_{0\delta}\right)\dot{\mathbf{r}}_{0\delta} - \left(\mathbf{r}_{0\delta} \cdot \dot{\mathbf{r}}_{0\delta}\right)\mathbf{r}_{0\delta}\right].\end{aligned} \quad (4.2.3{:}5)$$

Since the time derivative of distance $\dot{r}_{0\delta}$ is the component of $\dot{\mathbf{r}}_{0\delta}$ in the direction of $\mathbf{r}_{0\delta}$, it is possible to express $\dot{r}_{0\delta}$ in the form of a dot product

$$\dot{r}_{0\delta} = \frac{\mathbf{r}_{0\delta}}{r_{0\delta}} \cdot \dot{\mathbf{r}}_{0\delta}, \quad (4.2.3{:}6)$$

and, accordingly, equation (4.2.3:6) can be expressed as

$$\mathbf{r}_{0\delta} \cdot \dot{\mathbf{r}}_{0\delta} = r_{0\delta}\dot{r}_{0\delta}. \quad (4.2.3{:}7)$$

Equation (4.2.3:5) can now be expressed as

$$\mathbf{k}_{0\delta} \times \ddot{\mathbf{r}}_{0\delta} = -\mu\left(1 - r_c/r_{0\delta}\right)^3 \left[\frac{\dot{\mathbf{r}}_{0\delta}}{r_{0\delta}} - \frac{\dot{r}_{0\delta}\mathbf{r}_{0\delta}}{r_{0\delta}^2}\right], \quad (4.2.3{:}8)$$

where the expression in parentheses can be identified as the time derivative

$$\left[\frac{\dot{\mathbf{r}}_{0\delta}}{r_{0\delta}} - \frac{\dot{r}_{0\delta}\mathbf{r}_{0\delta}}{r_{0\delta}^2}\right] = \frac{d\left(\mathbf{r}_{0\delta}/r_{0\delta}\right)}{dt}, \quad (4.2.3{:}9)$$

and equation (4.2.3:8) can be expressed as

$$\mathbf{k}_{0\delta} \times \ddot{\mathbf{r}}_{0\delta} = -\frac{d\left(\mu \mathbf{r}_{0\delta}/r_{0\delta}\right)}{dt} + \mu A_r \frac{d\left(\mathbf{r}_{0\delta}/r_{0\delta}\right)}{dt}, \quad (4.2.3{:}10)$$

where

$$A_r = 1 - \left(1 - \frac{r_c}{r_{0\delta}}\right)^3 = \frac{3r_c}{r_{0\delta}} - 3\left(\frac{r_c}{r_{0\delta}}\right)^2 + \left(\frac{r_c}{r_{0\delta}}\right)^3. \quad (4.2.3{:}11)$$

As shown in (4.2.3:4), the time derivative of $\mathbf{k}_{0\delta}$ is zero. Accordingly, the vector product $\mathbf{k}_{0\delta} \times \ddot{\mathbf{r}}_{0\delta}$ can be expressed in the form

$$\mathbf{k}_{0\delta} \times \ddot{\mathbf{r}}_{0\delta} = \frac{d(\mathbf{k}_{0\delta} \times \dot{\mathbf{r}}_{0\delta})}{dt}. \tag{4.2.3:12}$$

Combining equations (4.2.3:10) and (4.2.3:12) gives

$$\frac{d(\mathbf{k}_{0\delta} \times \dot{\mathbf{r}}_{0\delta})}{dt} + \frac{d(\mu\,\mathbf{r}_{0\delta}/r_{0\delta})}{dt} = \mu A_r \frac{d(\mathbf{r}_{0\delta}/r_{0\delta})}{dt}, \tag{4.2.3:13}$$

which can be written in the form

$$\frac{d(\mathbf{k}_{0\delta} \times \dot{\mathbf{r}}_{0\delta} + \mu\,\mathbf{r}_{0\delta}/r_{0\delta})}{dt} = \mu A_r \frac{d(\mathbf{r}_{0\delta}/r_{0\delta})}{dt}. \tag{4.2.3:14}$$

The expression in parentheses on the left-hand side of the equation is equal to the eccentricity vector $-\mathbf{e}_{0\delta}\mu$, showing the direction of the perihelion or periastron radius in Kepler's orbital equation. Applying $\mathbf{e}_{0\delta}$ in equation (4.2.3:14), we get the time derivative

$$\frac{d\mathbf{e}_{0\delta}}{dt} = -A_r \frac{d(\mathbf{r}_{0\delta}/r_{0\delta})}{dt}, \tag{4.2.3:15}$$

which in Newtonian mechanics is equal to zero. Equation (4.2.3:15) implies that the eccentricity vector $\mathbf{e}_{0\delta}$ changes with time. Solving (4.2.3:15) gives

$$\frac{d\mathbf{e}_{0\delta}}{dt} = -A_r \left[\frac{\dot{\mathbf{r}}_{0\delta}}{r_{0\delta}} - \frac{\dot{r}_{0\delta}\mathbf{r}_{0\delta}}{r_{0\delta}^2} \right] = -A_r \left[\frac{d\mathbf{r}_{0\delta}/dt}{r_{0\delta}} - \frac{(dr_{0\delta}/dt)\mathbf{r}_{0\delta}}{r_{0\delta}^2} \right]. \tag{4.2.3:16}$$

In polar coordinates on the flat space plane, vector $d\mathbf{r}_{0\delta}$ can be expressed as

$$d\mathbf{r}_{0\delta} = r_{0\delta}d\varphi\,\hat{\mathbf{r}}_\perp + dr_{0\delta}\,\hat{\mathbf{r}}_\parallel, \tag{4.2.3:17}$$

where $\hat{\mathbf{r}}_\perp$ and $\hat{\mathbf{r}}_\parallel$ are the unit vectors perpendicular to $\mathbf{r}_{0\delta}$ and in the direction of $\mathbf{r}_{0\delta}$, respectively.

Substituting (4.2.3:17) into (4.2.3:16) gives

$$\frac{d\mathbf{e}_{0\delta}}{dt} = -A_r \left[\frac{(r_{0\delta}d\varphi\,\hat{\mathbf{r}}_\perp + dr_{0\delta}\,\hat{\mathbf{r}}_\parallel)/dt}{r_{0\delta}} - \frac{(dr_{0\delta}/dt)r_{0\delta}\hat{\mathbf{r}}_\parallel}{r_{0\delta}^2} \right], \tag{4.2.3:18}$$

and further

$$\frac{d\mathbf{e}_{0\delta}}{dt} = -A_r \left[\frac{d\varphi}{dt}\hat{\mathbf{r}}_\perp + \left(\frac{dr_{0\delta}}{r_{0\delta}\,dt} - \frac{dr_{0\delta}}{r_{0\delta}\,dt} \right)\hat{\mathbf{r}}_\parallel \right] = -A_r \frac{d\varphi}{dt}\hat{\mathbf{r}}_\perp. \tag{4.2.3:19}$$

As shown by (4.2.3:19), the change in $\mathbf{e}_{0\delta}$ occurs as a rotational change only, which means that the orbit conserves its eccentricity but is subject to a rotation of the main axis. Multiplying (4.2.3:19) by dt gives

$$d\mathbf{e}_{0\delta} = -A_r d\varphi\,\hat{\mathbf{r}}_\perp. \tag{4.2.3:20}$$

The differential rotation $d\psi_{0\delta}$ of the polar coordinate system that eliminates the differential change of the eccentricity vector $d\mathbf{e}_{0\delta}$ can be solved from the equation

Energy structures in space

$$d\mathbf{e}_{0\delta} = -A_r d\varphi\, \hat{\mathbf{r}}_\perp + d\psi_{0\delta} = 0, \qquad (4.2.3{:}21)$$

as

$$d\psi_{0\delta} = A_r d\varphi\, \hat{\mathbf{r}}_\perp, \qquad (4.2.3{:}22)$$

which, by substitution of equation (4.2.3:11) for A_r gives

$$d\psi_{0\delta} = A_r d\varphi = \left[1 - \left(1 - \frac{r_c}{r_{0\delta}}\right)^3\right] d\varphi \approx \frac{3r_c}{r_{0\delta}} d\varphi. \qquad (4.2.3{:}23)$$

In a coordinate system that rotates by angle $d\psi_{0\delta}$ in the direction of the orbital motion, the time derivative of $\mathbf{e}_{0\delta}$ is zero, which is the requirement of Kepler's orbital equation.

Applying Kepler's equation

$$r_{0\delta} = \frac{a(1-e^2)}{1+e\cos\varphi} \qquad (4.2.3{:}24)$$

for $r_{0\delta}$ in (4.2.3:23), we can express the rotation $d\psi_{0\delta}$ as

$$d\psi_{0\delta} \approx \frac{3r_c(1+e\cos\varphi)}{a(1-e^2)} d\varphi. \qquad (4.2.3{:}25)$$

Rotation $\Delta\psi_{0\delta}$ can be obtained by integrating (4.2.3:25)

$$\Delta\psi_{0\delta} \approx \frac{3r_c}{a(1-e^2)} \int_0^\varphi (1+e\cos\varphi)\, d\varphi = \frac{3r_c(\varphi + e\sin\varphi)}{a(1-e^2)}. \qquad (4.2.3{:}26)$$

According to equation (4.2.3:26), the coordinate system conserving Kepler's orbital equations rotates by angle $\Delta\psi_{0\delta}(\varphi)$ in the direction of the orbital motion. To express the orbital equation in the non-rotating polar coordinate system, we have to subtract angle $\Delta\psi_{0\delta}(\varphi)$ from the φ-coordinate as

$$r_{0\delta} = \frac{a(1-e^2)}{1+e\cos(\varphi - \Delta\psi_{0\delta})}, \qquad (4.2.3{:}27)$$

which is Kepler's equation supplemented with a perihelion advance of angle $\Delta\psi_{0\delta}(\varphi)$. Setting $\varphi = 2\pi$ in equation (4.2.3:26), the perihelion advance for a full revolution can be expressed as

$$\Delta\psi_{0\delta}(2\pi) = \frac{6\pi r_c}{a(1-e^2)}. \qquad (4.2.3{:}28)$$

By applying equations (4.1.6:9) and (4.2.2:4) in (4.2.3:28), the perihelion advance for a full revolution can be expressed as

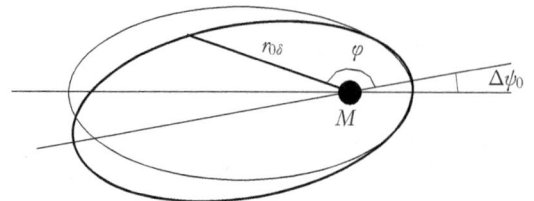

Figure 4.2.3-1. Perihelion advance results in the rotation of the main axis. For each full revolution, the rotation is $6\pi\, r_c/a(1-e^2)$.

$$\Delta\psi_{0\delta}(2\pi) = \frac{6\pi G(M+m)}{c^2 a(1-e^2)},\qquad(4.2.3{:}29)$$

which is the same result as given by the general theory of relativity for perihelion advance, Figure 4.2.3-1.

4.2.4 Kepler's energy integral

To complete our analysis of the orbit on the flat space plane, we now study the energy integral derived from the dot product of the velocity and the acceleration given in equation (4.2.2:5)

$$\dot{\mathbf{r}}_{0\delta}\cdot\ddot{\mathbf{r}}_{0\delta} = \dot{\mathbf{r}}_{0\delta}\cdot\mathbf{r}_{0\delta}\frac{-\mu(1-r_c/r_{0\delta})^3}{r_{0\delta}^3} = -\frac{\mu(1-r_c/r_{0\delta})^3}{r_{0\delta}^2}\dot{r}_{0\delta},\qquad(4.2.4{:}1)$$

which, by substituting equation (4.2.3:11) for $(1-r_c/r_{0\delta})^3$, can be expressed as

$$\dot{\mathbf{r}}_{0\delta}\cdot\ddot{\mathbf{r}}_{0\delta} = -\frac{\mu(1-A_r)}{r_{0\delta}^2}\dot{r}_{0\delta} = -\frac{\mu}{r_{0\delta}^2}\dot{r}_{0\delta} + \frac{\mu A_r}{r_{0\delta}^2}\dot{r}_{0\delta}.\qquad(4.2.4{:}2)$$

The first term on the right-hand side in equation (4.2.4:2) can be written as

$$-\frac{\mu}{r_{0\delta}^2}\dot{r}_{0\delta} = -\frac{\mu}{r_{0\delta}^2}\frac{dr_{0\delta}}{dt} = \frac{d(\mu/r_{0\delta})}{dt},\qquad(4.2.4{:}3)$$

and by substituting equation (4.2.4:3) into equation (4.2.4:2), we can write

$$\dot{\mathbf{r}}_{0\delta}\cdot\ddot{\mathbf{r}}_{0\delta} = \frac{d(\mu/r_{0\delta})}{dt} + \frac{\mu A_r}{r_{0\delta}^2}\dot{r}_{0\delta}.\qquad(4.2.4{:}4)$$

The dot product of the velocity and the acceleration can also be expressed as

$$\dot{\mathbf{r}}_{0\delta}\cdot\ddot{\mathbf{r}}_{0\delta} = \frac{d(1/2\,\dot{\mathbf{r}}_{0\delta}\cdot\dot{\mathbf{r}}_{0\delta})}{dt} = \frac{d(\dot{r}_{0\delta}^2/2)}{dt} = \frac{d(v_{r(0\delta)}^2/2)}{dt},\qquad(4.2.4{:}5)$$

where $\dot{\mathbf{r}}_{0\delta} = \mathbf{v}_{r(0\delta)}$ is the radial velocity on the flat space plane. Combining equations (4.2.4:4) and (4.2.4:5) gives

Energy structures in space

$$\frac{d\left(v_{r(0\delta)}^2/2 - \mu/r_{0\delta}\right)}{dt} = \frac{db}{dt} = \dot{b} = \frac{\mu A_r}{r_{0\delta}^2} \dot{r}_{0\delta}, \tag{4.2.4:6}$$

where b, in Kepler's formalism,

$$b = \frac{v_{r(0\delta)}^2}{2} - \frac{\mu}{r_{0\delta}}, \tag{4.2.4:7}$$

is referred to as the energy integral.

In the case of Newtonian mechanics, the time derivative of the energy integral is zero. In the DU, as shown by equation (4.2.4:6), the time derivative of b is not zero.

In Kepler's orbital equation

$$r_{0\delta} = \frac{k^2}{\mu(1+e\cos\varphi)} = \frac{a(1-e^2)}{(1+e\cos\varphi)}, \tag{4.2.4:8}$$

constants μ, e, b, and k are related as

$$k^2 = \frac{-\mu^2(1-e^2)}{2b} \quad ; \quad b = \frac{-\mu^2(1-e^2)}{2k^2}. \tag{4.2.4:9}$$

To determine the effect of the time-dependent b on the orbital equation, we solve for the time derivative of $k_{0\delta}^2$ (for stable mass centers μ is constant)

$$\ddot{k} = \frac{-\mu^2(1-e^2)}{2}\left(-\frac{1}{b^2}\right)\dot{b} = \frac{\mu^2(1-e^2)}{2b}\frac{1}{b}\dot{b} = -\frac{k_{0\delta}^2}{b}\dot{b}. \tag{4.2.4:10}$$

Substituting equation (4.2.4:6) for $\dot{b}$ into equation (4.2.4:10) gives

$$\ddot{k} = -\frac{k_{0\delta}^2}{b}\frac{\mu A_r}{r_{0\delta}^2}\dot{r}_{0\delta}, \tag{4.2.4:11}$$

and substituting equation (4.2.4:9) for b into equation (4.2.4:11) gives

$$\ddot{k} = -\frac{2k_{0\delta}^4}{-\mu(1-e^2)}\frac{A_r}{r_{0\delta}^2}\dot{r}_{0\delta}. \tag{4.2.4:12}$$

The time derivative $\dot{r}_{0\delta}$ in (4.2.4:12) can be solved from (4.2.4:8)

$$\dot{r}_{0\delta} = \frac{k^2}{\mu(1+e\cos\varphi)^2}e\sin\varphi\,\dot{\varphi}. \tag{4.2.4:13}$$

Substitution of equation (4.2.4:13) for $\dot{r}_{0\delta}$ in equation (4.2.4:12) and multiplication of the equation by dt gives

$$dk^2 = \frac{2k_{0\delta}^6}{\mu^2(1-e^2)(1+e\cos\varphi)^2}\frac{A_r}{r_{0\delta}^2}e\sin\varphi\,d\varphi. \tag{4.2.4:14}$$

From equation (4.2.4:8), we get

$$dr_{0\delta} = \frac{1}{\mu(1+e\cos\varphi)} dk^2. \qquad (4.2.4:15)$$

Substituting (4.2.4:14) for dk^2 in (4.2.4:15) gives

$$dr_{0\delta} = \frac{2k_{0\delta}^6}{\mu^3(1-e^2)(1+e\cos\varphi)^3} \frac{A_r}{r_{0\delta}^2} e\sin\varphi\, d\varphi, \qquad (4.2.4:16)$$

which can be developed further as

$$dr_{0\delta} = \frac{2r_{0\delta}^3}{(1-e^2)} \frac{A_r}{r_{0\delta}^2} e\sin\varphi\, d\varphi = \frac{2eA_r r_{0\delta}}{(1-e^2)} \sin\varphi\, d\varphi. \qquad (4.2.4:17)$$

Applying the first order approximation for $A_r \approx 3r_c/r_{0\delta}$, equation (4.2.4:17) can be expressed as

$$dr_{0\delta} = \frac{6er_c}{(1-e^2)} \sin\varphi\, d\varphi, \qquad (4.2.4:18)$$

and the total perturbation of distance $r_{0\delta}$ as (Figure 4.2.4-1)

$$\Delta r_{0\delta}(\varphi) = \frac{6er_c}{(1-e^2)} \int_0^\varphi \sin\varphi\, d\varphi = \frac{6er_c(1-\cos\varphi)}{(1-e^2)}. \qquad (4.2.4:19)$$

The increase of $r_{0\delta}$, $\Delta r_{0\delta}$, is zero at perihelion and achieves its maximum value at aphelion:

perihelion: $\Delta r_{0\delta}(0) = 0$ \qquad (4.2.4:20)

aphelion: $\Delta r_{0\delta}(\pi) = \dfrac{12e}{(1-e^2)} r_c$. \qquad (4.2.4:21)

Combining equations (4.2.3:27) and (4.2.4:19) gives the complete orbital equation of the flat space projection of the orbit

$$r_{0\delta} = \frac{a_{0\delta}(1-e^2)}{1+e\cos(\varphi-\Delta\psi_{0\delta})} + \frac{6er_c\left[1-\cos(\varphi-\Delta\psi_{0\delta})\right]}{(1-e^2)}. \qquad (4.2.4:22)$$

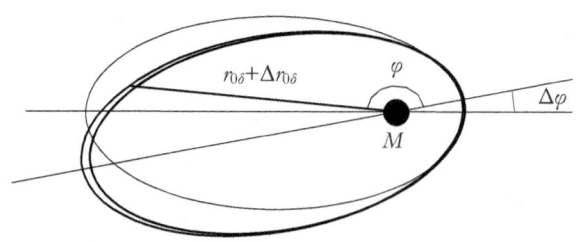

Figure 4.2.4-1. Kepler's orbit is perturbed by distance $\Delta r_{0\delta} = 6r_c e(1-\cos\varphi)/(1-e^2)$, equation (4.2.4-1).

Energy structures in space

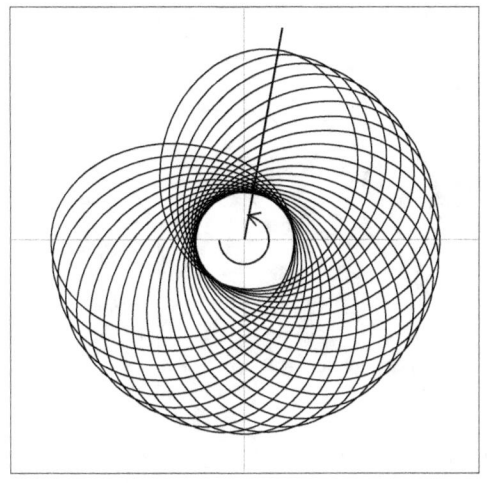

Figure 4.2.4-2. For $\delta = 4 \cdot 10^{-3}$ and $e = 0.6$, the rotation of the perihelion proceeds about 270° in 40 revolutions. The DU orbit conserves its shape but is slightly larger than Kepler's orbit, shown as the ellipse drawn with stronger line, with an arrow showing the orbiting direction. At perihelion, the distance from the orbit to the mass center is the same in the DU and Kepler's orbits.

Equation (4.2.4:22) is applicable in gravitational potentials $\delta \ll 1$ where the approximation $(1-\delta)^3 \approx (1-3\delta)$ is accurate enough. Figure 4.2.4-2 illustrates the development of the orbit according to equation (4.2.4:22).

4.2.5 The fourth dimension

The orbital coordinates are completed by adding the z-coordinate, which extends the orbital calculation made on the flat space plane to actual space curved in the fourth dimension.

With reference to equation (4.1.9:4), the z-coordinate, the distance from the central plane (in the flat space direction) intersecting the orbiting surface at $\varphi = \pm \pi/2$, can be expressed as

$$z(r_{0\delta}) = 2\sqrt{2r_c}\left[\sqrt{r_{0\delta}} - \sqrt{a_{0\delta}\left(1-e_{0\delta}^2\right)}\right], \qquad (4.2.5:1)$$

where $r_{0\delta}$ is the flat space distance from the center of the gravitational frame given in equation (4.2.4:22). The expression $a_{0\delta}(1-e_{0\delta}^2)$ in equation (4.2.5:1) is the value of $r_{0\delta}$ at $\varphi_{0\delta} = \pm \pi/2$, which is used as the reference value for the z-coordinate. Equations (4.2.4:22) and (4.2.5:1) give the 4-dimensional coordinates of an orbiting object as a function of angle $\varphi_{0\delta}$ determined relative to the perihelion direction in the flat space projection of the orbit, Figure 4.2.5-1.

The differential of a line element in the $z_{0\delta}$-direction can be expressed in terms of the differential in the $r_{0\delta}$-direction on the flat space plane and the tilting angle θ,

$$dz_{0\delta} = dR''_{0\delta} = \tan\theta \, dr_{0\delta} = B \, dr_{0\delta}, \qquad (4.2.5:2)$$

where [see equation (4.1.9:2)]

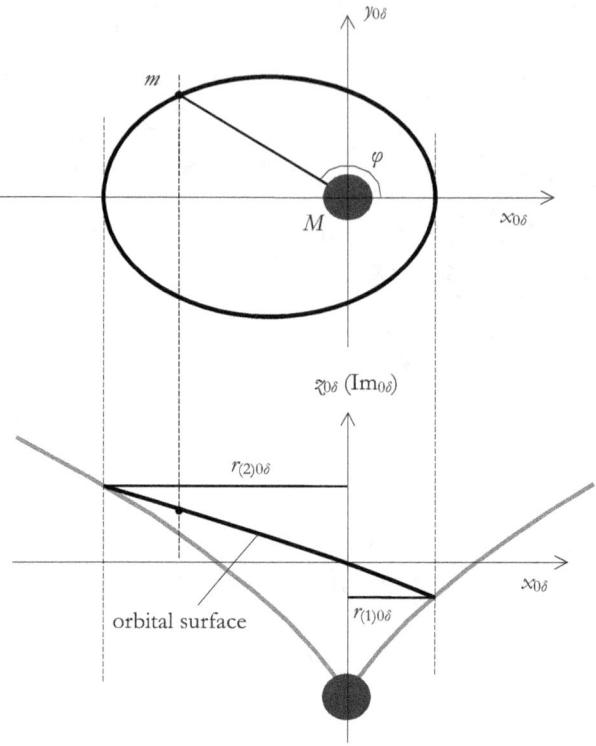

Figure 4.2.5-1. Projections of an elliptic orbit on the $x_{0\delta}-y_{0\delta}$ and $x_{0\delta}-z_{0\delta}$ planes in a gravitational frame around mass center M.

$$B = \tan\theta = \frac{\sqrt{1-(1-\delta)^2}}{(1-\delta)} = \frac{\sqrt{1-(1-r_c/r_{0\delta})^2}}{(1-r_c/r_{0\delta})}. \tag{4.2.5:3}$$

The distance differential $dr_{0\delta}$ in equation (4.2.5:2) can be obtained from the derivative of equation (4.2.4:22) as

$$dr_{0\delta} = A\,d\varphi_{0\delta}, \tag{4.2.5:4}$$

where

$$A = \left\{ \frac{a(1-e^2)}{\left[1+e\cos(\varphi-\Delta\psi)\right]^2} + \frac{6r_c}{(1-e^2)} \right\} e\sin(\varphi-\Delta\psi). \tag{4.2.5:5}$$

The line element of the orbit can be expressed in cylindrical coordinates as

$$d\mathbf{s} = dr\,\hat{\mathbf{u}}_r + r_{0\delta}d\varphi\,\hat{\mathbf{u}}_\varphi + dz\,\hat{\mathbf{u}}_z, \tag{4.2.5:6}$$

where $\hat{\mathbf{u}}$ is the unit vector in each coordinate direction.

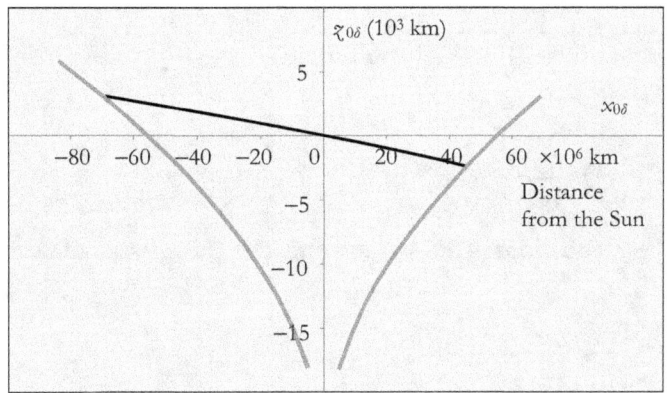

Figure 4.2.5-2. The $z_{0\delta} - x_{0\delta}$ profile of the orbit of Mercury. Note the different scales in the $z_{0\delta}$- and $x_{0\delta}$-directions.

The squared line element ds^2 of an orbit around a mass center can now be expressed as

$$ds^2 = \left[r_{0\delta}^2 + A^2 + A^2 B^2 \right] d\varphi^2, \quad (4.2.5\!:\!7)$$

where $r_{0\delta}$ is the flat space radius given in equation (4.2.4:22). The scalar value of the line element can now be expressed as

$$ds = \sqrt{r_{0\delta}^2 + A^2 \left(1 + B^2\right)}\, d\varphi. \quad (4.2.5\!:\!8)$$

The length of the path along the orbit from φ_1 to φ_2 can be obtained by integrating (4.2.5:8) as

$$s = \int_{\varphi_1}^{\varphi_2} \sqrt{r_{0\delta}^2 + A^2 \left(1 + B^2\right)}\, d\varphi. \quad (4.2.5\!:\!9)$$

Figure 4.2.5-2 shows the $x_{0\delta} - z_{0\delta}$ profile of the orbit of Mercury in the solar gravitational frame.

4.2.6 Effect of the expansion of space

The orbital elements a, e, k, and μ are related as

$$a\left(1 - e^2\right) = \frac{k^2}{\mu}. \quad (4.2.6\!:\!1)$$

The parameter k is the angular momentum per unit mass, which at the perihelion point can be expressed as

$$k = a\left(1 - e\right) v_p, \quad (4.2.6\!:\!2)$$

where v_p is the orbital velocity at the perihelion. By applying equations (4.1.1:8), (4.2.2:4), (4.2.6:1), and (4.2.6:2), the semi-major axis, a, can be expressed as

$$a = \frac{\mu(1+e)}{(1-e)} \frac{1}{v_p^2} r_c = \frac{(1+e)}{(1-e)} \frac{c_{0\delta}^2 r_c}{v_p^2} = \frac{(1+e)}{(1-e)} \frac{r_c}{\beta_{p(0\delta)}^2}, \qquad (4.2.6:3)$$

where

$$\mu = \frac{c_{0\delta}}{c_0} G(M+m) = \frac{c_{0\delta}}{c_0} c_0 c_{0\delta} r_c = c_{0\delta}^2 r_c. \qquad (4.2.6:4)$$

With reference to equations (4.1.6:10) and (4.2.2:4), the critical radius r_c can be expressed as

$$r_c = \delta r_{0\delta} = \frac{G(M+m)}{c_0 c_{0\delta}} = \frac{M+m}{M"} R"_{0\delta}. \qquad (4.2.6:5)$$

Substitution of equation (4.2.6:5) for r_c in equation (4.2.6:3) relates the semi-major axis to the imaginary radius of space

$$a = \frac{1}{\beta_{p(0\delta)}^2} \frac{1+e}{1-e} \frac{M+m}{M"} R"_{0\delta}. \qquad (4.2.6:6)$$

The conservation of energy in the cosmological expansion of space requires that β_p be conserved. Equations (4.2.6:5) and (4.2.6:6) confirm that r_c and the semi-major axis, a, increase in direct proportion to the imaginary radius $R"_{0\delta}$.

> Gravitationally bound local systems expand in direct proportion to the expansion of space.

4.2.7 Effect of the gravitational state in the parent frame

As shown by equation (4.2.6:6), the semi-major axis, a, increases in direct proportion to the imaginary radius $R"_{0\delta}$, which is the imaginary radius of the apparent homogeneous space of the local rotational system. When the local rotational system rotates in an elliptical orbit in its parent frame, the gravitational state of the local system and, thereby, the imaginary radius $R"_{0\delta}$, oscillates with the rotation in the parent frame. Solving for the radius of a local orbiting system from equation (4.1.1:8) gives

$$r_{0\delta} = \frac{M}{M"} \frac{R"_{0\delta}}{\delta} \quad ; \quad r_\delta = \frac{M}{M"} \frac{R"_\delta}{\delta}, \qquad (4.2.7:1)$$

which shows that conservation of the local gravitational factor, δ, makes $r_{0\delta}$ directly proportional to $R"_{0\delta}$, just as was concluded from equation (4.2.6:6).

The imaginary radius of the apparent homogeneous space of the local frame, $R"_{0\delta}$, is the local imaginary radius in the parent frame, which, according to equation (4.1.1:25), can be related to the imaginary radius of the apparent homogeneous space of the parent frame as

Energy structures in space

$$R''_{0\delta} = R''_{\delta P} = \frac{R''_{0\delta P}}{1-\delta_P} \quad ; \quad \delta_P = 1 - \frac{R''_{0\delta P}}{R''_{0\delta}}, \tag{4.2.7:2}$$

where δ_P is the gravitational factor of the orbiting system in the parent frame. When the imaginary radius of the apparent homogeneous space of the parent frame, $R''_{0\delta P}$, is constant, differentiation of equation (4.2.7:2) gives

$$d\delta_P = (1-\delta_P)\frac{dR''_{0\delta}}{R''_{0\delta}}. \tag{4.2.7:3}$$

With reference to equation (4.1.1:8), δ_P and its differential can be expressed as

$$\delta_P = \frac{M_P R''_{0\delta P}}{M''r_{0\delta P}} \quad \Rightarrow \quad d\delta_P = -\delta_P \frac{dr_{0\delta P}}{r_{0\delta P}}. \tag{4.2.7:4}$$

Combining equations (4.2.7:3) and (4.2.7:4) gives

$$\frac{dR''_{0\delta}}{R''_{0\delta}} = \frac{-\delta_P}{1-\delta_P}\frac{dr_{0\delta P}}{r_{0\delta P}} \approx -\delta_P \frac{dr_{0\delta P}}{r_{0\delta P}}, \tag{4.2.7:5}$$

which relates the change in the imaginary radius of the local frame to the change in the distance of the local frame from the central mass of the parent frame.

Assuming the gravitational factor in the local frame to be constant, differentiation of equation (4.2.7:1) gives

$$\frac{dr_{0\delta}}{r_{0\delta}} = \frac{dR''_{0\delta}}{R''_{0\delta}}, \tag{4.2.7:6}$$

which by substitution of equation (4.2.7:5) gives

$$\frac{dr_{0\delta}}{r_{0\delta}} \approx -\delta_P \frac{dr_{0\delta P}}{r_{0\delta P}} = -\frac{GM_P}{c^2 r_{0\delta P}}\frac{dr_{0\delta P}}{r_{0\delta P}} \approx -\frac{g_P}{c^2}\Delta r_P, \tag{4.2.7:7}$$

where g_P is the gravitational acceleration of the central mass of the parent frame at the local orbiting system.

The last form of equation (4.2.7:7) applies for small relative changes in $r_{0\delta}$, corresponding to the case where the eccentricity of the orbit of the local rotational system in the parent frame is small. For example, the eccentricity of the orbit of the Earth is $e = 0.0167$, which means that the annual change in the Earth to Moon distance can be calculated from the last form of equation (4.2.7:7).

An important message of equation (4.2.7:7) is that the orbital radius of a local system increases when the distance to the central mass of the parent frame decreases, Figure 4.2.7-1.

4.2.8 Local singularity in space

The velocity in a circular orbit around a mass center can be solved from the acceleration of the motion on the flat space plane and the acceleration due to the mass center given in equation (4.2.2:5)

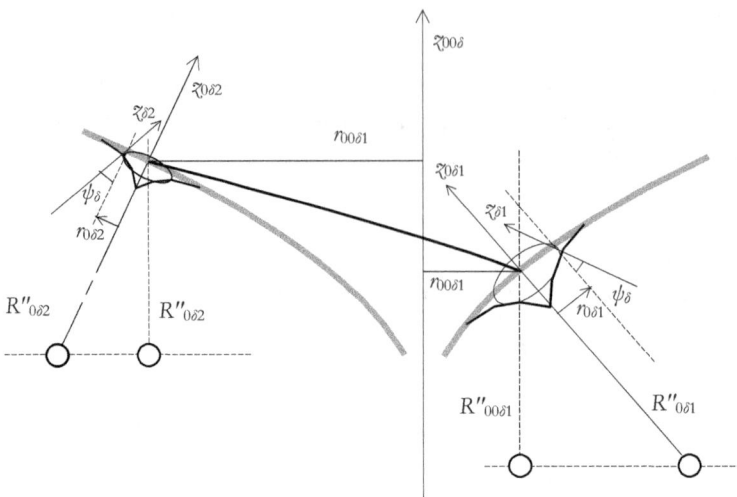

Figure 4.2.7-1. The orbital radius of a local rotational system increases when the local system comes closer to the central mass of its parent frame. The relative increase of the orbital radius $r_{0\delta}$ is directly proportional to the relative increase of the imaginary radius $R''_{0\delta}$ of the apparent homogeneous space of the local frame [see equation (4.2.7:6)].

$$\frac{v_{orb}^2}{r_{0\delta}}\hat{\mathbf{r}}_{0\delta} = \frac{GM(1-\delta)^3}{\chi_{0\delta}r_{0\delta}^2}\hat{\mathbf{r}}_{0\delta} = \frac{c_{0\delta}^2\delta(1-\delta)^3}{r_{0\delta}}\hat{\mathbf{r}}_{0\delta}, \qquad (4.2.8:1)$$

which gives

$$\beta_{orb(0\delta)}^2 = \frac{v_{orb}^2}{c_{0\delta}^2} = \beta_{orb(0\delta)}^2 = \delta(1-\delta)^3, \qquad (4.2.8:2)$$

and

$$\beta_{orb(0\delta)} = \sqrt{\delta(1-\delta)^3} = (1-\delta)\sqrt{\delta(1-\delta)}, \qquad (4.2.8:3)$$

or in terms of the local velocity of light $c_\delta = c_{0\delta}(1-\delta)$

$$\beta_{orb(\delta)} = \frac{v_{orb}}{c_\delta} = \sqrt{\delta(1-\delta)}. \qquad (4.2.8:4)$$

When related to the local velocity of light, the orbital velocity achieves its maximum $v_{orb} = 0.5\,c_\delta$ at $r_{0\delta} = 2r_c$ and goes to zero when $r_{0\delta} \Rightarrow r_c$ ($\delta \Rightarrow 1$).

As demonstrated by equation (4.2.8:4) and Figure 4.2.8-1, the local orbital velocity in a circular orbit near a local singularity is stable and approaches zero at the critical radius where the local velocity of light also approaches zero. This suggests that at orbits with $r_{0\delta} < 2 \cdot r_c$, a local singularity maintains the mass characteristic to the singularity.

Energy structures in space

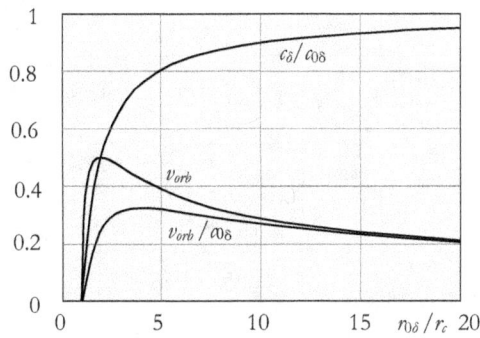

Figure 4.2.8-1. In extreme gravitational conditions ($r_{0\delta} \Rightarrow r_c$), the orbital velocity for a circular orbit goes to zero after passing the maximum $v_{orb(max(\delta))} = 0.5 \cdot c_\delta$ at $r_{0\delta} = 2r_c$ or, when related to the velocity of light in apparent homogeneous space $v_{orb(max(0\delta))} = 0.32 \cdot c_{0\delta}$, at $r_{0\delta} = 4r_c$.

The orbital period for circular orbits can be solved from (4.2.8:3) as

$$P = \frac{2\pi r_{0\delta}}{c_{0\delta} \beta_{orb(0\delta)}} = \frac{2\pi r_{0\delta}}{c_{0\delta} \sqrt{\frac{r_c}{r_{0\delta}}\left(1 - \frac{r_c}{r_{0\delta}}\right)^3}} = \frac{2\pi r_c}{c_{0\delta}} \left[\frac{r_c}{r_{0\delta}}\left(1 - \frac{r_c}{r_{0\delta}}\right)\right]^{-3/2} . \quad (4.2.8:5)$$

Derivation of (4.2.8:5) gives

$$\frac{dP}{dr} = \frac{3\pi}{c}\left(\frac{r_c}{r}\right)^{-\frac{1}{2}}\left(1 - \frac{r_c}{r}\right)^{-\frac{5}{2}}\left(1 - \frac{2r_c}{r}\right), \quad (4.2.8:6)$$

which goes to zero at $r = 2r_c$ corresponding to the minimum period of circular orbits (Figure 4.2.8-2)

$$P_{min} = \frac{2\pi \cdot 2r_c}{c\sqrt{\frac{1}{2}(1-\frac{1}{2})^3}} = \frac{16\pi r_c}{c} = \frac{16\pi GM}{c^3} . \quad (4.2.8:7)$$

The black hole at the center of the Milky Way, at compact radio source Sgr A*, has an estimated mass of about 3.6 times the solar mass, which means $M_{black\ hole} \approx 7.2 \cdot 10^{36}$ kg. When substituted for M in (4.2.8:7), the prediction for the minimum period in a circular orbit around the black hole is about 14.8 min, which is in line with the observed 16.8 ± 2 min period [42], Figure 4.2.8-2. Table 4.2.8-I compares important predictions in Schwarzschild space and DU space.

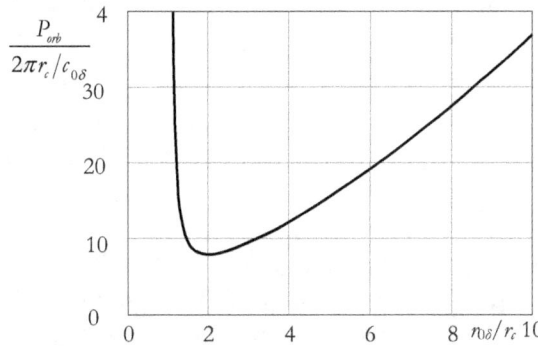

Figure 4.2.8-2. The orbital period for circular obits with radius $r_{0\delta}$ close to the critical radius r_c.

	Schwarzschild space	DU space
1) Velocity of free fall $\delta = GM/rc^2$	$\beta_{ff} = \sqrt{2\delta}(1-2\delta)$ (coordinate velocity)	$\beta_{ff(0\delta)} = (1-\delta)\sqrt{1-(1-\delta)^2}$ (eq. 4.1.6:15 & 4.1.6:12)
2) Orbital velocity at circular orbits	$\beta_{orb} = \dfrac{1-2\delta}{\sqrt{1/\delta - 3}}$ (coordinate velocity)	$\beta_{orb(0\delta)} = \sqrt{\delta(1-\delta)^3}$ (eq. 4.2.8:3)
3) Orbital period in Schwarzschild space (coordinate period) and in DU space	$P = \dfrac{2\pi r}{c}\sqrt{\dfrac{2}{\delta}}$, $r > 3 \cdot r_{c(Schwd)}$	$P = \dfrac{2\pi r_c}{c_{0\delta}}[\delta(1-\delta)]^{-3/2}$
4) Perihelion advance for a full revolution	$\Delta\psi(2\pi) = \dfrac{6\pi G(M+m)}{c^2 a(1-e^2)}$	$\Delta\psi(2\pi) = \dfrac{6\pi G(M+m)}{c^2 a(1-e^2)}$

Table 4.2.8-I. Predictions related to celestial mechanics in Schwarzschild space [41] and in DU space. In DU, space velocity β is the velocity relative to the velocity of light in the apparent homogeneous space of the local singularity, which corresponds to the coordinate velocity in Schwarzschild space.

The velocity of free fall, v_{ff}, reaches the local velocity of light at $r_{0\delta} \approx 3.414 \cdot r_c$ where the tilting angle of space is $\psi = 45°$.

In binary pulsars, the mass of the emitting neutron stars is typically about 1.5 times the mass of the Sun, corresponding to a critical radius of about $r_c \approx 2.3$ km. The estimated radius of typical neutron stars is about 8 km, which corresponds roughly to the distance $3.414 \cdot r_c$, where the velocity of free fall reaches the local velocity of light. Such a condition may be favorable for matter to radiation and elementary particle conversions, Figure 4.2.8-3.

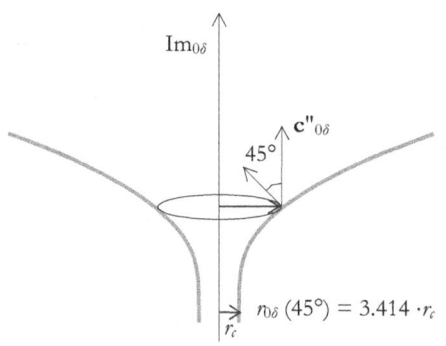

Figure 4.2.8-3. In a local singularity, space is tilted 90°. At the tilting angle 45° degrees, the velocity in free fall reaches the local velocity of light.

4.2.9 Orbital decay

In general relativity, orbiting bodies are predicted to emit gravitational radiation as a consequence of the changing quadrupole moment of orbiting systems. The energy released results in a decreasing orbital period that is strong enough to be observed in binary pulsar

Energy structures in space

systems. In the DU framework, the orbital decay of binary systems can be related to the rotation of the 4D angular momentum due to the periastron advance of eccentric orbits. In the DU solution, circular orbits are not subject to decay, but the prediction obtained for the decay of eccentric orbits is essentially the same as the corresponding GR prediction based on the quadrupole moment.

The effect of orbit plane rotation on the angular momentum of the orbit

Elliptic orbits are subject to periastron advance, which can be described as the rotation of the orbit plane. The tilting of space results in tilting of the obit plane relative to the flat space plane. Accordingly, periastron advance means rotation of the orbit plane around the normal, the z-coordinate direction, of non-tilted space. The rotation of the orbit plane means rotation of the 4D angular momentum of the orbit, Figure 4.2.9-1. The energy needed for the rotation is obtained from the decay of the orbital period.

According to DU, the z-coordinate (Im$_{0\delta}$-axis) of an orbiting object can be expressed as

$$z_{0\delta} = 2\sqrt{2r_c z}\left[\sqrt{r_{0\delta}} - \sqrt{a_{0\delta}\left(1-e_{0\delta}^2\right)}\right]. \qquad (4.2.9:1)$$

The difference in z-coordinate between apastron and heliastron can be expressed as

$$z_{aa} - z_{pa} = 2\sqrt{2r_c}\left[\left(\sqrt{a_{0\delta}(1+e_{0\delta})} - \sqrt{a_{0\delta}(1-e_{0\delta}^2)}\right) + \left(\sqrt{a_{0\delta}(1-e_{0\delta}^2)} - \sqrt{a_{0\delta}(1-e_{0\delta})}\right)\right]$$

$$= 2\sqrt{2r_c a_{0\delta}}\left[\sqrt{1+e_{0\delta}} - \sqrt{1-e_{0\delta}}\right] = 2\sqrt{2r_c a_{0\delta}} \cdot A_e, \qquad (4.2.9:2)$$

and the tilting angle

$$\gamma \approx \tan\gamma = \frac{z_{aa} - z_{pa}}{2a_{0\delta}} = \frac{2\sqrt{2r_c a_{0\delta}} \cdot A_{e1}}{2a_{0\delta}} = A_e\sqrt{2}\sqrt{\frac{r_c}{a_{0\delta}}} \approx \sin\gamma. \qquad (4.2.9:3)$$

The increase of angular momentum related to the rotation of the orbit plane around the $z_{0\delta}$ axis due to periastron advance is

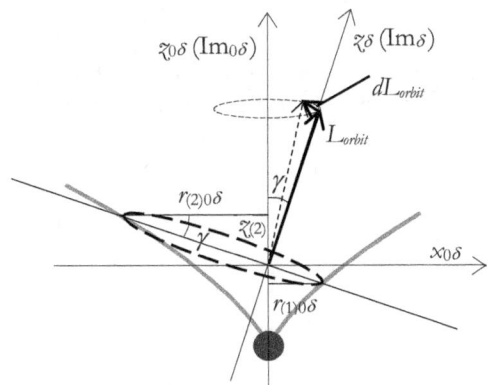

Figure 4.2.9-1. The 4D angular momentum L_{orbit} of an eccentric orbit, in the direction of the Im$_\delta$ axis of the orbital plane, rotates with the periastron advance of the obit. The energy released in the decay of the orbital period is assumed as the energy needed for the rotation of the angular momentum.

$$\frac{dL_r}{dt} = L_r \cdot \sin\gamma \cdot \frac{d\psi}{dt} \approx L_r A_e \sqrt{2} \sqrt{\frac{r_c}{a_{0\delta}}} \cdot \frac{\Delta\psi_{(2\pi)}}{P} \qquad (4.2.9{:}4)$$

where the angular velocity of the periastron advance is expressed in terms of the advance angle per a full cycle $\Delta\psi_{(2\pi)}$ with period P.

Angular momentum L_r can be expressed in terms of the angular momentum of the orbital motion L_o as the share of the advance angle ψ per cycle

$$L_r = \frac{\Delta\psi_{(2\pi)}}{2\pi} L_o . \qquad (4.2.9{:}5)$$

Substitution of equation (4.2.9:5) to equation (4.2.9:4) gives

$$\frac{dL_r}{dt} = \frac{\sqrt{2}}{2\pi} \frac{r_c^{1/2}}{a_{0\delta}^{1/2}} \cdot \frac{L_o}{P} A_e \Delta\psi_{(2\pi)}^2 , \qquad (4.2.9{:}6)$$

and

$$dL_r = \frac{L_o}{P} \cdot \frac{\sqrt{2}}{2\pi} \frac{r_c^{1/2}}{a_{0\delta}^{1/2}} A_e \Delta\psi_{(2\pi)}^2 dt . \qquad (4.2.9{:}7)$$

The period of a Keplerian orbit can be related to the semimajor axis a as

$$P^2 = \frac{4\pi^2 a^3}{G(M+m)} . \qquad (4.2.9{:}8)$$

The periastron advance of elliptic orbits for one period can be expressed

$$\Delta\psi_{(2\pi)} = \frac{6\pi G(M+m)}{c^2 a(1-e^2)} . \qquad (4.2.9{:}9)$$

Substitution of a, solved from equation (4.2.9:8), allows the expression of periastron advance in terms of the total mass and period m

$$\Delta\psi_{(2\pi)} = \frac{6\pi G(M+m)(2\pi)^{2/3}}{c^2 \left[G(M+m)\right]^{1/3} P^{2/3}(1-e^2)} = \frac{6\pi}{c^2(1-e^2)} \left(\frac{2\pi G}{P}\right)^{2/3} \left[G(M+m)\right]^{2/3} .$$

(4.2.9:10)

In the case of double pulsars, the period and the periastron advance are observed quantities, which means that equation (4.2.9:10) can be used for the determination of the total mass $(M+m)$ of the system. In terms of solar mass

$$M_s + m_s = \left(\frac{(1-e^2)\Delta\psi_{(2\pi)}}{6\pi}\right)^{3/2} \frac{Pc^3}{2\pi GM_\odot} . \qquad (4.2.9{:}11)$$

Keplerian orbit

To express L_o in terms of the orbital period P, we first apply the Keplerian relation m

$$L_o = \frac{2\pi a^2 m}{P}\sqrt{1-e^2} \ . \tag{4.2.9:12}$$

The period of a Keplerian orbit can be related to the semimajor axis a as

$$P^2 = \frac{4\pi^2 a^3}{G(M+m)} \ . \tag{4.2.9:13}$$

The semimajor axis a can be solved from equation (4.2.9:13) as

$$a = \frac{P^{1/2} L_o^{1/2}}{\left(1-e^2\right)^{1/4} \left(2\pi m\right)^{1/2}} \ . \tag{4.2.9:14}$$

Substitution of equation (4.2.9:14) to equation (4.2.9:13) we get

$$P^2 = \frac{4\pi^2}{G(M+m)} \frac{P^{3/2} L_o^{3/2}}{\left(1-e^2\right)^{3/4} \left(2\pi m\right)^{3/2}} \ , \tag{4.2.9:15}$$

and P is solved as

$$P^{2-3/2} = P^{1/2} = \frac{4\pi^2}{G(M+m)} \frac{L_o^{3/2}}{\left(1-e^2\right)^{3/4} \left(2\pi m\right)^{3/2}}$$

$$P = \frac{(2\pi)^4}{G^2(M+m)^2} \frac{L_o^3}{\left(1-e^2\right)^{3/2} \left(2\pi m\right)^3} = A \cdot L_o^3 \ , \tag{4.2.9:16}$$

where A is

$$A = \frac{2\pi}{G^2(M+m)^2 m^3} \frac{1}{\left(1-e^2\right)^{3/2}} \ . \tag{4.2.9:17}$$

Differentiation of (4.2.9:13) gives

$$dP = 3A \cdot L_o^2 \, dL_o = 3A \cdot L_o^3 \frac{dL_o}{L_o} = \frac{3P}{L_o} dL_o \tag{4.2.9:18}$$

Dependence of dP on dL in Keplerian orbit

Reduction of angular momentum dL_r by the rotation of the orbital plane was given by equation (8). Substitution of the negative of dL_r for dL_o in equation (4.2.9:18) gives the reduction of period required by the buildup of the angular momentum of the rotation of the orbital plane

$$dP = \frac{3P}{L_0} \frac{\sqrt{2}}{2\pi} \frac{r_c^{1/2}}{a_{0\delta}^{1/2}} \cdot \frac{L_o}{P} A_e \Delta \psi_{(2\pi)}^2 dt = \frac{3\sqrt{2}}{2\pi} \frac{r_c^{1/2}}{a_{0\delta}^{1/2}} A_e \Delta \psi_{(2\pi)}^2 dt \ , \tag{4.2.9:19}$$

and further, the time derivative of period P after substitution of (4.2.9:1) for $\Delta\psi$ in (4.2.9:19)

$$\frac{dP}{dt} = \frac{3\sqrt{2}}{2\pi} r_c^{1/2} A_e a_{0\delta}^{-1/2} \frac{36\pi^2 r_c^2}{a_{0\delta}^2 (1-e^2)^2} = \frac{54\pi\sqrt{2} \, r_c^{5/2} A_e}{(1-e^2)^2} a_{0\delta}^{-5/2}. \qquad (4.2.9\text{:}20)$$

Solving a from equation (4.2.9:13) gives

$$a = \frac{G^{1/3}(M+m)^{1/3}}{(2\pi)^{2/3}} P^{2/3} = \frac{c^{2/3} G^{1/3}(M+m)^{1/3}}{c^{2/3}(2\pi)^{2/3}} P^{2/3} = \frac{c^{2/3} r_c^{1/3}}{(2\pi)^{2/3}} P^{2/3}. \qquad (4.2.9\text{:}21)$$

Substitution of equation (4.2.9:21) for and $a_{0\delta}$ in equation (4.2.9:20) gives

$$\frac{dP}{dt} = \frac{54\pi\sqrt{2} \, r_c^{5/2} A_e}{(1-e^2)^2} a_{0\delta}^{-5/2} \frac{c^{-5/3} r_c^{-5/6}}{(2\pi)^{-5/3}} P^{-5/3}, \qquad (4.2.9\text{:}22)$$

which, after substitution of A_e from equation (4.2.9:2), can be written as

$$\frac{dP}{dt} = 54\pi\sqrt{2} \left(\frac{2\pi r_c}{cP}\right)^{5/3} \frac{\left[\sqrt{1+e_{0\delta}} - \sqrt{1-e_{0\delta}}\right]}{(1-e^2)^2}. \qquad (4.2.9\text{:}23)$$

GR prediction for the orbital decay

The GR prediction for the decay of the orbital period is given in [65,66]

$$\frac{dP}{dt} = \frac{(192/5) \cdot \pi G^{5/3}}{c^5} \left(\frac{P}{2\pi}\right)^{-5/3} \cdot \frac{1 + (73/24)e^2 + (37/96)e^4}{(1-e^2)^{7/2}} \cdot \frac{m_p m_c}{(m_p + m_c)^2} (m_p + m_c)^{5/3}, \qquad (4.2.9\text{:}24)$$

where the numerical constant $195/5 \cdot \pi \approx 123$, and the mass term is written in form $m_p m_c (m_p+m_c)^{1/3} = m_p m_c/(m_p+m_c)^2 \cdot (m_p+m_c)^{5/3}$.

By regrouping and substitution of $r_c = G(M+m)/c^2$ equation (4.2.9:23) obtains the form

$$\frac{dP}{dt} = \frac{54\pi\sqrt{2} \, G^{5/3}}{c^5} \left(\frac{P}{2\pi}\right)^{-5/3} \frac{\left[\sqrt{1+e_{0\delta}} - \sqrt{1-e_{0\delta}}\right]}{(1-e^2)^2} (M+m)^{5/3}, \qquad (4.2.9\text{:}25)$$

where the numerical constant $54 \cdot \pi\sqrt{2} \approx 240$. Equation (4.2.9:25) applies for $M \gg m$. By replacing the mass term based on the central mass ($M \gg m$) condition with the mass term for the binary star condition ($M \approx m$) used in the GR solution, equation (4.2.9:25) obtains the form

$$\frac{dP}{dt} \approx 240 \frac{G^{5/3}}{c^5} \left(\frac{P}{2\pi}\right)^{-5/3} \left(\frac{\left[\sqrt{1+e_{0\delta}} - \sqrt{1-e_{0\delta}}\right]}{(1-e^2)^2}\right) \cdot \frac{m_p m_c}{(m_p + m_c)^2} (m_p + m_c)^{5/3}. \qquad (4.2.9\text{:}26)$$

For a comparison with the GR equation, we take a factor of 2 from the factor 240 in (4.2.9:26) into the eccentricity factor, resulting in

Energy structures in space

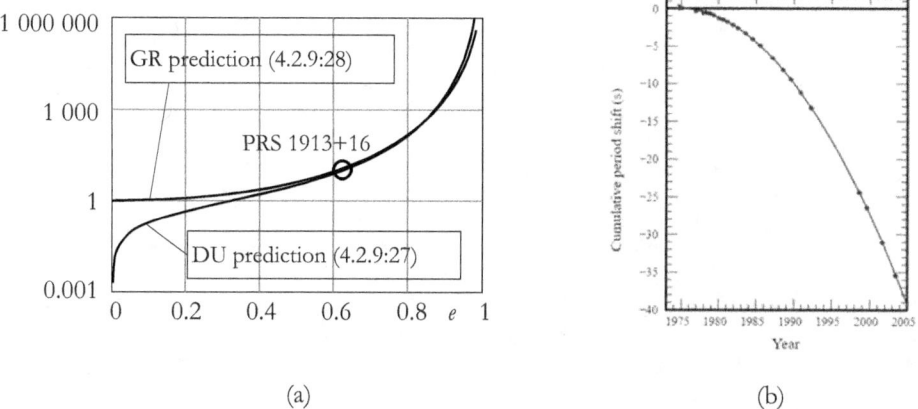

Figure 4.2.9-2. (a) The eccentricity factor of the decay of binary star orbit period. At the eccentricity $e = 0.616$ of the PSR 1913+16 orbit the eccentricity factor of the GR and DU for the orbit decay are essentially the same and lead to same prediction for the decay. (b) The predicted (solid curve) and observed orbital decay (dots) of PSR B1913+16 binary pulsar. Picture: *Wikimedia Commons*.

$$\frac{dP}{dt}_{(DU)} \approx 120 \cdot \frac{G^{5/3}}{c^5} \left(\frac{P}{2\pi}\right)^{-5/3} \left(2 \cdot \frac{\left[\sqrt{1+e_{0\delta}} - \sqrt{1-e_{0\delta}}\right]}{\left(1-e^2\right)^2}\right) \cdot \frac{m_p m_c}{\left(m_p + m_c\right)^2} \left(m_p + m_c\right)^{5/3},$$

(4.2.9:27)

which now can be compared with the GR equation (4.2.9:24):

$$\frac{dP}{dt}_{(GR)} \approx 123 \cdot \frac{G^{5/3}}{c^5} \left(\frac{P}{2\pi}\right)^{-5/3} \cdot \frac{1+(73/24)e^2 + (37/96)e^4}{\left(1-e^2\right)^{7/2}} \cdot \frac{m_p m_c}{\left(m_p + m_c\right)^2} \left(m_p + m_c\right)^{5/3}.$$

(4.2.9:28)

The eccentricity factors in equations (4.2.9:27) and (4.2.9:28) of GR and DU are compared in Figure 4.2.9-2. In the DU prediction, the eccentricity factor goes to zero at $e=0$, which means that there is no decay for circular orbits. The GR prediction shows decay also for circular orbits.

5. Mass, mass objects, and electromagnetic radiation

In the DU framework, the descriptions of mass objects, electromagnetism, and atomic structures can all be based on mass as a wavelike substance. Such a unification means revisiting the basis and conclusions of Planck's equation. We do not need to consider Planck's equation as a heuristic finding violating classical electromagnetism, but a consequence of Maxwell's equations solved for the emission of a single cycle of a harmonic oscillator. The unified perspective of mass and radiation allows the description of mass objects as resonant mass wave structures, with results essentially the same as those obtained by quantum mechanics.

While relativity in the DU is expressed in terms of locally available rest energy, the effects of gravitation and motion are directly reflected in the energy states of atomic objects, and thereby in the characteristic emission and absorption frequencies.

The linkage of Planck's equation to Maxwell's equation has exceedingly important consequences:

- *The solution reveals the embedding of the velocity of light in Planck's constant.*

- *The removal of the velocity of light from the Planck constant produces the "intrinsic Planck" constant, h_0, with dimensions of mass-meter [kgm].*

- *The renewed Planck's equation demonstrates the linkage of mass and the wavelength of radiation, by enabling the definition of the wavelength equivalence of mass and the mass equivalence of wavelength, respectively.*

- *The intrinsic Planck constant can be expressed in terms of fundamental electrical constants, the unit charge, and the vacuum permeability.*

- *The linkage between the fine structure constant and any other physical constant is removed. As a consequence, the fine structure constant appears as a purely numerical factor.*

- *A quantum of electromagnetic radiation receives a precise expression: A quantum of radiation is the energy of a cycle of radiation emitted by a single electron oscillation in the emitting object.*

- *The linkage of mass and wavelength allows the description of mass objects as resonant mass wave structures.*

5.1 The mass equivalence of radiation

5.1.1 Quantum of radiation

The Planck equation

In the early 1900's, the German physicist Max Planck concluded that if radiation in a cavity is in equilibrium with the atoms of the walls, there must be a correspondence between the energy distribution in the radiation and the energy state of the atoms emitting and absorbing the radiation. He described atoms as harmonic oscillators with specific frequencies and assumed that each oscillator absorbs or emits radiation energy only in doses proportional to the frequency of the oscillator. Mathematically, Planck expressed the idea with an equation stating that the energy in a single emission or absorption process is proportional to the frequency as

$$E = hf \qquad (5.1.1:1)$$

where h is the Planck constant, assumed to be the same for all oscillators. The message of Planck's equation was, and still is, accepted as a law of nature in contradiction with classical electromagnetism and Maxwell's equations.

In fact, the Planck equation is not in contradiction with classical electrodynamics once we specify the meaning of a single emission or absorption process as a cycle of oscillation of a unit charge in a harmonic oscillator. Obviously, the emission/absorption counterpart of such an oscillation cycle is a cycle of electromagnetic radiation.

In order to find the solution, it is essential to relate the length of the dipole to the wavelength emitted – in the case of atomic oscillators, the effective length of the dipole is not related to the atomic diameter, but to the distance a point-like emitter moves in the fourth dimension in a cycle of emission. In the DU framework, such a distance is equal to the wavelength, i.e., a point emitter can be regarded as a one-wavelength dipole in the fourth dimension. In fact, such a conclusion is not too strange in the SR/GR framework either; for a point emitter at rest, the spacetime line-element in $dt = 1/f$ is $ds = cdt = c/f = \lambda$.

The energy described by the Planck equation (5.1.1:1) should be understood as the energy of one cycle of radiation emitted or absorbed by a harmonic oscillator per one unit charge oscillation. In his Nobel Prize lecture in 1920, Max Planck stated:

"Either the quantum of action was a fictional quantity, then the whole deduction of the radiation law was, in the main, illusory and represented nothing more than an empty non-significant play on formulae, or the derivation of the radiation law was based on a sound physical conception. In this case, the quantum of action must play a fundamental role in physics, and here was something entirely new, never before heard of, which seemed called upon to basically revise all our physical thinking, built as this was, since the establishment of the infinitesimal calculus by Leibniz and Newton, upon the acceptance of the continuity of all causative connections[67]."

From the DU perspective, the Planck equation has a solid basis in classical electrodynamics. However, the concept of "a quantum of action" may be misleading – a revised interpretation of the Planck equation is obtained by removing the embedded velocity of light from the Planck constant. Such a revision reveals the *intrinsic Planck constant* with dimensions of mass-meter [kg·m], and the Planck equation, as the energy of a cycle of electromagnetic radiation emitted by an atomic emitter by a single electron transition, obtains the form

$$E = h_0 c_0 f = \frac{h_0}{\lambda} c_0 c. \qquad (5.1.1{:}2)$$

In equation (5.1.1:2), the quantity h_0/λ has the dimension of mass [kg], which allows it to be regarded as the mass equivalence of radiation. The concept of mass equivalence of radiation is of high value in a unified description of mass objects and radiation – the mass equivalence returns the energy of a cycle into the same form as the rest energy of a mass object. The concept of mass equivalence also applies in reverse – the wavelength equivalence of mass objects obtains the form of (5.1.1:2) by applying the *wavelength equivalence of mass*, which for the rest mass is equal to the Compton wavelength.

Maxwell's equations: solution of one cycle of radiation

Moving electric charges result in electromagnetic radiation through the buildup of changing electric and magnetic fields as described by Maxwell's equations.

The electric and magnetic fields produced by an oscillating electric dipole at a distance r ($r/z_0 > 2z_0/\lambda$) can be expressed as

$$\mathbf{E} = \frac{\Pi_0 \omega^2 \sin\theta}{4\pi\varepsilon_0 r c^2} \sin(kr - \omega t)\,\hat{\mathbf{r}}_\theta, \qquad (5.1.1{:}3)$$

and

$$\mathbf{B} = \frac{1}{c}\mathbf{E}\,\hat{\mathbf{r}}_\varphi = \frac{\Pi_0 \omega^2 \sin\theta}{4\pi\varepsilon_0 r c^3} \sin(kr - \omega t)\,\hat{\mathbf{r}}_\varphi, \qquad (5.1.1{:}4)$$

where θ is the angle between the dipole and the distance vectors and

$$\Pi_0 = Nez_0, \qquad (5.1.1{:}5)$$

is the peak value of the dipole momentum, where N is the number of unit charges, e, oscillating in a dipole of effective length z_0. Both field vectors, $\mathbf{E}$ and $\mathbf{B}$, are perpendicular to the distance vector $\mathbf{r}$.

The Poynting vector, showing the direction of the energy flow, has the direction of $\mathbf{r}$, Figure 5.1.1-1. The energy density of radiation can be expressed as E

$$E = \varepsilon_0 E^2 = \frac{\Pi_0^2 \chi \mu_0 \omega^4 \sin^2\theta}{16\pi^2 r^2 c^2} \sin^2(kr - \omega t), \qquad (5.1.1{:}6)$$

where the vacuum permittivity ε_0 is replaced with the vacuum permeability μ_0

$$\mu_0 = \frac{1}{\varepsilon_0 c_0 c}. \qquad (5.1.1{:}7)$$

The factor χ in (5.1.1:6) is the frame conversion factor $\chi = c_0/c$ defined in equation (4.1.4:13). The average energy density of radiation is

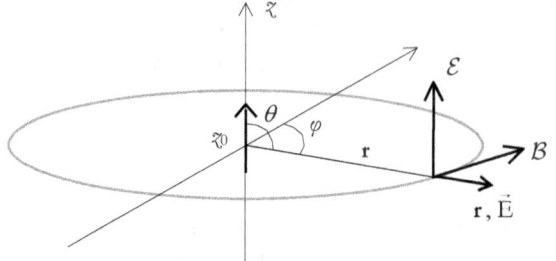

Figure 5.1.1-1. An electric dipole in the direction of the z-axis results in maximum radiation density in the normal plane of the dipole, $\theta = \pi/2$.

$$E_{ave} = \frac{1}{2}E = \frac{E_0}{2\pi}\int_0^{2\pi} \sin^2(kr - \omega t)d(\omega t) = \frac{\Pi_0^2 \chi \mu_0 \omega^4}{32\pi^2 r^2 c^2}\sin^2\theta. \qquad (5.1.1:8)$$

The average energy flow from the dipole is

$$\left\langle \frac{dE}{dt}\right\rangle = P = c\int_{sphre}\frac{\Pi_0^2 \chi \mu_0 \omega^4}{32\pi^2 r^2 c^2}\sin^2\theta\, d\theta = \frac{\Pi_0^2 \chi \mu_0 \omega^4}{32\pi^2 r^2 c}\int_{sphre}\sin^2\theta\, d\theta. \qquad (5.1.1:9)$$

With the substitution of equation (5.1.1:5) for Π_0, $\omega = 2\pi f = 2\pi c/\lambda$, and $\chi = c_0/c$, equation (5.1.1:9), the energy flow of one cycle of radiation can be expressed as

$$E_\lambda = \frac{P}{f} = \frac{N^2 e^2 z_0^2 \chi \mu_0 16\pi^4 f^4}{32\pi^2 r^2 c\, f}\frac{2}{3}4\pi r^2 = N^2\left(\frac{z_0}{\lambda}\right)^2\frac{2}{3}\left(2\pi^3 e^2 \mu_0 c_0\right)f. \qquad (5.1.1:10)$$

In equation (5.1.1:10), N^2 is the intensity factor related to the number of electrons oscillating in the dipole, the ratio (z_0/λ) relates the dipole length to the wavelength emitted, and the factor 2/3 is the ratio of average energy in a cycle emitted by the dipole to the energy in a cycle emitted by a hypothetical isotropic dipole. The factor $(2\pi^3 e^2 \mu_0 c_0)$ has the dimensions of momentum–length, like Planck's constant h, and the numerical value $5.997 \cdot 10^{-34} = h/1.1049$ [kgm²/s], assuming that $c_0 \approx c$.

Due to the motion of space in the fourth dimension at velocity c, a point source at rest in local space moves a distance $r_4 = c \cdot T = \lambda$ in the fourth dimension. An atomic emitter/absorber can be studied as a point source, as a one-wavelength dipole in the fourth dimension. As a first approximation, the emission/absorption of such a source has the form of equation (5.1.1:10) with $z_0 = \lambda$ and factor χ_λ relating the energy of a cycle to the energy of a cycle of a hypothetical isotropic one-wavelength dipole

$$E_\lambda = N^2 \chi_\lambda \left(2\pi^3 e^2 \mu_0 c_0\right)f = N^2 h \cdot f = N^2 h_0 \cdot c_0 \cdot f = N^2 \frac{h_0}{\lambda}c_0 c. \qquad (5.1.1:11)$$

By relating equation (5.1.1:11), with $N = 1$, to the Planck equation, we can find out that the value of factor χ_λ is close to one, $\chi_\lambda = 1.1049$. The Planck constant h can now be expressed in terms of fundamental physical constants e and μ_0 as

$$h = \chi_\lambda \cdot 2\pi^3 e^2 \mu_0 c_0 = 1.104905316 \cdot 2\pi^3 e^2 \mu_0 c_0. \qquad (5.1.1:12)$$

Mass, mass objects and electromagnetic radiation 169

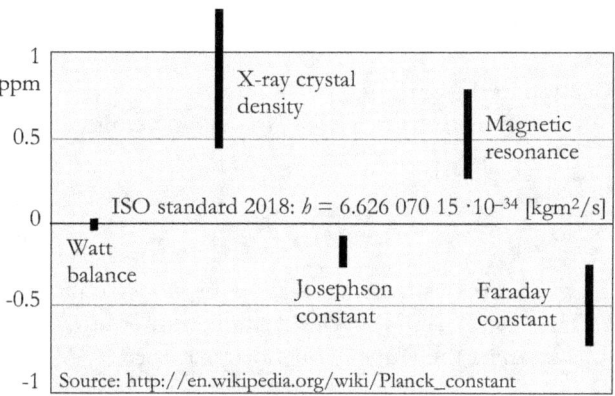

Figure 5.1.1-2. Determination of the Planck constant with five different methods: Watt balance, X-ray crystal density, Josephson constant, Magnetic resonance, and Faraday constant. The estimated accuracy of each method is shown by the vertical bars in the figure. The CODATA 2006 value of the Planck constant is fixed to the value obtained with the Watt balance. All measured values lie within about a one ppm range, which is the level of deviation we may assume resulting from a different effect of the c_0/c ratio in different methods.

The physical basis of the factor $\chi_\lambda = 1.1049$ has not been solved analytically. It can be regarded as the geometrical factor of a point emitter as an antenna in the fourth dimension. One of the factors in χ_λ is the ratio c_0/c. The difference between c_0 and c is estimated to be of the order of 1ppm. The c_0/c ratio does not explain constant χ_λ, but it may result in a different effect in different methods used to determine the exact value of the Planck constant, Figure 5.1.1-2.

The intrinsic Planck constant

Equations (5.1.1:11-13) reveal the physical basis of the Planck equation and relate the Planck constant to primary electrical constants. They also show that the velocity of light $c_0 \cong c$ is a hidden factor in the Planck constant.

In the last two forms of equation (5.1.1:11), the velocity of light is removed from the Planck constant by introducing the *intrinsic Planck constant* h_0

$$h_0 = \frac{h}{c} = \chi_\lambda \cdot 2\pi^3 e^2 \mu_0 = 2.210219 \cdot 10^{-42} \quad [\text{kg·m}]. \tag{5.1.1:13}$$

The intrinsic Planck constant has dimensions of [kg·m]; accordingly, the quantity h_0/λ has dimensions of mass [kg]. For the emission of a single electron oscillation by a *Planck source*, equation (5.1.1:11) obtains the form of the Planck equation

$$E_{\lambda 0} = hf = c_0 h_0 f = c_0 \frac{h_0}{\lambda} c = c_0 \cdot m_{\lambda 0} c, \tag{5.1.1:14}$$

where $m_{\lambda 0}$ is the *unit mass equivalence of a cycle of radiation* of a Planck emitter

$$m_{\lambda 0} = \frac{h_0}{\lambda}, \tag{5.1.1:15}$$

per a single electron transition in the emitter.

For parallel transitions of N electrons in a cycle, the energy emitted by a Planck source is

$$E_{\lambda(N)} = N^2 hf = N^2 c_0 h_0 f = N^2 c_0 \frac{h_0}{\lambda} c = N^2 c_0 \cdot m_{0\lambda} c = c_0 \cdot m_\lambda c, \tag{5.1.1:16}$$

where $m_\lambda = N^2 h_0/\lambda = N^2 m_{0\lambda}$ expresses the total mass equivalence emitted in a cycle by N electrons in the source. The energy of N_1 cycles of radiation emitted by single electron transitions has the original form of the Planck equation proposed by Max Planck

$$E_{(N_1 \cdot \lambda)} = N_1 \cdot hf \quad \left(= N_1 \cdot c_0 \frac{h_0}{\lambda} c \right). \tag{5.1.1:17}$$

The derivation of equations (5.1.1:10-17) corresponds closely to the original idea of a quantum of radiation suggested by Max Planck about 1900. Max Planck assumed that atoms on the walls of a blackbody cavity behave like harmonic oscillators with different characteristic frequencies. Such oscillators work like narrow-band antennas emitting and absorbing radiation corresponding to the oscillator's frequency. As the smallest dose of radiation, he postulated a quantum of radiation, which, in the light of equation (5.1.1:10), means a single electron transition in the emitter.

Physical meaning of a quantum

In full agreement with Max Planck's original idea, a quantum of radiation is related to energy exchange between radiation and the receiving or sending oscillator (antenna). Atomic emitters and absorbers are regarded as resonators sensitive to the radiation with the nominal frequency of the resonator.

An antenna is not selective to the energy of radiation but to the wavelength of radiation. The energy of radiation is subject to the intensity as given in equation (5.1.1:16). The minimum energy emitted into one cycle of radiation is the quantum of radiation due to a single electron transition in the antenna as defined in equation (5.1.1:14).

Absorption of a quantum of radiation requires that 1) the wavelength of the wave to be absorbed is matched to the nominal wavelength of the antenna, and 2) the energy of the wave within the effective area of the absorber (antenna) is at least the energy of a quantum, i.e., the energy required to result in a single electron transition in the absorber:

1) $\lambda_{absorber} = \lambda_{radiation}$

2) $E_\lambda \left(A_{\mathit{eff}} \right) = E_\lambda \left(G \lambda^2 / 2\pi \right) \geq E_{\lambda 0} = c_0 \dfrac{h_0}{\lambda} c = c_0 \cdot m_{\lambda 0} c.$ \tag{5.1.1:18}

For a dipole, in the direction of the normal plane, the effective area in 2) is

$$A_{\mathit{eff}} = \frac{3}{2} \frac{\lambda^2}{4\pi}, \tag{5.1.1:19}$$

which is equal to a circular area with a diameter

$$d_{A(\mathit{eff})} = \frac{\lambda}{2\pi}\sqrt{\frac{3}{2}} \approx 0.19 \cdot \lambda. \qquad (5.1.1{:}20)$$

As shown by equation (5.1.1:10), the Planck equation is not in contradiction with the classical theory of electromagnetism and Maxwell's equations. Essential for such a conclusion is that the quantum of radiation is understood as the energy emitted or absorbed by a single electron transition in a cycle.

Applying the intrinsic Planck constant, the momentum of a quantum of radiation with wavelength λ can be expressed as

$$p_{0\lambda} = \hbar_0 f = \frac{\hbar_0}{\lambda} c = \hbar_0 k \cdot c = m_{0\lambda} c, \qquad (5.1.1{:}21)$$

where $\hbar_0 = h_0/2\pi$ and $k = 2\pi/\lambda$ is the wavenumber corresponding to wavelength λ.

A quantum of electromagnetic radiation is defined as a cycle of radiation emitted by a quantum emitter. An atom emitting electromagnetic radiation has the properties of a quantum emitter or a Planck source.

Equation (5.1.1:21) defines the momentum of a radiation quantum in terms of the mass equivalence of a cycle of radiation, $m_0\lambda = h_0/\lambda$. An implication of equation (5.1.1:21) is that the momentum of a radiation quantum cannot be defined or determined in a distance less than a wavelength. To obtain full information about the substance available for the expression of momentum, we need to observe the full wavelength of radiation.

The intensity factor

Applying the concept of a quantum for the emission of a standard dipole, equation (5.1.1:10) can be rewritten in the form

$$E_\lambda = N^2 \left(\frac{z_0}{\lambda}\right)^2 \frac{2}{3}\left(2\pi^3 e^2 \mu_0 c_0\right) f = N^2 \left(\frac{z_0}{\lambda}\right)^2 \frac{2}{3}\frac{1}{\chi_\lambda}\frac{\hbar_0}{\lambda} c_0 c$$

$$= I_\lambda \frac{\hbar_0}{\lambda} c_0 c = m_\lambda c_0 c, \qquad (5.1.1{:}22)$$

where I_λ is the *intensity factor*, and m_λ is the mass equivalence of a radiation cycle emitted by N electrons oscillating in a dipole with effective length z_0. Generally, the intensity factor and the mass equivalence of radiation emitted are expressed as

$$I_\lambda = N^2 \left(\frac{z_0}{\lambda}\right)^2 \frac{A}{\chi_\lambda}$$

$$m_\lambda = I_\lambda \frac{\hbar_0}{\lambda}, \qquad (5.1.1{:}23)$$

where A is a geometrical factor characteristic of the type of antenna ($A = 2/3$ for the dipole described by equation (5.1.1:10)). Equations (5.1.1:22) and (5.1.1:23) apply to any antenna emitting or receiving electromagnetic radiation.

5.1.2 The fine structure constant and the Coulomb energy

The fine structure constant

The fine structure constant a is traditionally defined as

$$a \equiv \frac{e^2 \mu_0 c}{2h} = \frac{e^2 \mu_0}{2h_0}. \tag{5.1.2:1}$$

Substitution of equation (5.1.1:14) for h_0 in equation (5.1.2:1) gives the fine structure constant in the form

$$a = \frac{e^2 \mu_0}{2 \cdot \chi_\lambda \cdot 2\pi^3 e^2 \mu_0} = \frac{1}{4\pi^3 \chi_\lambda} \simeq 7.2973525376 \cdot 10^{-3} \simeq \frac{1}{137.0360}. \tag{5.1.2:2}$$

Equation (5.1.2:2) shows the very fundamental nature of a as a purely numerical factor without any relationship to physical constants.

The fine structure constant a is a dimensionless factor independent of any dimensioned physical constant (5.1.2:2).

The Coulomb energy

The traditional form of Coulomb energy of point-like charges q_1 and q_2 at a distance r from each other is

$$E_{EM} = \frac{q_1 q_2}{4\pi \varepsilon_0 r} = N_1 N_2 \frac{e^2 \mu_0}{4\pi r} c_0 c, \tag{5.1.2:3}$$

where, in the last form, charges q_1 and q_2 are expressed in term of unit charges as $N_1 e$ and $N_2 e$, and the vacuum permittivity ε_0 in terms of μ_0 (equation (5.1.1:7)).

In equation (5.1.2:3), the factor $N_1 N_2 e^2 \mu_0 / 4\pi$ has the dimension of mass. Substitution of equation (5.1.2:1) for $e^2 \mu_0$ in equation (5.1.2:3) obtains the form

$$E_{EM} = N_1 N_2 \frac{e^2 \mu_0}{4\pi r} c_0 c = N_1 N_2 a \frac{h_0}{r} c_0 c = N_1 N_2 a \frac{h_0}{L_r} c_0 c = m_{EM} c_0 c, \tag{5.1.2:4}$$

where L_r is the circumference of a circle with radius r, i.e., the length of an equipotential orbit around the accompanying charge. Equation (5.1.2:4) reveals the mass equivalence of the Coulomb energy of point-like charges $N_1 e$ and $N_2 e$ at a distance r from each other

$$m_{EM} = N_1 N_2 \frac{e^2 \mu_0}{4\pi r} = N_1 N_2 a \frac{h_0}{r} = N_1 N_2 a \frac{h_0}{L_r}. \tag{5.1.2:5}$$

For unit charges at a distance r from each other, the mass equivalence is

$$m_{EM(0)} = \frac{e^2 \mu_0}{4\pi r} = a \frac{h_0}{r} = a \frac{h_0}{L_r}. \tag{5.1.2:6}$$

Box 5.1.2-A

Due to the motion of space, objects at rest in space move at the velocity of light in the fourth dimension. The action of the imaginary motion on electrical charges at rest in space can be regarded as an electromagnetic interaction between them, formally identical with the Coulomb force.

The electromagnetic force created between charges q_1 and q_2 can be derived by applying the conventional expression of magnetic force $\mathbf{F}^\square_{EM}$ as

$$\mathbf{F}^\square_{EM} = q_1 (i\mathbf{c} \times \mathbf{B}^\square), \qquad (5.1.2:A1)$$

where $i\mathbf{c}$ is the imaginary velocity of q_1 and q_2, and $\mathbf{B}^\square$ is the magnetic flux density [Vs/m²] generated by the motion of q_2 at distance $\mathbf{r}$. $\mathbf{B}^\square$ can be expressed as (see Figure 5.1.2-A1)

$$\mathbf{B}^\square = \frac{q_2 \mu_0}{4\pi r^2} (i\mathbf{c} \times \hat{\mathbf{r}}). \qquad (5.1.2:A2)$$

In equation (5.1.2:A2), μ_0 is the permeability of the vacuum (i.e., space), r is the distance between q_1 and q_2, and $\hat{\mathbf{r}}$ is a unit vector in the direction of $\mathbf{r}$. Since the space direction is perpendicular to the imaginary direction, the magnetic force $\mathbf{F}^\square_{EM}$ between charges q_1 and q_2 can be expressed with the aid of equations (5.1.2:A1) and (5.1.2:A2) as

$$\begin{aligned}\mathbf{F}^\square_{EM} &= \frac{q_1 q_2 \mu_0}{4\pi r^2} [i\mathbf{c} \times (i\mathbf{c} \times \hat{\mathbf{r}})] = \frac{q_1 q_2 \mu_0}{4\pi r^2} [(i\mathbf{c} \cdot \hat{\mathbf{r}}) i\mathbf{c} - (i\mathbf{c} \cdot i\mathbf{c}) \hat{\mathbf{r}}] \\ &= -\frac{q_1 q_2 \mu_0}{4\pi r^2} (ic)^2 \hat{\mathbf{r}} = \frac{q_1 q_2 \mu_0}{4\pi r^2} c^2 \hat{\mathbf{r}}.\end{aligned} \qquad (5.1.2:A3)$$

The derivation of equation (5.1.2:A3) shows that the electromagnetic force generated by the imaginary velocity of space is opposite in sign to the electromagnetic force generated by parallel motion of charges in space. Electrical currents flowing in the same direction in parallel conductors result in an attractive force, whereas currents in opposite directions result in a repulsive force between the conductors. In ion beams, the attractive effect is observed as constriction of the discharge (the pinch effect). Due to the square of the imaginary unit $i^2 = -1$ in equation (5.1.2:A3), the expression for the effect of the motion of space on electrical charges obtains a form identical to the Coulomb law.

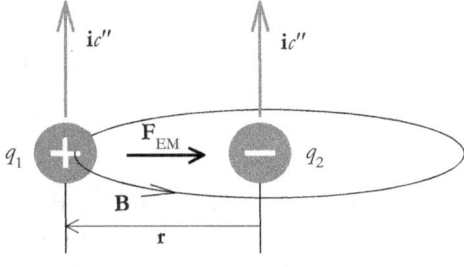

Figure 5.1.2-A1. The electrostatic interaction (Coulomb force) between electrical charges at rest in space can be described as a magnetic interaction due to the imaginary motion of space.

The energy released by a Coulomb system, for example, in an accelerator, can be expressed in terms of the release of mass

$$\Delta E_{EM} = E_{EM(1)} - E_{EM(2)} = \left(m_{EM(1)} - m_{EM(2)}\right) c_0 c = \Delta m_{EM} c_0 c , \qquad (5.1.2:7)$$

that appears as the mass contribution of the kinetic energy of the accelerated object (see equation 4.1.2:5). Traditionally, Coulomb energy is derived from the static Coulomb force postulated for charges at rest in space. Formally, the motion of space at velocity c in the fourth dimension creates a magnetic force between charges at rest in space (Box 5.1.2-A).

Energy carried by electric and magnetic fields

In expanding space, the vacuum impedance decreases in direct proportion to the decreasing velocity of light

$$Z = \frac{E}{H} = \sqrt{\frac{\mu_0}{\varepsilon_0}} = \mu_0 c , \qquad (5.1.2:8)$$

where E is the electric field, and H is the magnetic field.

Despite the change in the ratio between electric and magnetic fields in electromagnetic waves, the energies carried by the electric and magnetic fields remain equal. The energy density of an electromagnetic wave is

$$E = \frac{1}{2}\left(\varepsilon_0 E^2 + \mu_0 H^2\right). \qquad (5.1.2:9)$$

Substitution of equation (5.1.2:8) for E and H in (5.1.2:9), and $\varepsilon_0 = 1/\mu_0 c^2$ gives the energy density of an electromagnetic wave in terms of the magnetic field

$$E = \frac{1}{2}\left(\frac{1}{\mu_0 c^2} \mu_0^2 c^2 H^2 + \mu_0 H^2\right) = \frac{1}{2}\left(\mu_0 H^2 + \mu_0 H^2\right) = \mu_0 H^2 , \qquad (5.1.2:10)$$

and the electric field

$$E = \frac{1}{2}\left(\varepsilon_0 E^2 + \mu_0 \frac{E^2}{\mu_0^2 c^2}\right) = \frac{1}{2}\left(\varepsilon_0 E^2 + \varepsilon_0 E^2\right) = \varepsilon_0 E^2 , \qquad (5.1.2:11)$$

respectively.

5.1.3 Wavelength equivalence of mass

The Compton wavelength

Applying the concept of mass equivalence, the momentum of electromagnetic radiation obtains a form equal to that of the rest momentum and rest energy of mass objects. Equations (5.1.1:22) and (5.1.1:15) show the momentum and energy of a quantum of radiation in the form of the rest momentum

$$p_\lambda = m_\lambda c , \qquad (5.1.3:1)$$

and the rest energy

$$E_\lambda = c_0 m_\lambda c . \tag{5.1.3:2}$$

The difference, however, is that the momentum of electromagnetic radiation appears in the direction of the propagation of the radiation in space direction only, whereas the rest momentum of matter appears in the fourth dimension.

The concept of mass equivalence of radiation can be extended to its inverse quantity, the *wavelength equivalence of mass*

$$\lambda_m = \frac{h_0}{m} \quad \text{and} \quad k_m = \frac{2\pi}{\lambda} = \frac{m}{\hbar_0} , \tag{5.1.3:3}$$

where $\hbar_0$ is the *intrinsic reduced Planck constant* $\hbar_0 = h_0/2\pi$. The rest energy of mass m can be expressed as

$$E_{rest} = c_0 |\mathbf{p}_{rest}| = c_0 mc = c_0 \frac{h_0}{\lambda_m} c = c_0 \hbar_0 k_m c . \tag{5.1.3:4}$$

The wavelength and wavenumber equivalences of mass m in (5.1.3:4) can be identified as the Compton wavelength and wavenumber

$$\lambda_{Compton} \equiv \frac{h}{mc} = \frac{h_0}{m} = \lambda_m \qquad k_{Compton} \equiv \frac{mc}{\hbar} = \frac{m}{\hbar_0} = k_m . \tag{5.1.3:5}$$

Wave presentation of the energy four vector

The energy-momentum four-vector is traditionally written in the form

$$E^2_{m(tot)} = c^2 (mc)^2 + c^2 p^2 . \tag{5.1.3:6}$$

In the DU framework, the total energy of a mass object m, moving at velocity β in the local energy frame, is presented as a complex function [see equation (4.1.2:11)]

$$E_{m(tot)} = E^{\square} = c_0 |p^{\square}| = c_0 |m_\beta \beta c + i\, mc| = c_0^2 \sqrt{(mc)^2 + (m_\beta \beta c)^2} , \tag{5.1.3:7}$$

where m_β is the mass contributing to the real component of the momentum [see equation (4.1.2:10)]

$$m_\beta = m + \Delta m = m/\sqrt{1-\beta^2} = \hbar_0 k_{m(\beta)} = \hbar_0 k_m /\sqrt{1-\beta^2} , \tag{5.1.3:8}$$

where the last two forms apply the wave number equivalence of mass as defined in equation (5.1.3:3). The wave number presentation of the total energy of (5.1.3:7) obtains the complex form

$$E^{\square}_{m(tot)} = c_0 \hbar_0 k_{m(\beta)} \beta c + i c_0 \hbar_0 k_m c = c_0 \hbar_0 k_{m(\beta)} c (\sin\varphi + i \cos\varphi) , \tag{5.1.3:9}$$

or in algebraic form

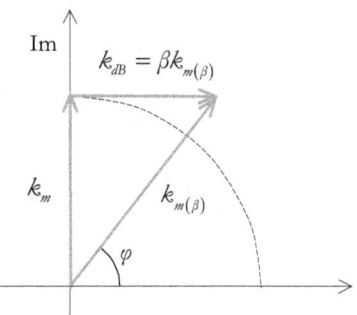

Figure 5.1.3-1. Complex wave number presentation of the energy-momentum four-vector.

$$E_{m(tot)} = c_0 \hbar_0 k_m c \sqrt{1+\left(\frac{k_{dB}}{k_m}\right)^2} = E_{rest}\sqrt{1+\left(\frac{k_{dB}}{k_m}\right)^2} \,. \quad (5.1.3{:}10)$$

Division of equation (5.1.3:9) by c_0 gives the complex presentation of the total momentum

$$p^\square = \hbar_0 k_{m(\beta)} \beta c + i\hbar_0 k_m c = \hbar_0 k_{m(\beta)} c\left(\cos\varphi + i\sin\varphi\right), \quad (5.1.3{:}11)$$

and further dividing by $(\hbar_0 c)$ returns the complex presentation of the wave number of the total mass m_β, Figure 5.1.3-1

$$\beta k_{m(\beta)} + i k_m = k_{m(\beta)}\left(\cos\varphi + i\sin\varphi\right) = \frac{k_m}{\sqrt{1-\beta^2}}\left(\cos\varphi + i\sin\varphi\right). \quad (5.1.3{:}12)$$

In equation (5.1.3:12), the quantity $\beta k_{m(\beta)}$ can be identified as the de Broglie wave number

$$k_{dB} = \beta k_{m(\beta)} = \frac{2\pi}{\lambda_{dB}} = \frac{\beta c \cdot m_\beta}{\hbar} = \frac{\beta m_\beta}{\hbar_0}\,. \quad (5.1.3{:}13)$$

The real component of the complex momentum in (5.1.3:11) can be expressed in the forms

$$p' = \hbar_0 k_{m(\beta)} \cdot \beta c = \hbar_0 k_{dB} \cdot c\,, \quad (5.1.3{:}14)$$

where

1) the first form describes a mass wave with wave number k_β propagating at velocity βc, and
2) the second form describes a mass wave with de Broglie wave number k_{dB} propagating at velocity c.

The physical meanings of the two interpretations are discussed in Section 5.3.

There are no classical "mass particles" in the Dynamic Universe. A mass object in DU space can be described as a standing wave structure characterized by the Compton wavelength. The momentum of a mass object can be expressed in terms of a wave front with wavelength λ_β of (5.3.4:3) propagating along with the object at velocity βc in space (see Section 5.3.4).

Resonant mass wave in a potential well

In a potential well, i.e., in a closed 1-dimensional space of length a, harmonic waves may propagate in both directions, i.e., the wave configuration is the sum of the waves along x and $-x$ directions

$$\psi = \psi_0 \sin(\omega t + kx) + \psi'_0 \sin(\omega t - kx). \qquad (5.1.3{:}15)$$

As a requirement of the boundary conditions at $x = 0$ and $x = a$, the amplitude of the wave has to be zero. The boundary condition at $x = 0$ means

$$\psi_{x=0} = (\psi_0 + \psi'_0)\sin \omega t = 0 \quad \text{i.e.} \quad \psi'_0 = -\psi_0. \qquad (5.1.3{:}16)$$

Substitution of (5.1.3:16) to (5.1.3:15) gives

$$\psi = \psi_0 \left[\sin(\omega t + ka) - \sin(\omega t - ka)\right] = 2\psi_0 \cos\omega t \sin kx. \qquad (5.1.3{:}17)$$

To fulfill the boundary condition at $x = a$, $\sin(kx)$ must be zero at $x = a$, i.e., $kx = n\cdot\pi/a$, resulting in

$$\psi = 2\psi_0 \cos\omega t \sin\frac{n\pi}{a}. \qquad (5.1.3{:}18)$$

In the case of a mass object in a one-dimensional potential well, the wave number in the direction of the real axis across the potential well has to fulfill equation (5.1.3:18)

$$k_r = \frac{n\pi}{a}. \qquad (5.1.3{:}19)$$

The momentum of the object consists of half-wave momenta propagating in opposite directions, which means that the net momentum is zero in the potential. Substitution of the wave number k_r into (5.1.3:19) for the expression of kinetic energy obtained by combining equations (5.1.3:10) and (4.1.2:12), gives the energy levels available in the potential well, Figure 5.1.3-2,

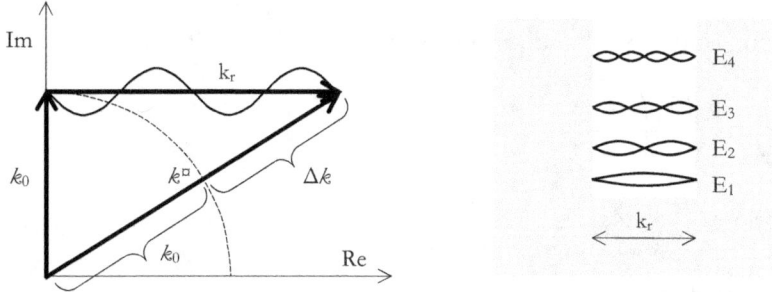

Figure 5.1.3-2. Resonant mass wave (wave number k_r) as the real component of the complex momentum $p_r = \hbar_0 k_r c$.

$$E_n = \Delta k \cdot \hbar_0 c_0 c = E_{rest}\left(\sqrt{1+\left(\frac{n\pi}{a}\bigg/k_m\right)^2}-1\right). \tag{5.1.3:20}$$

Substitution of the rest mass wave number $k_m = m/\hbar_0$ (=Compton wave number) into (5.1.3:20) we get

$$E_n = c_0 mc\left[\sqrt{1+\left(\frac{n\pi}{a}\bigg/\frac{m}{\hbar_0}\right)^2}-1\right] \approx \left(\frac{n\pi}{a}\right)^2 \frac{\hbar_0^2}{2m}c^2. \tag{5.1.3:21}$$

The first form of (5.1.3:21) is the "relativistic solution" solution, and the last form is the first-order approximation equal to the result obtained from the Schrödinger equation.

5.1.4 Hydrogen-like atoms

Principal energy states

Applying the concept of a mass wave, the base energy states of an electron in hydrogen-like atoms can be solved by assuming a resonance condition of the de Broglie wave in a Coulomb equipotential orbit around the nucleus. With reference to equation (5.1.2:4), the Coulomb energy of Z electrons at distance r from the nucleus is

$$E_{Coulomb} = -Za\frac{\hbar_0}{2\pi r}c_0 c = -Za\frac{\hbar_0}{r}c_0 c. \tag{5.1.4:1}$$

For a resonance condition, the de Broglie wavelength $n\lambda_{dB} = 2\pi r$, which is equal to the wave number boundary condition

$$k_{dB} = \frac{n}{r}. \tag{5.1.4:2}$$

With reference to equation (5.1.3:10) for the total energy of motion, the energy of an electron as the sum of kinetic energy and Coulomb energy in a Coulomb equipotential orbit with radius r is

$$E_n = E_{kin} + E_{Coulomb}. \tag{5.1.4:3}$$

Substitution of (5.1.3:20) and (5.1.4:1) for E_{kin} and $E_{Coulomb}$ in (5.1.4:3) gives

$$E_n = \hbar_0 k_m c_0 c\left[\sqrt{1+\left(\frac{n}{k_m r}\right)^2}-1-\frac{Za}{k_m r}\right]. \tag{5.1.4:4}$$

The solution of (5.1.4:4) is illustrated in Figure 5.1.4-1; for each value of n, the total energy E_n is a continuous function of r. The "quantized" energy states are energy minima of E_n for each value of n.

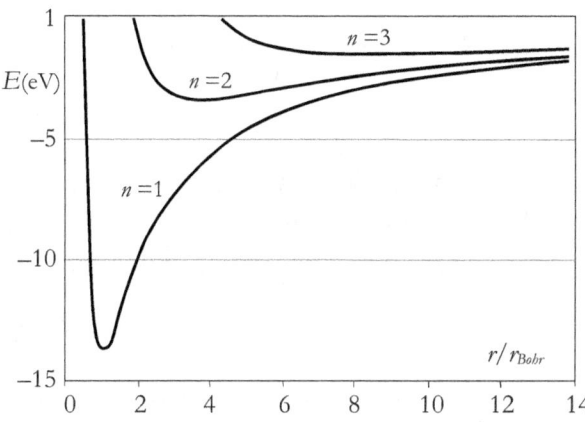

Figure 5.1.4-1. Total energy of electron in hydrogen-like atoms for principal quantum number $n = 1$, $n = 2$, $n = 3$ according to equation (5.1.4:4). Orbital radii of the energy minima are $r/r_{Bohr}=1$, $r/r_{Bohr}=4$, and $r/r_{Bohr}=9$, respectively.

To find the radius for minimum energy, we determine the zero of the derivative of (5.1.4:4)

$$\frac{dE_n}{dr} = \frac{\hbar_0 c_0 c}{r^2}\left[-\frac{n^2}{r}\bigg/\sqrt{k_m^2+\left(\frac{n}{r}\right)^2} + Za\right] = 0. \qquad (5.1.4:5)$$

The solutions of (5.1.4:5) are

1) $r = \infty$

2) $Za - \dfrac{n^2}{r}\bigg/\sqrt{k_m^2+\left(\dfrac{n}{r}\right)^2} = Za - \dfrac{n^2}{\sqrt{r^2 k_0^2 + n^2}} = 0.$ (5.1.4:6)

The radii for minimum energy E_n solved from (5.1.4:6) are

$$r_n = \frac{n^2}{Zak_m}\sqrt{1-\left(\frac{Za}{n}\right)^2}, \qquad (5.1.4:7)$$

where the factor in front of the square root, for $n = 1$, is equal to the classical Bohr radius. The classical notation of Bohr radius is obtained by substitutions of the fine structure constant a and the Compton wave number k_m into the front factor of (5.1.4:7).

$$E_{Z,n} = -mc_0 c\left[1-\sqrt{1-\left(\frac{Za}{n}\right)^2}\right] \approx -\left(\frac{Z}{n}\right)^2 \frac{a^2}{2}mc^2, \qquad (5.1.4:8)$$

where the first-order approximation is equal to the result obtained from the standard solution based on Schrödinger's equation. The first order "relativistic correction" applied to the standard solution is equal to the second order term in the serial approximation of the exact form equation (5.1.4:8).

To find the additional quantum numbers and the fine structure states, the wave equation should be solved for spherical harmonics. Such an analysis is left outside the scope of this treatise.

The effects of gravitation and motion

With reference to (4.1.4:9), the electron rest mass *m* in equation (5.1.4:8) is

$$m_e = m_{e(0)} \prod_{i=1}^{j} \sqrt{1-\beta_i^2} \,, \qquad (5.1.4:9)$$

where $m_{e(0)}$ is the electron mass at rest in a hypothetical homogeneous space, and *j* means the electron moving in the nucleus frame. With reference to equation (4.1.4:10), the velocity of light in equation (5.1.4:8) is

$$c = c_0 \prod_{i=1}^{j} (1-\delta_i) \,. \qquad (5.1.4:10)$$

Substitution of equations (5.1.4:9) and (5.1.4:10) into the last form of equation (5.1.4:7) gives the principal energy states of hydrogen-like atoms in the form

$$E_{Z,n} = \frac{Z^2}{n^2} \frac{\alpha^2}{2} m_{e(0)} c_0^2 \prod_{i=1}^{j} (1-\delta_i) \sqrt{1-\beta_i^2} \,, \qquad (5.1.4:11)$$

showing the dependence of the energy states of an atom on the state of motion and gravitation of the atom.

The energy difference between two energy states is

$$\Delta E_{(n1,n2)} = Z^2 \left[\frac{1}{n_1^2} - \frac{1}{n_2^2} \right] \frac{\alpha^2}{2} m_{e(0)} c_0^2 \prod_{i=1}^{j} (1-\delta_i) \sqrt{1-\beta_i^2} \,. \qquad (5.1.4:12)$$

Differences between the energy states of electrons determine the characteristic emission and absorption energies of atoms. Accordingly, equation (5.1.4:12) shows the dependence of the characteristic emission and absorption energies on the gravitational state and motion of the atom in the local energy frame and in the parent frames.

Characteristic absorption and emission frequencies

Applying equation (5.1.4:12), the characteristic emission and absorption frequency corresponding to the energy transition $\Delta E_{(n1,n2)}$ can be expressed as

$$f_{(n1,n2)} = \frac{\Delta E_{(n1,n2)}}{h_0 c_0} = f_{0(n1,n2)} \prod_{i=1}^{j} (1-\delta_i) \sqrt{1-\beta_i^2} \,, \qquad (5.1.4:13)$$

where $f_{0(n1,n2)}$ is the frequency of the transition for an atom at rest in a hypothetical homogeneous space

$$f_{0(n1,n2)} = Z^2 \left[\frac{1}{n_1^2} - \frac{1}{n_2^2} \right] \frac{\alpha^2}{2h_0} m_{e(0)} c_0 \,. \qquad (5.1.4:14)$$

The velocity of the expansion of space, $c_0 = c_4$, is a function of the time from the singularity. Substitution of equation (3.3.3:8) for c_0 in equation (5.1.3:8) gives frequency $f_{0(n1,n2)}$ in the form

Mass, mass objects and electromagnetic radiation

$$f_{0(n1,n2)} = Z^2 \left[\frac{1}{n_1^2} - \frac{1}{n_2^2} \right] \frac{a^2 m_{e(0)}}{2 h_0} \left(\frac{2}{3} GM'' \right)^{1/3} t^{-1/3}, \qquad (5.1.4:15)$$

which expresses frequency $f_{0(n1,n2)}$ in terms of the age of expanding space, the gravitational constant, and the total mass in space.

The emission wavelength corresponding to the emission frequency of equation (5.1.4:13) and the energy transition $\Delta E_{(n1,n2)}$ is

$$\lambda_{(n1,n2)} = \frac{c}{f_{(n1,n2)}} = \frac{c_0}{f_{0(n1,n2)}} \frac{\prod_{i=1}^{n}(1-\delta_i)}{\prod_{i=1}^{n}(1-\delta_i)\sqrt{1-\beta_i^2}} = \frac{\lambda_{0(n1,n2)}}{\prod_{i=1}^{n}\sqrt{1-\beta_i^2}}, \qquad (5.1.4:16)$$

where

$$\lambda_{0(n1,n2)} = \frac{c_0}{f_{0(n1,n2)}}, \qquad (5.1.4:17)$$

is the wavelength of radiation emitted by the energy transition $\Delta E_{(n1,n2)}$ of the atom at rest in hypothetical homogeneous space. Substitution of equation (5.1.3:8) for $f_{0(n1,n2)}$ in equation (5.1.4:17) gives

$$\lambda_{0(n1,n2)} = \frac{2 h_0}{Z^2 \left[1/n_1^2 - 1/n_2^2 \right] a^2 m_{e(0)}}. \qquad (5.1.4:18)$$

Applying the standard solution of the Bohr radius (the approximate value of equation (5.1.4:7)) and equation (5.1.3:3), we can express the radius of the hydrogen atom as

$$a_0 = \frac{h_0^2}{\pi \mu_0 e^2 m_e} = \frac{a_{0(0)}}{\prod_{i=1}^{n}\sqrt{1-\beta_i^2}}, \qquad (5.1.4:19)$$

where $a_{0(0)}$ is the Bohr radius of a hydrogen atom at rest in a hypothetical homogeneous space

$$a_{0(0)} = \frac{h_0^2}{\pi \mu_0 e^2 m_{e(0)}} = \frac{h_0}{2\pi a m_{e(0)}}. \qquad (5.1.4:20)$$

As shown by equations (5.1.4:18) and (5.1.4:19), both the emission wavelength and the atomic radius are functions of the velocity of the atom in the local energy frame and the velocities of the local frame and the parent frames. The emission wavelength and the atomic radius, however, are not functions of the gravitational state, the local velocity of light, or the expansion velocity of space.

When h_0 (solved in terms of a from equation (5.1.4:19)), and the Bohr radius $a_{0(0)}$ (solved from equation (5.1.4:20)) are substituted into equation (5.1.4:18), equation (5.1.4:16) can be expressed in the form

$$\lambda_{(n1,n2)} = \frac{4\pi\, a_{0(0)}}{aZ^2 \left[1/n_1^2 - 1/n_2^2\right] \prod_{i=1}^{n} \sqrt{1-\beta_i^2}} = \frac{4\pi a_0}{aZ^2 \left[1/n_1^2 - 1/n_2^2\right]}, \qquad (5.1.4:21)$$

which shows that the wavelength emitted is directly proportional to the Bohr radius of the atom.

In fact, the last form of equation (5.1.4:21) is just another form of Balmer's formula, which does not require any assumptions tied to the DU model. Equation (5.1.4:21) also means that, like the dimensions of an atom, the characteristic emission and absorption wavelengths of an atom are unchanged in the course of the expansion of space but are dependent on the velocity of the emitter and absorber in their local and parent frames.

> DU predicts an increase in the size of atoms (in three dimensions) due to motion, instead of the length contraction in the direction of motion predicted by the special theory of relativity.

The effects of motion and gravitation on the wavelengths and frequencies of atoms can be extended to electromagnetic resonators and lasers of macroscopic dimensions. The increase in atomic size with motion means that the dimensions of resonators coupled to moving oscillators increase in direct proportion to the increase in the wavelength of the electromagnetic wave produced by the oscillator.

The characteristic frequency of an atomic oscillator, unlike the wavelength, is subject to change with a changing velocity of light and the expansion of space. With reference to equations (5.1.4:15) and (3.3.3:8), the characteristic frequency $f_{(t)}$ at time t from the singularity of space, when the 4-radius of space is $R_{4(t)}$, can be expressed as

$$f_{(t)} = f_{(t0)} \frac{c_{0(t)}}{c_{0(t0)}} = f_{(t0)} \left(\frac{R_{4(t0)}}{R_{4(t)}}\right)^{1/2} = f_{(t0)} \left(\frac{t_0}{t}\right)^{1/3}, \qquad (5.1.4:22)$$

where $f_{(t0)}$ is the frequency when the 4-radius of space is $R_{4(t0)}$, the velocity of light is $c_{0(t0)}$, and the time from the singularity is t_0.

5.2 Effect of gravitation and motion on clocks and radiation

5.2.1 Effect of gravitation and motion on clocks and radiation

Applying equation (5.1.4:13), the frequencies of two identical atomic oscillators moving at velocities β_A and β_B in gravitational states δ_A and δ_B in a gravitational frame can be expressed as

$$f_A = f_{0\delta}(1-\delta_A)\sqrt{1-\beta_A^2}, \qquad (5.2.1:1)$$

and

$$f_B = f_{0\delta}(1-\delta_B)\sqrt{1-\beta_B^2}, \qquad (5.2.1:2)$$

where $f_{0\delta}$ is the frequency of the oscillators at rest in the apparent homogeneous space of the local gravitational frame

$$f_{0\delta} = f_0 \prod_{i=1}^{n-1}\left[(1-\delta_i)\sqrt{1-\beta_i^2}\right], \qquad (5.2.1:3)$$

where frames $i = 1…n-1$ are the parent frames of the local gravitational frame n.

Combining equations (5.2.1:1) and (5.2.1:2) allows the ratio of the frequencies f_B and f_A to be expressed as

$$\frac{f_B}{f_A} = \frac{(1-\delta_B)\sqrt{1-\beta_B^2}}{(1-\delta_A)\sqrt{1-\beta_A^2}}, \qquad (5.2.1:4)$$

and the relative frequency difference $\Delta f/f_A = (f_B-f_A)/f_A$ as

$$\frac{\Delta f}{f_A} = \frac{(1-\delta_B)\sqrt{1-\beta_B^2}}{(1-\delta_A)\sqrt{1-\beta_A^2}} - 1. \qquad (5.2.1:5)$$

Substituting equation (4.1.1:30) for δ_A and δ_B in equation (5.2.1:5), we get

$$\frac{\Delta f}{f_A} = \frac{(1-GM/r_B c_0 c)\sqrt{1-\beta_B^2}}{(1-GM/r_A c_0 c)\sqrt{1-\beta_A^2}} - 1. \qquad (5.2.1:6)$$

When $\beta_A, \beta_B \ll 1$ and $\delta_A, \delta_B \ll 1$, then also $c_A \approx c_B \approx c$, and equation (5.2.1:6) can be approximated as

$$\frac{\Delta f}{f_A} = \frac{GM}{c^2}\left[\frac{1}{r_A} - \frac{1}{r_B}\right] - \frac{1}{2}\left(\beta_B^2 - \beta_A^2\right), \qquad (5.2.1:7)$$

where the first term is the gravitational shift, and the second term is the shift due to the motions. When $|r_B - r_A|/r_A \ll 1$, equation (5.2.1:7) can be expressed as

$$\frac{\Delta f}{f_A} = \frac{gh}{c^2} - \frac{1}{2}\left(\beta_B^2 - \beta_A^2\right), \qquad (5.2.1:8)$$

where $h = r_B - r_A$ is the difference in altitude in the gravitational frame and g is the gravitational acceleration at distance $r = r_A \approx r_B$ from mass center M

$$g = \frac{GM}{r^2}. \qquad (5.2.1:9)$$

Equations (5.2.1:5–9) express the shift in the frequencies of atomic oscillators in different states of gravitation and motion. The equations are essentially the same as the expressions for the gravitational shift and the effect of motion on atomic oscillators in the general theory of relativity. The validity of the equations has been confirmed in numerous experiments (see Chapter 7).

Instead of explaining the effects of motion and gravitation on an atomic clock as a frequency shift, the theory of relativity explains them in terms of proper time, as a change in the flow of time for an object in motion and a different state of gravitation relative to the observer.

Based on equations (5.2.1:1–3), a general expression for the ratio of the frequencies of two identical atomic oscillators is

$$\frac{f_B}{f_A} = \frac{\prod_{j=1}^{n}(1-\delta_{Bi})\sqrt{1-\beta_{Bi}^2}}{\prod_{i=1}^{m}(1-\delta_{Ai})\sqrt{1-\beta_{Ai}^2}}, \qquad (5.2.1:10)$$

where δ_{Ai}, δ_{Bi}, and β_{Ai}, β_{Bi} describe the states of gravitation and motion in the local energy frame and in the nested parent frames relevant to oscillators A and B.

In general relativity, the combined effect of motion and gravitation on the "proper frequency" of atomic oscillators in a local gravitational frame is given by the equation

$$f_{\delta,\beta} = f_{0,0}\sqrt{1 - 2\delta - \beta^2}, \qquad (5.2.1:11)$$

where

$$\delta = \frac{GM}{rc^2}. \qquad (5.2.1:12)$$

Equation (5.2.1:11) of general relativity corresponds to equation (5.2.1:1) in the Dynamic Universe. The difference between the GR and DU equations appears only in the 4th order terms in the series approximations of equations (5.2.1:1) and (5.2.1:11)

$$f_{\delta,\beta(DU)} = f_{0,0}(1-\delta)\sqrt{1-\beta^2} \approx f_{0,0}\left(1 - \delta - \frac{1}{2}\beta^2 - \frac{1}{8}\beta^4 + \frac{1}{2}\delta\beta^2\right), \qquad (5.2.1:13)$$

and

$$f_{\delta,\beta(GR)} = f_{0,0}\sqrt{1-2\delta-\beta^2} \approx f_{0,0}\left(1 - \delta - \frac{1}{2}\beta^2 - \frac{1}{8}\beta^4 - \frac{1}{2}\delta\beta^2 - \frac{1}{2}\delta^2\right). \qquad (5.2.1:14)$$

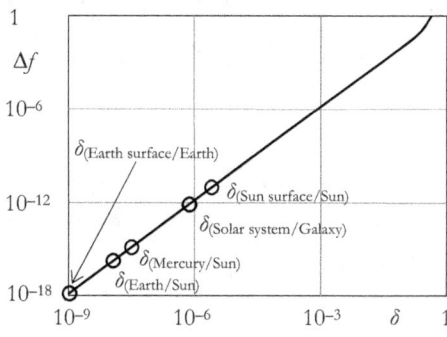

Figure 5.2.1-1(a). The difference in the DU and GR predictions of the gravitational correction of atomic oscillators in different gravitational states. On the surface of the Earth $\delta \approx 10^{-9}$ and the difference in the two predictions appear in the 18:th decimal.

Figure 5.2.1-1(b). The difference in the DU and GR predictions of the frequency of atomic oscillators at extreme conditions when $\delta = \beta^2 \to 1$. Such condition may appear close to a black hole in space. The GR and DU predictions in the figure are based on equations (5.2.1:11) and (5.2.1:13), respectively.

The difference between the DU and GR frequencies in equations (5.2.1:13) and (5.2.1:14) is

$$\Delta f_{\delta,\beta(DU-GR)} \approx \delta\beta^2 + \tfrac{1}{2}\delta^2. \tag{5.2.1:15}$$

The difference given by equation (5.2.1:15) is too small to be detected with clocks in Earth satellites or spacecraft in the solar gravitational frame, Figure 5.2.1-1(a). The difference, however, is essential in extreme conditions where δ and β approach unity, Figure 5.2.1-1(b).

5.2.2 Gravitational shift of electromagnetic radiation

As discussed in the previous section, the frequency of an atomic oscillator is a function of its gravitational state. The frequency of oscillation is reduced as the δ-factor characterizing the gravitational state increases.

When an atomic oscillator at rest in δ_A-state emits radiation at the oscillation frequency f_A, the frequency received by an object at rest in δ_B state is the same, f_A. In a steady state, because of the absolute time, the same number of cycles emitted in a time interval will also be received.

The wavelength of the signal sent from the object at rest in the δ_A-state can be expressed in terms of the frequency, f_A, and the local velocity of light, c_A, as

$$\lambda_A = \frac{c_A}{f_A}. \tag{5.2.2:1}$$

With reference to equations (4.1.1:23) and (5.2.1:1), equation (5.2.2:1) can be expressed as

$$\lambda_A = \frac{c_A}{f_A} = \frac{c_{0\delta}(1-\delta_A)}{f_{0\delta}(1-\delta_A)} = \frac{c_{0\delta}}{f_{0\delta}}, \qquad (5.2.2:2)$$

which shows that, because the oscillation frequency and the local velocity of light depend similarly on the gravitational state, the wavelength emitted is independent of the gravitational state of the emitting object in the gravitational frame in question. Accordingly, the wavelength of the radiation sent by an object at rest in δ_B-state is

$$\lambda_B = \lambda_A = \frac{c_{0\delta}}{f_{0\delta}}. \qquad (5.2.2:3)$$

When radiation sent by an object at rest in δ_A-state is received by an object at rest in δ_B-state, the frequency received is f_A. The velocity of light in the δ_B-state is c_B. Thus, the wavelength received is

$$\lambda_{A(B)} = \frac{c_B}{f_A}. \qquad (5.2.2:4)$$

Substituting equation (5.2.2:1) for f_A in equation (5.2.2:4), $\lambda_{A(B)}$ can be expressed as

$$\lambda_{A(B)} = \frac{c_B}{c_A}\lambda_A, \qquad (5.2.2:5)$$

and by further applying equations (5.2.2:2) and (5.2.2:3), we get

$$\lambda_{A(B)} = \frac{c_B}{c_A}\lambda_A = \frac{c_B}{c_A}\lambda_B = \frac{f_B}{f_A}\lambda_A = \frac{f_B}{f_A}\lambda_B. \qquad (5.2.2:6)$$

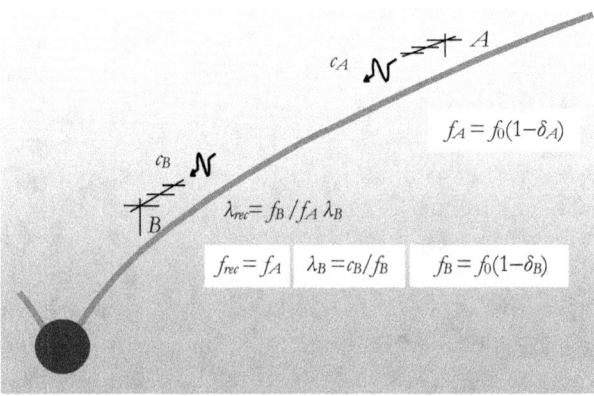

Figure 5.2.2-1. The velocity of light is lower close to a mass center, $c_B < c_A$, which results in a decrease in the wavelength of electromagnetic radiation transmitted from A to B. Accordingly, the signal received at B is blueshifted relative to the reference wavelength observed in radiation emitted by a similar object in the δ_B-state. The frequency of the radiation is unchanged during the transmission.

That is, the wavelength sent by the oscillator in the δ_A-state is changed by a factor equal to the inverse of the ratio of the corresponding frequencies in the two gravitational states, Figure 5.2.2-1.

Equation (5.2.2:6) expresses the gravitational redshift or blueshift of electromagnetic radiation. The frequency of electromagnetic radiation does not change when the radiation travels from one gravitational state to another. However, the wavelength of the radiation is shifted due to the different velocity of light in different gravitational states.

The DU model makes a clear distinction between the gravitational effects on the frequency and wavelength of atomic oscillators and the gravitational effects on the frequency and wavelength of electromagnetic radiation.

The DU predictions of the gravitational shifts of the frequencies and wavelengths of atomic oscillators and electromagnetic radiation are in complete agreement with experiments (see Chapter 7).

The characteristic frequency of an oscillator is directly proportional to the local velocity of light in the gravitational state of the oscillator.

The characteristic wavelength of electromagnetic radiation sent by an oscillator is independent of the gravitational state in which the oscillator is located.

The gravitational red or blue shift of electromagnetic radiation is the shift of the wavelength of the radiation due to the difference in the velocity of light at different gravitational states. No change in the frequency of the radiation occurs during propagation.

5.2.3 The Doppler effect of electromagnetic radiation

Doppler effect in a local gravitational frame

The Doppler effect of electromagnetic radiation is derived analogously to the Doppler effect of any wave motion emitted by a source in motion relative to the state of rest in the propagation frame and received by an observer also moving relative to the state of rest in the propagation frame. If the source and the receiver both are objects in the same gravitational state in a local gravitational frame, the propagation velocity is the local imaginary velocity of space in the frame.

The motion of the source in the local gravitational frame affects both the characteristic frequency of the source and the wavelength emitted in different directions. The shortening of the wavelength, the Doppler effect, is governed by the distance the source moves during a cycle in the direction of the wave emitted. If the velocity of the source is $\mathbf{v}_A$, the change in the wavelength of radiation emitted in the direction $\mathbf{r}$ is

$$\Delta\lambda_\mathbf{r} = T\mathbf{v}_A \cdot \hat{\mathbf{r}} = \frac{v_A}{c}\lambda_{A(\beta)}\hat{\mathbf{v}}_A \cdot \hat{\mathbf{r}} = \frac{\lambda_{A(0)}\beta_{A(\mathbf{r})}}{\sqrt{1-\beta_A^2}}, \qquad (5.2.3:1)$$

where $T = 1/f$ is the cycle time, $\lambda_{A(\beta)}$ is the characteristic wavelength of the source moving at velocity β_A in the δ-state, $\hat{\mathbf{v}}_A$ and $\hat{\mathbf{r}}$ are the unit vectors in the directions of $\mathbf{v}_A$ and $\mathbf{r}$,

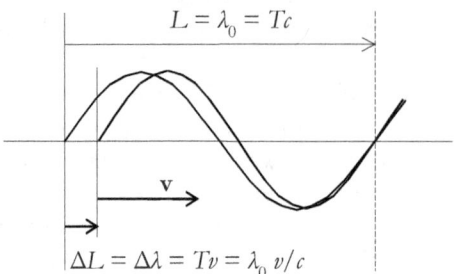

Figure 5.2.3-1. The wavelength of electromagnetic radiation emitted by a moving source is shortened in the direction of the motion by the distance moved by the source during the cycle time, $\Delta\lambda = \lambda_0\, v/c$.

respectively, and $\beta_{A(r)} = \boldsymbol{\beta}_A \cdot \hat{\mathbf{r}}$ is the component of velocity $\boldsymbol{\beta}_A$ in the direction of $\mathbf{r}$, Figure 5.2.3-1.

The wavelength emitted in direction $\mathbf{r}$ is

$$\lambda_{rA} = \frac{\lambda_{A(0)}}{\sqrt{1-\beta_A^2}}\left(1-\beta_{A(r)}\right) = \lambda_{A(\beta)}\left(1-\beta_{A(r)}\right). \tag{5.2.3:2}$$

By substituting (5.2.3:2) into (5.1.1:22), the momentum of the radiation emitted in the $\mathbf{r}$ direction by a source with velocity $\beta_{A(r)}$ in the direction of the emission is

$$\mathbf{P}_{(\text{rad})rA} = \frac{h_0}{\lambda_{rA}} c\hat{\mathbf{r}} = \frac{h_0}{\lambda_{A(0)}} \frac{\sqrt{1-\beta_A^2}}{\left(1-\beta_{A(r)}\right)} c\hat{\mathbf{r}}. \tag{5.2.3:3}$$

The momentum of radiation observed in the $\mathbf{r}$ direction by a receiver moving at velocity $\mathbf{v}_B$ in the δ-state is

$$\mathbf{P}_{(\text{rad})rA(B)} = \frac{h_0}{\lambda_{rA}}\left(c - \mathbf{v}_B \cdot \hat{\mathbf{r}}\right)\hat{\mathbf{r}} = \frac{h_0}{\lambda_{rA}}\left(1-\beta_{B(r)}\right)c\hat{\mathbf{r}} = \frac{h_0}{\lambda_{rA}} \mathbf{c}', \tag{5.2.3:4}$$

where

$$\beta_{B(r)} = \frac{\mathbf{v}_B \cdot \hat{\mathbf{r}}}{c} = \boldsymbol{\beta}_B \cdot \hat{\mathbf{r}}, \tag{5.2.3:5}$$

is the component of velocity $\boldsymbol{\beta}_B$ in the direction of $\mathbf{r}$ and

$$\mathbf{c}' = \left(c - \mathbf{v}_B \cdot \hat{\mathbf{r}}\right)\hat{\mathbf{r}} = c\left(1-\beta_{B(r)}\right)\hat{\mathbf{r}} \tag{5.2.3:6}$$

is the effective velocity at which the radiation is received in the direction of $\mathbf{r}$ (the velocity of light minus the velocity of the receiver in the local gravitational frame).

Substituting equation (5.2.3:3) for h_0/λ_{rA} in equation (5.2.3:4) we get

$$\mathbf{P}_{(\text{rad})rA(B)} = \frac{h_0\sqrt{1-\beta_A^2}}{\lambda_{A(0)}} \frac{\left(1-\beta_{B(r)}\right)}{\left(1-\beta_{A(r)}\right)} c\hat{\mathbf{r}}. \tag{5.2.3:7}$$

The wavelength of radiation from two identical emitters at rest in the same gravitational frame is the same, $\lambda_{A(0)} = \lambda_{B(0)} = \lambda_0$. The wavelength of radiation from a reference oscillator moving with the receiver at velocity β_B is

$$\lambda_{B(\beta)} = \lambda_B = \frac{\lambda_{B(0)}}{\sqrt{1-\beta_B^2}} = \frac{\lambda_{A(0)}}{\sqrt{1-\beta_A^2}} = \frac{\lambda_0}{\sqrt{1-\beta_B^2}}. \qquad (5.2.3{:}8)$$

Substituting equation (5.2.3:8) into equation (5.2.3:7) gives the observed momentum in terms of the wavelength λ_B of the reference oscillator in the same δ-state as

$$\mathbf{P}_{(\text{rad})rA(B)} = \frac{h_0}{\lambda_B}\frac{\sqrt{1-\beta_A^2}}{\sqrt{1-\beta_B^2}}\frac{\left(1-\beta_{B(r)}\right)}{\left(1-\beta_{A(r)}\right)} c\hat{\mathbf{r}}. \qquad (5.2.3{:}9)$$

By applying equation (5.1.1:22) for $h_0/\lambda_B \cdot c$, equation (5.2.3:9) can be expressed in terms of the frequency of the radiation observed as

$$f_{A(B)} = f_B \frac{\sqrt{1-\beta_A^2}}{\sqrt{1-\beta_B^2}}\frac{\left(1-\beta_{B(r)}\right)}{\left(1-\beta_{A(r)}\right)}, \qquad (5.2.3{:}10)$$

which combines the effect of the Doppler shift and the effects of the different velocities of the source and reference oscillators on the frequency of each oscillator, Figure 5.2.3-2.

If the source and the receiver are in different gravitational states δ_A and δ_B, equation (5.2.3:10) needs to be supplemented with the effect of gravitation in accordance with equations (5.2.1:1) and (5.2.1:2), as

$$f_{A(B)} = f_B \frac{(1-\delta_A)\sqrt{1-\beta_A^2}}{(1-\delta_B)\sqrt{1-\beta_B^2}}\frac{\left(1-\beta_{B(r)}\right)}{\left(1-\beta_{A(r)}\right)}. \qquad (5.2.3{:}11)$$

Substituting equation (5.2.1:4) for f_B in equation (5.2.3:11) gives

$$f_{A(B)} = f_A \frac{\left(1-\beta_{B(r)}\right)}{\left(1-\beta_{A(r)}\right)}, \qquad (5.2.3{:}12)$$

where f_A and f_B are the frequencies of the source and the reference oscillators moving with the receiver

$$f_A = f_{0\delta}(1-\delta_A)\sqrt{1-\beta_A^2} \quad ; \quad f_B = f_{0\delta}(1-\delta_B)\sqrt{1-\beta_B^2}, \qquad (5.2.3{:}13)$$

where $f_{0\delta}$ is the frequency of the oscillators at rest in the apparent homogeneous space of the local gravitational frame.

Equations (5.2.3:11) and (5.2.3:12) can be expressed in terms of wavelengths related to velocity c as

$$\lambda_{A(B)} = \lambda_B \frac{(1-\delta_B)\sqrt{1-\beta_B^2}}{(1-\delta_A)\sqrt{1-\beta_A^2}}\frac{\left(1-\beta_{A(r)}\right)}{\left(1-\beta_{B(r)}\right)}, \qquad (5.2.3{:}14)$$

and

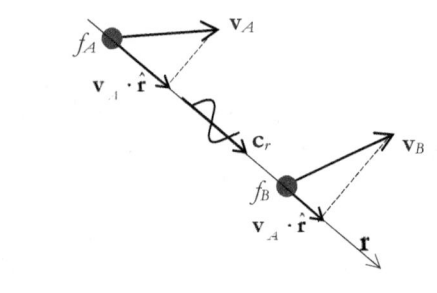

Figure 5.2.3-2. The Doppler effect combines the effects of the velocities of the source and the receiver in the direction of the signal path.

$$\lambda_{A(B)} = \lambda_A \frac{\left(1-\beta_{A(\mathbf{r})}\right)}{\left(1-\beta_{B(\mathbf{r})}\right)}, \quad (5.2.3{:}15)$$

where

$$\lambda_A = \frac{\lambda_{A(0)}}{\sqrt{1-\beta_A^2}} \quad ; \quad \lambda_B = \frac{\lambda_{B(0)}}{\sqrt{1-\beta_B^2}}, \quad (5.2.3{:}16)$$

where $\lambda_{A(0)} = \lambda_{B(0)} = \lambda_0$ is the wavelength of the oscillators at rest, i.e., the wavelength emitted by the oscillators at rest in the local gravitational frame.

Equations (5.2.3:12) and (5.2.3:15) are essentially identical with the classical Doppler equations, just as equations (5.2.3:11) and (5.2.3:14) correspond to the Doppler equations derived from the general theory of relativity. In the terminology of the theory of relativity, the effect of motion on the oscillators is referred to as the "time dilation term", or the "transversal or secondary Doppler effect", and the gravitational effect is referred to as the gravitational red- or blueshift [68,69].

Doppler effect in nested energy frames

If the source and the receiver are in different energy frames, the frequencies of the corresponding oscillators are calculated from equation (5.2.1:10). The simplest approach to calculate the effects of the motions of the source and the receiver within their parent frames is to follow the same procedure as we did for the source and an object in the same frame by considering each frame as an object in its parent frame. On the source side, the wavelength is reduced in each step from the local frame towards the "root" parent frame and finally to hypothetical homogeneous space. With reference to equation (5.2.3:2), the wavelength emitted to hypothetical homogeneous space step by step through the chain of nested energy frames can be deduced as

$$\begin{aligned}\lambda_{A[n-1](\mathbf{r})} &= \lambda_{A[n]}\left(1-\beta_{A[n](\mathbf{r})}\right) \\ \lambda_{A[n-2](\mathbf{r})} &= \lambda_{A[n-1](\mathbf{r})}\left(1-\beta_{A[n-1](\mathbf{r})}\right) = \lambda_{A[n]}\left(1-\beta_{A[n](\mathbf{r})}\right)\left(1-\beta_{A[n-1](\mathbf{r})}\right) \\ &\cdots \\ \lambda_{A[0](\mathbf{r})} &= \lambda_{A[n]}\prod_{i=1}^{n}\left(1-\beta_{A[i](\mathbf{r})}\right). \end{aligned} \quad (5.2.3{:}17)$$

Mass, mass objects and electromagnetic radiation

At the receiver, with reference to equation (5.2.3:6), in the nth frame, the effective velocity of the receiving signal propagating in hypothetical homogeneous space is

$$\mathbf{c}_1' = c_0\left(1-\beta_{[1]B(r)}\right)\hat{\mathbf{r}}$$

$$\mathbf{c}_2' = c_1'\left(1-\beta_{[2]B(r)}\right)\hat{\mathbf{r}} = c_0\left(1-\beta_{[1]B(r)}\right)\left(1-\beta_{[2]B(r)}\right)\hat{\mathbf{r}} \qquad (5.2.3{:}18)$$

$$\ldots$$

$$\mathbf{c}_m' = c_0\prod_{j=1}^{m}\left(1-\beta_{[j]B(r)}\right)\hat{\mathbf{r}},$$

and the momentum observed in a signal with wavelength $\lambda_{A[0](r)}$ in hypothetical homogeneous space is

$$\mathbf{P}_{A[0]([j]B)r} = \frac{h_0}{\lambda_{rA[0]}}c_0\prod_{j=1}^{m}\left(1-\beta_{[j]B(r)}\right)\hat{\mathbf{r}} = \frac{h_0}{\lambda_{rA[0]}}\mathbf{c}_m'. \qquad (5.2.3{:}19)$$

Substitution of equation (5.2.3:17) for $\lambda_{rA[0]}$ in equation (5.2.3:19) gives the momentum as

$$\mathbf{P}_{A[0]([j]B)r} = \frac{h_0 c}{\lambda_{rA[n]}}\frac{\prod_{j=1}^{m}\left(1-\beta_{[j]B(r)}\right)}{\prod_{i=1}^{n}\left(1-\beta_{A[i](r)}\right)}\hat{\mathbf{r}} = h_0 f_{A(B)}\,\hat{\mathbf{r}}, \qquad (5.2.3{:}20)$$

where frequency $f_{A(B)}$ is the Doppler shifted frequency of a signal emitted by source A and received by B

$$f_{A(B)} = f_A\frac{\prod_{j=1}^{m}\left(1-\beta_{jB(r)}\right)}{\prod_{i=1}^{n}\left(1-\beta_{iA(r)}\right)}, \qquad (5.2.3{:}21)$$

where $f_A = h_0/\lambda_{rA[n]}$ is the frequency of the source in its local frame $A[n]$.

Equation (5.2.3:21) can be written in the form

$$f_{A(B)} = f_A\frac{\prod_{j=1}^{k}\left(1-\beta_{jB(r)}\right)\prod_{j=k+1}^{n}\left(1-\beta_{jB(r)}\right)}{\prod_{i=1}^{k}\left(1-\beta_{iA(r)}\right)\prod_{i=k+1}^{m}\left(1-\beta_{iA(r)}\right)} = f_A\frac{\prod_{j=k+1}^{n}\left(1-\beta_{jB(r)}\right)}{\prod_{i=k+1}^{m}\left(1-\beta_{iA(r)}\right)}, \qquad (5.2.3{:}22)$$

which demonstrates the elimination of the effects of the "root" parent frames 1 to k common to both the source and the receiver.

In equation (5.2.3:22), the k:th frame is the first root frame serving as the reference at rest for the transmission of a signal from the source to the receiver. As shown by equation (5.2.3:22), the effects of the motions of frames 1 to k on the Doppler effect are cancelled, Figure 5.2.3-3.

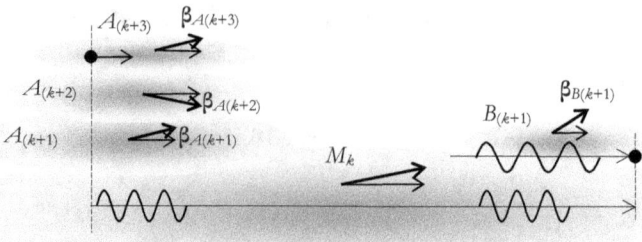

Figure 5.2.3-3. Transmission of electromagnetic radiation from the source at rest in frame $A_{(k+3)}$ to the receiver at rest in frame $B_{(k+1)}$. The motions of frames $A_{(k+1)}$ ... $A_{(k+3)}$ result in a change of the wavelength in radiation propagating in the M_k frame.

$$f_{A(B)} = f_B \frac{\prod_{j=k+1}^{n}\left(1-\beta_{Bj(r)}\right)\prod_{i=k+1}^{m}\left(1-\delta_{Ai}\right)\sqrt{1-\beta_{Ai}^2}}{\prod_{i=k+1}^{m}\left(1-\beta_{iA(r)}\right)\prod_{j=k+1}^{n}\left(1-\delta_{Bj}\right)\sqrt{1-\beta_{Bj}^2}}. \qquad (5.2.3{:}23)$$

In the k-frame, the momentum of the radiation is

$$\mathbf{P}_{A[0](k)\mathrm{r}} = \frac{h_0}{\lambda_{A[k]}} c_k \hat{\mathbf{r}}, \qquad (5.2.3{:}24)$$

and in the $B(k+1)$ frame

$$\mathbf{P}_{A[0][B(k+1)]\mathrm{r}} = \frac{h_0}{\lambda_{A[k]\mathrm{r}}} c_k \left(1-\beta_{[k+1]B(r)}\right)\hat{\mathbf{r}} = \frac{h_0 c_k}{\lambda_{A[k]} \big/ \left(1-\beta_{[k+1]B(r)}\right)}\hat{\mathbf{r}}. \qquad (5.2.3{:}25)$$

All the nested frames should be understood to be co-existing. Capturing of radiation from one frame to another changes the frequency and the reference at rest, but it does not change the physical propagation velocity of the radiation in the root frame. The reduction of momentum in equation (5.2.3:25) is a consequence of an increase in wavelength $\lambda_{k+1} = \lambda_k / (1-\beta_{[k+1]B(r)})$, which means a reduction of the mass equivalence of radiation due to the motion of the receiver in the propagation frame k. The reduction of the momentum can also be interpreted as a reduction of velocity $c_{k+1} = c_k(1-\beta_{[k+1]B(r)})$ in the B_{k+1} frame due to a kinematic component resulting from the motion of the receiver frame.

When received by a receiver B at rest in frame $k+1$, the frequency observed is

$$f_{A(B)} = \frac{c_k}{\lambda_{A[k]\mathrm{r}}}\left(1-\beta_{[k+1]B(r)}\right), \qquad (5.2.3{:}26)$$

where

$$\lambda_{A[k](r)} = \lambda_{A[k+3]} \prod_{i=k+1}^{k+3}\left(1-\beta_{A[i](r)}\right). \qquad (5.2.3{:}27)$$

Substituting equation (5.2.3:27) for $\lambda_{A[k](r)}$ in equation (5.2.3:26), frequency $f_{A(B)}$ obtains the form

$$f_{A(B)} = \frac{c_k}{\lambda_{A[k+3]}} \frac{\left(1-\beta_{[k+1]B(r)}\right)}{\prod_{i=k+1}^{k+3}\left(1-\beta_{A[i](r)}\right)} = f_A \frac{\left(1-\beta_{[k+1]B(r)}\right)}{\prod_{i=k+1}^{k+3}\left(1-\beta_{A[i](r)}\right)}. \qquad (5.2.3:28)$$

The frequency in equation (5.2.3:28) is the same as that obtained by applying equation (5.2.3:22), which was derived by regarding the source and receiver frame as moving objects in the (root) parent frame.

5.3 Localized energy objects

5.3.1 Momentum of radiation from a moving emitter

Emission from a point source

Emission of electromagnetic energy from a point source can be described as a turn of the imaginary energy of the emitter into the energy of electromagnetic radiation propagating in a spatial direction. By combining equations (5.1.1:22) and (5.1.1:24), the momentum of electromagnetic radiation emitted by a dipole in one cycle at rest in a local frame can be expressed as

$$\mathbf{p}_\lambda = \int_{sphere} \frac{N^2 I_\lambda}{A} \frac{h_0}{\lambda} c\hat{\mathbf{r}} \, dA = \int_{sphere} \frac{m_\lambda c}{A} \hat{\mathbf{r}} \, dA = 0, \qquad (5.3.1:1)$$

where the factors N and X_λ are the intensity and geometry factors of the emitter. The momentum vector integrated over all emission directions is zero, whereas the total substance of electromagnetic energy, the mass equivalence of a cycle of radiation, is equal to the mass equivalence of the electromagnetic energy of one oscillation cycle in the emitter (see equation (5.1.1:24)), Figure 5.3.1:1.

$$m_\lambda = N^2 \left(\frac{z_0}{\lambda}\right)^2 \frac{A}{\chi_\lambda} \frac{h_0}{\lambda} = I_\lambda \frac{h_0}{\lambda} = m_{EM}. \qquad (5.3.1:2)$$

The integrated absolute value of the momenta emitted to space directions is equal to the imaginary momentum of the emitter released by the emitter.

Emission from a plane emitter

In a symmetric bidirectional plane emitter at rest, the vector sum of the momenta of radiation emitted in opposite directions is zero, whereas the scalar sum of the momenta is equal to the scalar value of the momentum related to the electromagnetic energy released by the transmitter

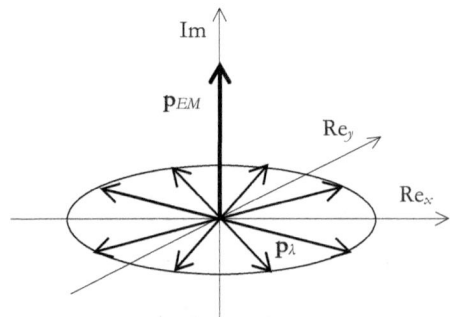

Figure 5.3.1-1. The momentum $\mathbf{p}_{EM}$ of the electromagnetic energy in an emitter at rest in the parent frame appears in the imaginary direction. The momentum of the emitted wave has its momentum $\mathbf{p}_\lambda$ in the direction of the emission in space. An emission event can be described as a turn of the imaginary momentum by 90° into the momentum in space.

$$\mathbf{P}_{\lambda(tot)} = \mathbf{P}_{\lambda(\rightarrow)} + \mathbf{P}_{\lambda(\leftarrow)} = 0 \quad ; \quad \left|\mathbf{p}_{\lambda(\rightarrow)}\right| + \left|\mathbf{p}_{\lambda(\leftarrow)}\right| = \left|\mathbf{p}_{EM}\right|, \tag{5.3.1:3}$$

where

$$\mathbf{P}_{EM} = \mathbf{i}\frac{E_{EM}}{c_0} \tag{5.3.1:4}$$

is the momentum of a cycle of electromagnetic energy being converted into a cycle of radiation at the emission by turning the momentum into spatial directions.

When a plane emitter moves at velocity $\mathbf{v} = \beta\mathbf{c}$ in the direction perpendicular to the emitter plane, the wavelength of radiation emitted in the direction of motion is reduced, and the wavelength of radiation emitted in the opposite direction is increased due to the Doppler effect. With reference to equations (5.2.3:3), (5.3.1:1), and (5.3.1:2), the corresponding momenta of a cycle of radiation in each direction are

$$\mathbf{P}_{(\rightarrow)} = \tfrac{1}{2}m_{\lambda,rest(0)}\frac{\sqrt{1-\beta^2}}{(1-\beta)}\mathbf{c} = \frac{\tfrac{1}{2}m_{\lambda,rest(\beta)}}{(1-\beta)}\mathbf{c} \tag{5.3.1:5}$$

in the direction of $\hat{\mathbf{c}}$ and

$$\mathbf{P}_{(\leftarrow)} = -\tfrac{1}{2}m_{\lambda,rest(0)}\frac{\sqrt{1-\beta^2}}{(1+\beta)}\mathbf{c} = -\frac{\tfrac{1}{2}m_{\lambda,rest(\beta)}}{(1+\beta)}\mathbf{c} \tag{5.3.1:6}$$

in the direction opposite to $\mathbf{c}$, Figure 5.3.1-2.

In equations (5.3.1:5) and (5.3.1:6) $m_{\lambda,rest(0)}$ means the mass equivalence of radiation emitted by the emitter at rest in the parent frame

$$2\cdot\tfrac{1}{2}m_{\lambda,rest(0)} = m_{\lambda,rest(0)} = m_{EM(0)}, \tag{5.3.1:7}$$

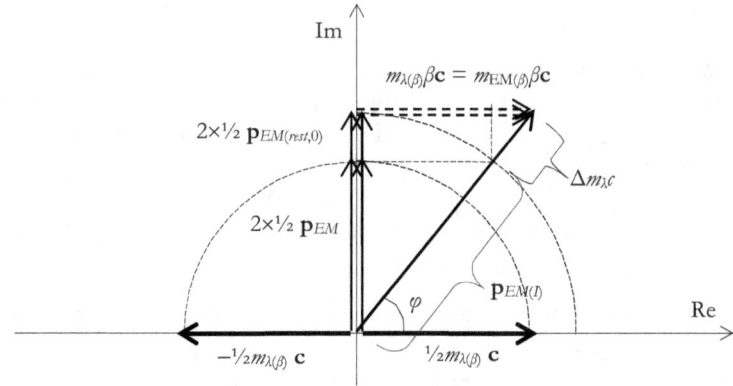

Figure 5.3.1-2. The rest momentum $\mathbf{p}_{EM(0)}$ and the internal momentum $\mathbf{p}_{EM(I)}$ of electromagnetic energy in a plane transmitter moving at velocity β in a parent frame. The momentum of plane waves emitted in the emitter frame (moving with the emitter) is the internal momentum of the emitter. In the direction of motion of the emitter, the total momentum of the electromagnetic energy, $\mathbf{p}_{net(\beta)}$, is formally the momentum related to the "relativistic mass" equivalence of the radiation emitted.

and $m_{\lambda,rest(\beta)}$ is the mass equivalence of radiation emitted in the emitter frame by the emitter moving at velocity β in the parent frame

$$m_{\lambda,rest(\beta)} = m_{rest(0)}\sqrt{1-\beta^2} = m_{EM(0)}\sqrt{1-\beta^2}. \tag{5.3.1:8}$$

Multiplication of the numerator and denominator in equations (5.3.1:5) and (5.3.1:6) by a factor $\sqrt{1-\beta^2}$ gives

$$\mathbf{P}_{(\rightarrow)} = \tfrac{1}{2} m_{\lambda,rest(0)} \frac{1-\beta^2}{(1-\beta)\sqrt{1-\beta^2}} \mathbf{c} = \tfrac{1}{2} \frac{m_{\lambda(0)}}{\sqrt{1-\beta^2}}(1+\beta)\mathbf{c}, \tag{5.3.1:9}$$

and

$$\mathbf{P}_{(\leftarrow)} = -\tfrac{1}{2} \frac{m_{\lambda(0)}}{\sqrt{1-\beta^2}}(1-\beta)\mathbf{c}. \tag{5.3.1:10}$$

By combining equations (5.3.1:9) and (5.3.1:10), the net momentum of radiation emitted by a moving plane emitter can now be expressed as

$$\begin{aligned}\mathbf{P}_{net(\beta)} &= \mathbf{P}_{(\rightarrow)} + \mathbf{P}_{(\leftarrow)} = \frac{m_{\lambda(0)}}{\sqrt{1-\beta^2}}\beta\mathbf{c} = m_{\lambda(\beta)}\beta\mathbf{c} \\ &= \left(m_{\lambda(0)} + \Delta m_\lambda\right)\beta\mathbf{c}.\end{aligned} \tag{5.3.1:11}$$

The mass equivalence of electromagnetic radiation $m_{\lambda(\beta)}$ due to the motion of the emitter in equation (5.3.1:11) has the form of the relativistic mass of any energy object with rest mass $m_{\lambda(0)}$ put into motion at velocity $\beta \cdot c$ in space

$$m_{\lambda(\beta)} = \frac{m_{\lambda(0)}}{\sqrt{1-\beta^2}} = N^2 X_\lambda \frac{h_0/\lambda_{(0)}}{\sqrt{1-\beta^2}}. \tag{5.3.1:12}$$

The additional mass equivalence Δm_λ in (5.3.1:11) is the mass increase needed to put the emitter into motion (see Section 4.1.2). As shown by equation (5.3.1:5), the rest mass equivalence of the moving energy object is reduced by the motion, like in the case of mass objects moving in their parent frame, as

$$m_{\lambda,rest(\beta)} = m_{\lambda,rest(0)}\sqrt{1-\beta^2} = N^2 X_\lambda \frac{h_0}{\lambda_{(0)}}\sqrt{1-\beta^2}. \tag{5.3.1:13}$$

In the frame moving with the emitter, the momenta of the opposite waves are

$$\mathbf{P}_{rest(\rightarrow)} = \tfrac{1}{2} m_{\lambda,rest(\beta)} \mathbf{c} = \tfrac{1}{2} m_{\lambda,rest(0)}\sqrt{1-\beta^2}\,\mathbf{c} = \tfrac{1}{2} N^2 \frac{h_0}{\lambda_{rest(\beta)}} \mathbf{c}, \tag{5.3.1:14}$$

and

$$\mathbf{P}_{rest(\leftarrow)} = -\tfrac{1}{2} m_{\lambda,rest(\beta)} \mathbf{c} = -\tfrac{1}{2} m_{\lambda(0)}\sqrt{1-\beta^2}\,\mathbf{c} = -\tfrac{1}{2} N^2 \frac{h_0}{\lambda_{rest(\beta)}} \mathbf{c}, \tag{5.3.1:15}$$

Mass, mass objects and electromagnetic radiation

with net momentum

$$\mathbf{P}_{rest(\beta)net} = \mathbf{P}_{rest(\to)} + \mathbf{P}_{rest(\leftarrow)} = 0. \qquad (5.3.1{:}16)$$

The net momentum is zero in the moving frame, also when the emitter is moving in its parent frame. Like in the case of mass objects, the state of rest within the moving frame is obtained against a reduction in the absolute values of the rest momenta. The sum of the absolute values of the opposite momenta can be expressed as the rest momentum of the emitter in the imaginary direction

$$\mathbf{P}_{rest(\beta)} = \mathbf{i}\, m_{\lambda,rest(\beta)} c = \mathbf{i}\, m_{\lambda,rest(0)} c \sqrt{1-\beta^2} = \mathbf{i}\, I_\lambda \frac{h_0}{\lambda_{rest(\beta)}} c. \qquad (5.3.1{:}17)$$

When $\beta = 0$, the emitter frame is indistinguishable from the parent frame, and the rest momentum in the emitter frame and in the parent frame is

$$\mathbf{P}_{rest(0)} = \mathbf{i}\, m_{\lambda,rest(0)} c = \mathbf{i}\, I_\lambda \frac{h_0}{\lambda_{rest(0)}} c, \qquad (5.3.1{:}18)$$

or with $I_\lambda = 1$ as it is for an ideal quantum emitter for a single unit charge oscillation

$$\mathbf{P}_{rest(0)} = \mathbf{i}\, m_{0\lambda,rest(0)} c = \mathbf{i}\, \frac{h_0}{\lambda_{rest(0)}} c. \qquad (5.3.1{:}19)$$

In the emitter frame, the emission of electromagnetic radiation can be described as the turn of the rest momentum of the emitter in the imaginary direction into the momentum of radiation in space directions.

5.3.2 Resonator as an energy object

The conclusions drawn regarding the momenta of plane waves are of special interest when applied to a one-dimensional resonator with plane waves propagating in opposite directions. A resonator creates a closed energy object, capturing the radiation into the frame of the emitter, feeding the resonator.

As taught by classical wave mechanics, a resonant superposition of waves in opposite directions produces a standing wave

$$A = 2A_0 \sin 2\pi \frac{r}{\lambda} \cos 2\pi f\, t = 2A_0 \sin kr \cos \omega t, \qquad (5.3.2{:}1)$$

with nodes at $r = n \cdot \lambda/2$. Like the momenta of waves emitted in opposite directions by a plane emitter given in equation (5.3.1:16), the momenta in a resonator have zero vector sum but non-zero scalar sum

$$\mathbf{P}_{int(\leftrightarrow)} = \mathbf{P}_{\lambda(\to)} + \mathbf{P}_{\lambda(\leftarrow)} = 0 \; ; \quad \left|\mathbf{P}_{\lambda(\to)}\right| + \left|\mathbf{P}_{\lambda(\leftarrow)}\right| = p_{\lambda(\leftrightarrow)} = p_{rest(\beta)}, \qquad (5.3.2{:}2)$$

where $p_{\lambda(\leftrightarrow)} = p_{\lambda,rest(\beta)}$ is the sum of the absolute values of the momenta in opposite directions, which is the rest momentum of the electromagnetic energy in the resonator, Figure 5.3.2-1.

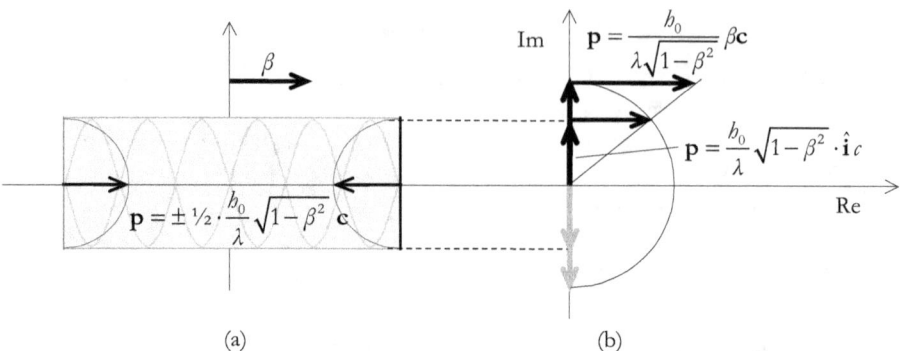

Figure 5.3.2-1. (a) An electromagnetic resonator can be studied as an energy object or a closed energy system with rest mass equal to the sum of the mass equivalences of the waves in opposite directions.

Due to the nature of the fourth dimension as the symmetry sum of vector quantities in space, the total momentum of a resonator can be described as momentum in the fourth dimension. Such a conclusion can also be drawn from the study of the central force effect in the fourth dimension due to motion at the velocity of light in space (see Section 4.1.8).

When the resonator is in motion in the direction of its longitudinal axis, the sum of the momenta of the opposite waves observed in the parent frame is the wave carrying the momentum of the resonator (5.3.1:11)

$$\mathbf{P}_{net(\beta)} = \mathbf{P}_{\lambda(\rightarrow)} + \mathbf{P}_{\lambda(\leftarrow)} = m_{\lambda(\beta)}\beta\mathbf{c} = \frac{m_{\lambda(0)}}{\sqrt{1-\beta^2}}\beta\mathbf{c} \;, \tag{5.3.2:3}$$

where βc is the velocity of the resonator in its parent frame, and mass $m_{\lambda(\beta)}$ means the mass equivalence of the electromagnetic energy in the resonator as an energy object with rest mass equivalence $m_{\lambda(0)}$.

The momentum (5.3.2:3) of the moving resonator in the parent frame can be expressed as

$$\mathbf{P}_{net(\beta)} = m_{\lambda(\beta)}\beta\mathbf{c} = \frac{m_{\lambda(0)}}{\sqrt{1-\beta^2}}\beta\mathbf{c} = \frac{h_0}{\lambda_{(0)}\sqrt{1-\beta^2}}\beta\mathbf{c} = \frac{h_0}{\lambda_{(\beta)}}\beta\mathbf{c} \;, \tag{5.3.2:4}$$

where $\lambda_{(\beta)}$ is the wavelength related to the net momentum of the resonator moving at velocity βc in the local frame, Figure 5.3.2-2.

In a resonator moving at velocity βc in the direction of the longitudinal axis, the internal frequency $f_{I(\beta)}$ of radiation can be interpreted as the frequency of radiation with external wavelength

$$\lambda_{ext} = \lambda_{I(0)} = \lambda_{I(\beta)}(1-\beta_r) \tag{5.3.2:5}$$

propagating at velocity $c(1-\beta)$ in the parent frame, or the frequency of radiation with the internal wavelength $\lambda_{I(\beta)}$ propagating at velocity c in the frame moving with the resonator at velocity βc

Mass, mass objects and electromagnetic radiation

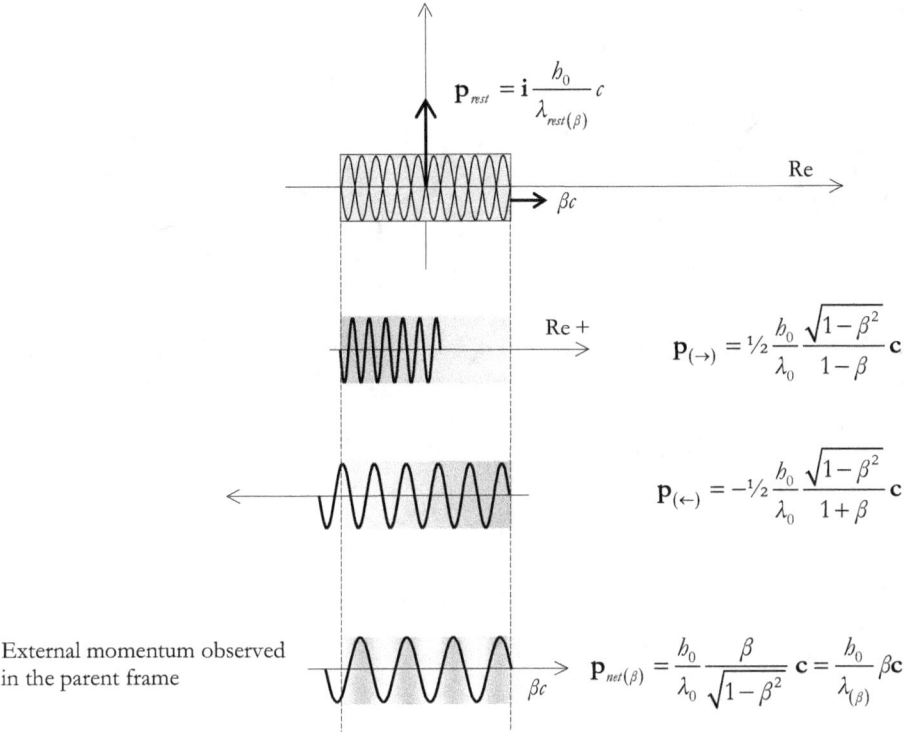

Figure 5.3.2-2. The sum of the absolute values of the momenta within a resonator is described as the rest momentum in the imaginary direction. The net momentum of the resonator in the direction of the motion in the parent frame is the sum of the Doppler-shifted front and back waves. The sum wave propagates at velocity $\beta\mathbf{c}$ in parallel with the resonator in the parent frame.

$$f_{I(\beta)} = \frac{c(1-\beta_r)}{\lambda_{I(0)}} = \frac{c(1-\beta_r)}{\lambda_{I(\beta)}(1-\beta_r)} = \frac{c}{\lambda_{I(\beta)}}, \qquad (5.3.2{:}6)$$

where $\lambda_{I(0)}$ is the internal wavelength of the resonator at rest in the local frame.

The phase velocity c relevant to the internal wavelength λ_I and internal frequency f_I is independent of the velocity of the resonator frame and equal to the local imaginary velocity of light

$$c_\varphi = f_{I(\beta)} \cdot \lambda_{I(\beta)} = c = c_0 \prod_i (1-\delta_i) \ . \qquad (5.3.2{:}7)$$

When studied as a closed energy system, momenta in opposite directions in a resonator result in radiation pressure at the reflectors. In a physical resonator, the recoil due to the radiation pressure at the opposite ends of the resonator is compensated through tension and an excited state in the chemical bonds between atoms in the resonator body.

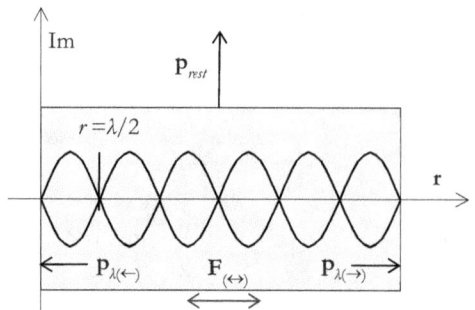

Figure 5.5.2-3. A resonator can be described as an energy object with the mass equivalence of the electromagnetic radiation in the standing wave. The radiation pressure inside the resonator results in a tension in the mechanical structure of the resonator.

A resonator as an energy frame, or energy object, comprises radiation as the carrier of the rest energy, and the resonator body that defines the physical dimensions of the energy object, Figure 5.3.2-3.

Waves carrying the opposite momenta of equation (5.3.2:2) in the resonator frame can be expressed in the form

$$A = A_0 \{\sin[\omega_I t + k_I r] - \sin[\omega_I t - k_I r]\}, \quad (5.3.2:8)$$

resulting in a standing wave through superposition

$$A = 2A_0 \sin 2\pi \frac{r}{\lambda_I} \cos 2\pi f_I t = 2A_0 \sin k_I r \cos \omega_I t \quad (5.3.2:9)$$

with zero amplitude nodes at

$$r = \tfrac{1}{2}n\lambda_I = L \quad (5.3.2:10)$$

Including the effect of all the parent frames, the wavelength and frequency of a resonator are

$$\lambda_{(\delta,\beta)} = \lambda_0 \Big/ \prod_{i=0}^{n}\left[\sqrt{1-\beta_i^2}\right], \quad (5.3.2:11)$$

and

$$f_{(\delta,\beta)} = f_0 \prod_{i=0}^{n}\left[(1-\delta_i)\sqrt{1-\beta_i^2}\right], \quad (5.3.2:12)$$

where λ_0 and f_0 are the wavelength and frequency of the emitter at rest in hypothetical homogeneous space (see Section 5.1.4 for derivation).

With reference to equation (5.1.4:19), atomic dimensions are functions of the velocity of the atom in the local energy frame. Accordingly, the length of the resonator L in equation (5.3.2:10) is subject to "swelling" due to the velocity of the resonator in its parent frames. Thus

$$L = L_0 \Big/ \prod_{i=0}^{n}\left[\sqrt{1-\beta_i^2}\right], \quad (5.3.2:13)$$

where L_0 is the length of the resonator at rest in hypothetical homogeneous space.

As shown by equations (5.3.2:11) and (5.3.2:13), the internal wavelengths and dimensions of a resonator increase equally along with the velocity of the resonator. Accordingly, the resonance condition and the number of nodes in a standing wave in a resonator are independent of the velocity of the resonator in the local frame, as well as independent of the velocities of the local frame in all the parent frames. Substituting equations (5.3.2:11) and (5.3.2:13) into equation (5.3.2:10), the number of half-waves in a resonator can be expressed as

$$n = \frac{2L}{\lambda_{I(\beta)}} = \frac{2L_0}{\lambda_{I(0)}} \frac{\prod_{i=0}^{n}\sqrt{1-\beta_i^2}}{\prod_{i=0}^{n}\sqrt{1-\beta_i^2}} = \frac{2L_0}{\lambda_{I(0)}}.$$ (5.3.2:14)

The resonance condition in equation (5.3.2:14) is independent of the direction of the resonator relative to its velocity in the local frame and in the parent frames. As shown by equations (5.3.2:11) and (5.3.2:12), the internal wavelength and the internal frequency are also independent of the direction of the resonator relative to its velocity in the parent frame.

5.3.3 Momentum of spherical emitter

In previous chapters, the momentum of radiation was studied in the case of plane waves from planar sources and in a one-dimensional resonator. The net momentum of radiation from a bidirectional planar emitter is zero when the emission direction is perpendicular to the motion of the emitter, and it has the form of the momentum of any mass object when the emission occurs in the direction of the motion.

In the case of an isotropic spherical source like a stellar radiation source on a macroscopic scale, the momentum of radiation from a surface differential dA can be expressed as

$$d\mathbf{p} = \frac{m_{EM}c\sqrt{1-\beta^2}}{4\pi r^2} \frac{2\pi r \sin\varphi \, r d\varphi}{1-\beta\cos\varphi} \hat{\mathbf{r}}_\varphi = \frac{m_{EM}c\sqrt{1-\beta^2}}{2} \frac{\sin\varphi}{1-\beta\cos\varphi} d\varphi \, \hat{\mathbf{r}}_\varphi ,$$ (5.3.3:1)

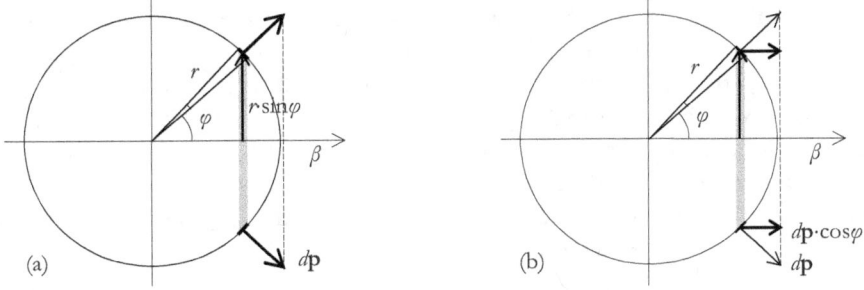

Figure 5.3.3-1. (a) Calculation of the momentum of radiation emitted by an isotropic spherical source. (b) Calculation of momentum in the direction of velocity β.

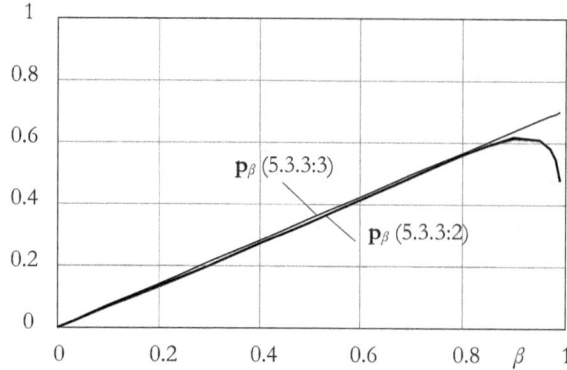

Figure 5.3.3-2. The momentum of radiation emitted by an isotropic spherical source according to equations (5.3.3:2) and (5.3.3:3).

where $dA = 2\pi r^2 \sin\varphi \, d\varphi$ is a spherical surface differential with its symmetry axis in the direction of velocity β, Figure 5.3.3-1.

Due to the symmetry, only the component of momentum $d\mathbf{p}$ in the direction of β, $d\mathbf{p}_\beta = d\mathbf{p} \cos\varphi$, contributes to the total momentum, which is obtained by integration

$$\mathbf{p}_\beta = \frac{m_{EM} c \sqrt{1-\beta^2}}{2} \int_0^\pi \frac{\sin\varphi \cos\varphi}{1-\beta \cos\varphi} d\varphi \, \hat{\mathbf{r}}_\beta, \qquad (5.3.3:2)$$

where β is the velocity of the emitter in the local frame. The integral in equation (5.3.3:2) cannot be solved in a closed form, but in a wide range of β ($0 < \beta < 0.85$) is close to a linear function of β, Figure 5.3.3-2

$$\mathbf{p}_\beta \approx \frac{m_{EM} \beta c}{\sqrt{2}} \hat{\mathbf{r}}_\beta. \qquad (5.3.3:3)$$

5.3.4 Mass object as a standing wave structure

A mass object in the DU framework can be described as a standing wave structure or resonator. A mass object at rest in an energy frame in space is described as a resonator hosting a standing wave at the Compton wavelength, which is the wavelength equivalence of the rest mass of the object. As discussed in Section 5.3.2, in the case of electromagnetic resonators, the momentum of a standing wave structure in space is zero due to the cancellation of the opposite momenta of the waves in opposite directions, but appears as momentum in the fourth dimension, just like the rest momentum of a mass object.

A standing wave structure for describing a mass object possesses complete symmetry in all space directions. Such a structure can be demonstrated with a one-dimensional resonator in any space direction, Figure 5.3.4-1.

When a "mass wave resonator" is put into motion at velocity βc in space, the rest wave number k_{rest} is reduced [see equations (4.1.3:10) and (5.3.1:8)] as

$$\mathbf{P}_{rest(\beta)} = \mathbf{i} \, mc\sqrt{1-\beta^2} = \mathbf{i} \, \hbar_0 k_{rest(0)} c \sqrt{1-\beta^2}, \qquad (5.3.4:1)$$

and the momentum in the direction of motion in space becomes (see equation (4.1.2:14))

Mass, mass objects and electromagnetic radiation

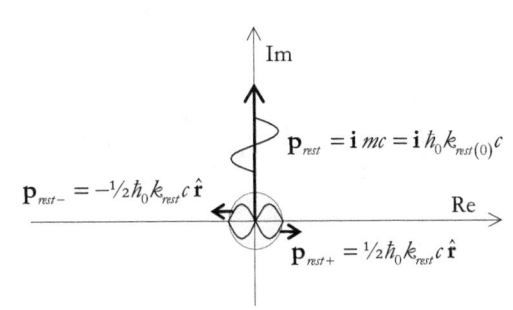

Figure 5.3.4-1. A mass object at rest in space is described as a one-dimensional resonator with momentum in the fourth dimension.

$$\mathbf{p}_{\hat{r}} = \frac{mv}{\sqrt{1-\beta^2}}\hat{\mathbf{r}} = \frac{\hbar_0 k_{rest(0)}}{\sqrt{1-\beta^2}}\beta c\,\hat{\mathbf{r}} = \hbar_0 k_{(\beta)}\beta c\,\hat{\mathbf{r}}, \qquad (5.3.4:2)$$

which is equal to the net momentum, as the sum of the Doppler-shifted front wave and back wave of the opposite rest waves in a one-dimensional resonator, Figure 5.3.4-2

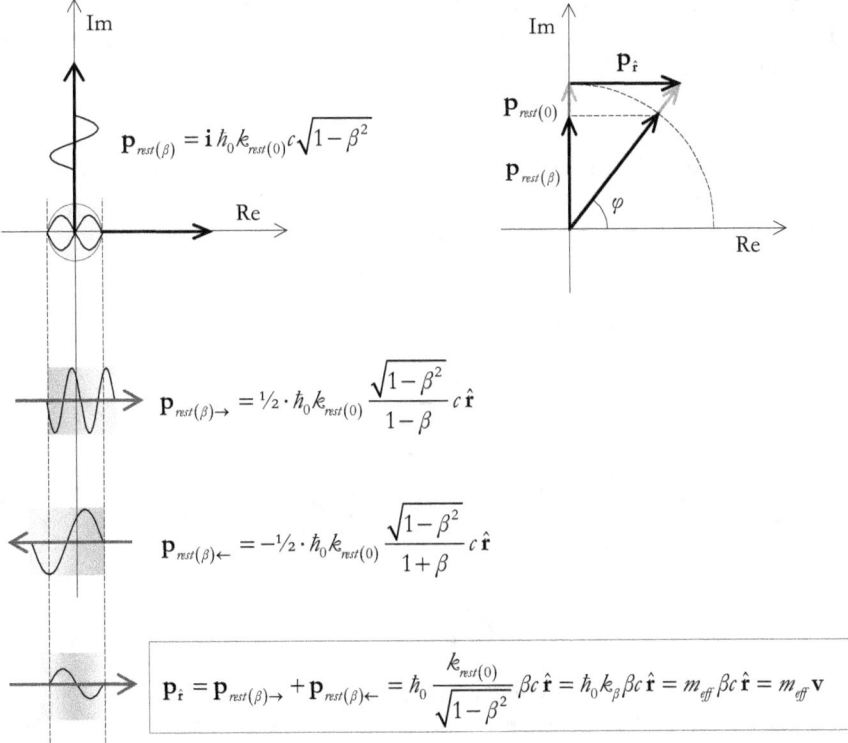

Figure 5.3.4-2. A mass object moving at velocity βc in space has its rest momentum in the imaginary direction and its momentum in space in the direction of motion. The momentum in space is the net momentum of the Doppler-shifted front wave back wave observed in the frame where the object is moving at velocity βc.

$$\mathbf{P}_{\hat{r}} = \tfrac{1}{2}\hbar_0 \frac{k_{rest(\beta)}}{1-\beta} c\hat{r} - \tfrac{1}{2}\hbar_0 \frac{k_{rest(\beta)}}{1+\beta} c\hat{r} = \hbar_0 k_{(\beta)} \beta c\hat{r} ,\qquad(5.3.4\text{:}3)$$

The resulting wave number $k_{(\beta)}$ in (5.3.4:3) is the wave number equivalence of the relativistic mass as shown in the derivation in equations (5.3.1:9–11) and (5.3.2:3). A message of equation (5.3.4:3) is that a moving object has the nature of a mass wave moving at velocity βc in a local frame.

5.3.5 The double slit experiment

An important difference between the wave presentation of the momentum of a mass object in standard quantum mechanics and in the DU comes from the removal of c from the Planck constant discussed in Section 5.1.2. In standard quantum mechanics, the momentum of a mass object is expressed in terms of de Broglie wavelength $\lambda = h/p$.

$$\mathbf{P}_r = \frac{h}{\lambda_{dB}} \hat{r} = \hbar k_{dB} \hat{r} .\qquad(5.3.5\text{:}1)$$

Equation (5.3.5:1) assumes implicitly that the propagation velocity of the de Broglie wave is the velocity of light, which is illustrated by replacing the Planck constant h with the intrinsic Planck constant h_0 or $\hbar_0$,

$$\mathbf{P}_r = \hbar_0 k_{dB} \cdot c\hat{r} .\qquad(5.3.5\text{:}2)$$

The mass wave with wave number $k_{(\beta)}$ in equation (5.3.4:3) creates the momentum of the object with mass $m+\Delta m$ propagating at velocity βc. In the DU framework, the wave presentation of the momentum $\mathbf{p}_r$ in (5.3.5:2) is expressed

$$\mathbf{P}_r = \hbar_0 k_{dB} \cdot c\hat{r} = \hbar_0 k_{(\beta)} \cdot \beta c\hat{r} ,\qquad(5.3.5\text{:}3)$$

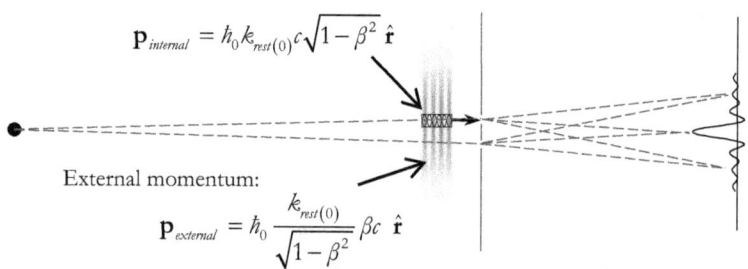

Figure 5.3.5-1. A mass object as a standing wave structure (drawn in the direction of the real axis). The momentum of the object moving at velocity βc is the external momentum as the sum of the Doppler shifted front and back waves, which can be described as the momentum of a wave front propagating in the local frame in parallel with the propagating mass object. The interference pattern observed in the double slit experiment demonstrates the momentum as a wave front by resulting in deflection of the propagation path observed as an interference pattern between wave fronts passing the two different slits.

which can be equally interpreted as the result of a wave with de Broglie wave number propagating at the velocity of light or a wave with wave number $k_{(\beta)}$ propagating at the velocity of the moving object. The wave number $k_{(\beta)}$ is related to the de Broglie wavelength as

$$k_{(\beta)} = \beta \cdot k_{dB}. \tag{5.3.5:4}$$

The DU interpretation of the momentum of a moving mass object as the momentum of a wave front is of special interest for understanding the double slit experiment [70]. Figure 5.3.5-1 illustrates the momentum of an object as a co-moving wave front in the vicinity of the moving object.

5.3.6 Planck units in the DU framework

In the DU framework, where $\hbar_0 = h/c$, the Planck mass and the Planck distance obtain the forms

Planck mass $\qquad m_0 = \sqrt{\dfrac{c_0^2}{G} \hbar_0} = 5.4556 \cdot 10^{-8} \qquad$ [kg] $\qquad (5.3.6:1)$

Planck distance $\qquad r_0 = \sqrt{\dfrac{G}{c_0^2} \hbar_0} = 4.05 \cdot 10^{-35} \qquad$ [m]. $\qquad (5.3.6:2)$

Multiplication of (5.3.6:1) with (5.3.6:2) gives

$$m_0 = \frac{\hbar_0}{r_0}, \tag{5.3.6:3}$$

showing the Planck distance r_0 as the wavelength equivalence of the Planck mass

$$r_0 = \lambda_{(m_0)} = \frac{\hbar_0}{m_0}. \tag{5.3.6:4}$$

Equation (5.3.6:4) is of special interest as the basis for the buildup of elementary particles as sub-harmonics of the wavelength equivalence of the Planck mass, as suggested by Ari Lehto [71].

In the DU framework, the ratio G/c_0^2 can be expressed in terms of the mass equivalence and the 4-radius of spherically closed space as

$$\frac{c_0^2}{G} = \frac{M''}{R_4}. \tag{5.3.6:5}$$

Substitution of (5.3.6:5) into (5.3.6:1) and (5.3.6:2) relates the Planck mass and the Planck distance to M'' and R_4 as

$$m_0 = \frac{r_0}{R_4} M'', \tag{5.3.6:6}$$

and

$$r_0 = \frac{m_0}{M''} R_4, \qquad (5.3.6{:}7)$$

respectively. Each one of the equations shows that

$$\frac{m_0}{M''} = \frac{r_0}{R_4} = U, \qquad (5.3.6{:}8)$$

where the common ratio is denoted as U. Substituting U back into (5.3.6:6) and (5.3.6:7), we get

$$m_0 = U \cdot M'' \qquad (5.3.6{:}9)$$

and

$$r_0 = U \cdot R_4, \qquad (5.3.6{:}10)$$

which shows that the Planck mass is related to the mass equivalence of the whole space in the same way as the Planck distance is related to the 4-radius of space.

5.4 Propagation of electromagnetic radiation in local frames

Dents in space around mass centers and the reduction of the velocity of light in the dents affect the propagation time and direction of light and radio signals passing a mass center. Prediction for the bending of light in the DU framework is essentially the same as they are in the framework of general relativity. The prediction for the Shapiro delay in the DU is slightly different from that in general relativity. In the cases confirmed with measurements, however, the predictions of the two theories are equal.

5.4.1 Shapiro delay in a local gravitational frame

The propagation of light near mass centers is affected both by the reduction of the velocity of light and by the lengthening of the propagation path due to the geometry of the fourth dimension. And since the two factors work in the same direction, the traveling time of light increases when light passes a mass center in space.

The propagation time of light from location A to location B in a gravitational frame can be generally expressed by the equation

$$T_{A,B} = \int_A^B dt = \int_A^B \frac{dx}{c}, \qquad (5.4.1:1)$$

where the local velocity of light c and the radial component of the distance differential dx are functions of the distance from the local mass center in the gravitational frame studied, Figure 5.4.1-1. For studying the effect of the curvature of space and the associated change in the velocity of light, it is useful to express the propagation time in terms of a hypothetical propagation time in hypothetical homogeneous space, followed by a correction due to the curvature of local space and the associated change in the velocity of light.

$$T_{A,B} = \int_A^B \left[\frac{dx_{0\delta}}{c_{0\delta}} + d\left(\frac{dx}{c}\right) \right] = T_{(A,B)0\delta} + \Delta T_{A,B} = T_{(A,B)0\delta} + \int_A^B d\left(\frac{dx}{c}\right). \qquad (5.4.1:2)$$

The first term in (5.4.1:2) is the propagation time based on the flat space distance from A to B, and the velocity of light in the hypothetical homogeneous space of the local gravitational frame. The second term in (5.4.1:2) shows the increase in propagation time resulting from the increased distance due to the curvature of space between locations A and B, and the local velocity of light along the propagation path. To solve the second term, we first write it in the form

$$\Delta T_{A,B} = \int_A^B \Delta\left(\frac{dx}{c}\right) = \int_A^B \left[\frac{\Delta(dx)}{c} - \left(\frac{\Delta c}{c^2} dx\right) \right], \qquad (5.4.1:3)$$

where

$$\Delta(dx) = dx - dx_{0\delta}, \qquad (5.4.1:4)$$

and

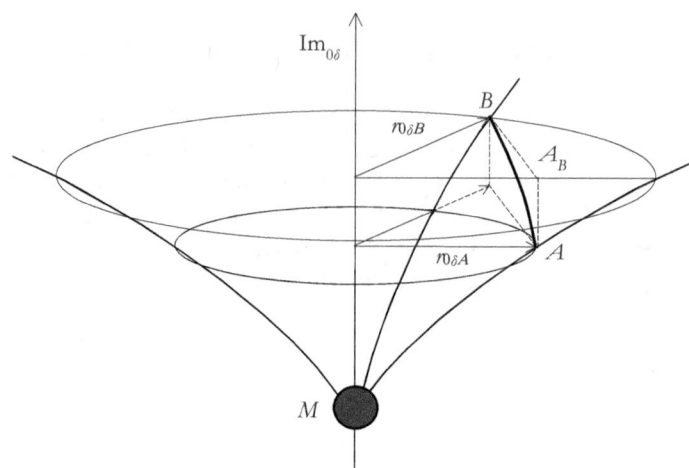

Figure 5.4.1-1. The light path AB from location A to location B follows the shape of the dent in space as a geodesic line in the gravitational frame of mass center M. Point A is at flat space distance $r_{0\delta A}$, and point B is at flat space distance $r_{0\delta B}$ from mass center M. Point A_B is the flat space projection of point A on the flat space plane crossing point B. Line A_BB is the distance between A and B as it would be without the dent. The velocity of light in the dent is reduced in proportion to $1/n_\delta$. i.e., the velocity of light at A is lower than the velocity of light at B. The distance AB_A is the projection of the path AB on the flat space plane.

$$\frac{\Delta c}{c} \approx \frac{\Delta c}{c_{0\delta}} = \frac{c - c_{0\delta}}{c_{0\delta}} = -\delta. \tag{5.4.1:5}$$

Substitution of (5.4.1:4) and (5.4.1:5) into (5.4.1:3) gives

$$\Delta T_{A,B} = \frac{1}{c_{0\delta}} \int_A^B \left[(dx - dx_{0\delta}) + \delta \cdot dx \right]. \tag{5.4.1:6}$$

To solve the distance term in the integrand in equation (5.4.1:6), the distance differential dx in the actual propagation path in tilted space is expressed

$$dx = \sqrt{dx_{0\delta}^2 + dz^2}, \tag{5.4.1:7}$$

where dz is the differential in the direction of the $\mathrm{Im}_{0\delta}$ perpendicular to the flat space plane

$$dz = dr \sin \psi = dr_{0\delta} \tan \psi, \tag{5.4.1:8}$$

where $dr_{0\delta}$ is the flat space projection of distance differential dr, Figure 5.4.1-2(a).

Angle a is the angle between the flat space projection of the path and the tangential direction perpendicular to radius $r_{0\delta}$, which allows the expression of $dr_{0\delta}$ as (Fig. 5.4.1-2(b,c))

$$dr_{0\delta} = dx_{0\delta} \sin a. \tag{5.4.1:9}$$

Substitution of (5.4.1:4) into (5.4.1:3) gives

Figure 5.4.1-2.

(a)

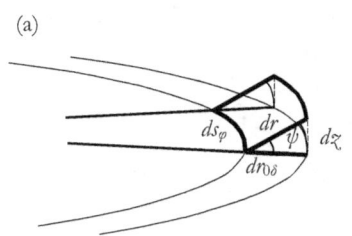

(a) Effect of the curvature of space on the distance differential dr. There is no lengthening in the distance differential ds_φ in the direction perpendicular to the radius.

(b)

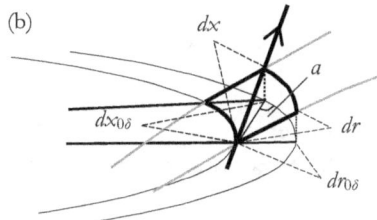

(b) Distance differential dx in the direction of a light beam at an angle a to the tangential direction on the flat space plane.

(c)

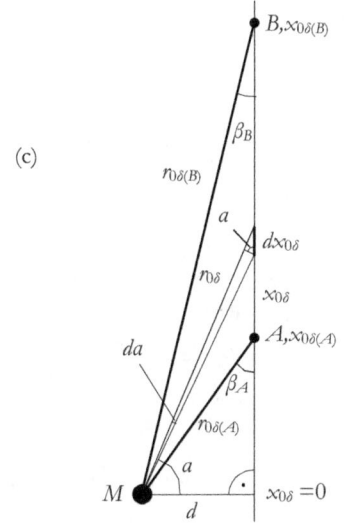

(c) Notation of distances in the analysis of the signal delay between points A and B in a gravitational frame around mass M. Distances $x_{0\delta(A)}$ and $x_{0\delta(B)}$ are the flat space distances from A and B to point $x = 0$, which is the shortest distance between line AB and mass center M.

Observe that

$$\sin a = \frac{r_{0\delta}}{x_{0\delta}} = \frac{dr_{0\delta}}{dx_{0\delta}}$$

and

$$\cos a = \frac{r_{0\delta} da}{dx_{0\delta}}$$

$$dx = \sqrt{dx_{0\delta}^2 + dr_{0\delta}^2 \tan^2\psi} = dx_{0\delta}\sqrt{1+\sin^2 a\left(\frac{1-\cos^2\psi}{\cos^2\psi}\right)}. \qquad (5.4.1{:}10)$$

Assuming $\psi \ll 1$, which means that distance $r_{0\delta} \gg r_c$ and $\cos\psi \approx 1$, the cosine term can be approximated as

$$\frac{1-\cos^2\psi}{\cos^2\psi} = \frac{(1-\cos\psi)(1+\cos\psi)}{\cos^2\psi} \approx 2(1-\cos\psi) = 2\delta, \qquad (5.4.1{:}11)$$

and (5.4.1:10) becomes

$$dx = dx_{0\delta}\sqrt{1+2\delta\sin^2 a} \approx dx_{0\delta}\left(1+\delta\sin^2 a\right). \qquad (5.4.1{:}12)$$

Substitution of (5.4.1:12) and into (5.4.1:6) gives

$$\Delta T_{A,B} = \frac{1}{c_{0\delta}}\int_A^B\left[\delta\sin^2 a + \delta\left(1+\delta\sin^2 a\right)\right]dx_{0\delta}$$

$$\approx \frac{1}{c_{0\delta}}\int_A^B \delta\left(\sin^2 a + 1\right)dx_{0\delta}, \qquad (5.4.1{:}13)$$

where the last form is obtained by neglecting the $\delta^2\sin^2 a$ term ($\delta \ll 1$).

The distance differential $dx_{0\delta}$ in (5.4.1:13) can be expressed in terms of $r_{0\delta}$ and the angle differential da, Fig. 5.4.1-2(c)

$$dx_{0\delta} = \frac{r_{0\delta}da}{\cos a}. \qquad (5.4.1{:}14)$$

Substitution of (5.4.1:14) for $dx_{0\delta}$ and (4.1.1:30) for δ in (5.4.1:13) results

$$\Delta T_{(A-B)} = \frac{GM}{c_{0\delta}^3}\int_A^B \frac{1}{r_{0\delta}}\left(\sin^2 a + 1\right)\frac{r_{0\delta}da}{\cos a} \approx \frac{GM}{c_{0\delta}^3}\int_A^B\left(\frac{1}{\cos a} - \cos a + \frac{1}{\cos a}\right)da$$

$$\approx \frac{GM}{c_{0\delta}^3}\left[2\int_A^B \frac{da}{\cos a} - \int_A^B \cos a\, da\right], \qquad (5.4.1{:}15)$$

where the velocity of light is $c_B \approx c_{0\delta}$. From equation (5.4.1:14), we get

$$\frac{da}{\cos a} = \frac{dx_{0\delta}}{r_{0\delta}} = \frac{dx_{0\delta}}{\sqrt{d_{0\delta}^2 + x_{0\delta}^2}} \approx \frac{dx}{\sqrt{d^2 + x^2}}, \qquad (5.4.1{:}16)$$

where $d_{0\delta} \approx d$ is the shortest distance between M and the light propagation path (Fig. 5.4.1-2(c)). Substitution of (5.4.1:16) into the first integral in (5.4.1:15) results in

$$\Delta T_{A,B} = \frac{GM}{c_{0\delta}^3}\left[2\int_A^B \frac{dx}{\sqrt{d^2+x^2}} - \int_A^B \cos a\, da\right]$$

$$= \frac{GM}{c_{0\delta}^3}\left\{2\ln\left[\frac{x_B + \sqrt{d^2+x_B^2}}{x_A + \sqrt{d^2+x_A^2}}\right] - \left[\sin a_B - \sin a_A\right]\right\}. \qquad (5.4.1{:}17)$$

The algebraic form of sin a is (Fig. 5.4.1-2(c))

$$\sin a = \frac{r_{0\delta}}{x_{0\delta}} \approx \frac{r}{x}, \qquad (5.4.1{:}18)$$

which allows the expression of (5.4.1:17) in the form

$$\Delta T_{A,B} = \frac{GM}{c_{0\delta}^3}\left\{2\ln\left[\frac{x_B+r_B}{x_A+r_A}\right] - \left[\frac{x_B}{r_B}-\frac{x_A}{r_A}\right]\right\}. \qquad (5.4.1{:}19)$$

The last term, which is missing from the corresponding GR prediction, is due to the fact that distance is increased only in the radial direction; the tangential component in the light path is unaffected by the dent in space.

Equation (5.4.1:19) applies for $x_A \geq 0$ and $x_B > 0$. When $d = 0$, line AB passes through M and $r_A = x_A$ and $r_B = x_B$, i.e., the propagation path is in the radial direction, and since $c_{0\delta} \approx c$, equation (5.4.1:19) obtains the form

$$\Delta T_{A,B} = \frac{2GM}{c^3}\ln\left[\frac{r_B}{r_A}\right], \qquad (5.4.1{:}20)$$

which is identical with the corresponding prediction in general relativity (because there is no tangential component in the propagation path). For $x_A < 0$, we apply equation (5.4.1:19) separately from 0 to x_B and from $-x_A$ to 0 (Figure 5.4.1-3)

$$\Delta T_{A,B} = \frac{GM}{c_{0\delta}^3}\left\{2\ln\left[\frac{x_B+r_B}{d}\right] - \left[\frac{x_B}{r_B}\right] + 2\ln\left[\frac{x_A+r_A}{d}\right] - \left[\frac{x_A}{r_A}\right]\right\}, \qquad (5.4.1{:}21)$$

or

$$\Delta T_{A,B} = \frac{GM}{c_{0\delta}^3}\left\{2\ln\left[\frac{(x_B+r_B)(x_A+r_A)}{d^2}\right] - \left[\frac{x_B}{r_B}+\frac{x_A}{r_A}\right]\right\}. \qquad (5.4.1{:}22)$$

For cosmological observations, it is convenient to express equation (5.4.1:22) in terms of angles β_A and β_B as illustrated in Figure 5.4.1-3

$$\begin{aligned}x_A &= r_A\cos\beta_A\\ x_B &= r_B\cos\beta_B\\ d &= r_A\sqrt{1-\cos^2\beta_A} = r_B\sqrt{1-\cos^2\beta_B}.\end{aligned} \qquad (5.4.1{:}23)$$

Substitution of (5.4.1:23) for x_A, x_B, and d in (5.4.1:22) gives

$$\begin{aligned}\Delta T_{A,B} &= \frac{GM}{c_{0\delta}^3}\left\{2\ln\left[\sqrt{\frac{1+\cos\beta_A}{1-\cos\beta_A}}\sqrt{\frac{1+\cos\beta_B}{1-\cos\beta_B}}\right] - \left(\cos\beta_A+\cos\beta_A\right)\right\}\\ &= \frac{GM}{c_{0\delta}^3}\left\{2\ln\left[\cot\frac{\beta_A}{2}\cot\frac{\beta_B}{2}\right] - \left(\cos\beta_A+\cos\beta_A\right)\right\}.\end{aligned} \qquad (5.4.1{:}24)$$

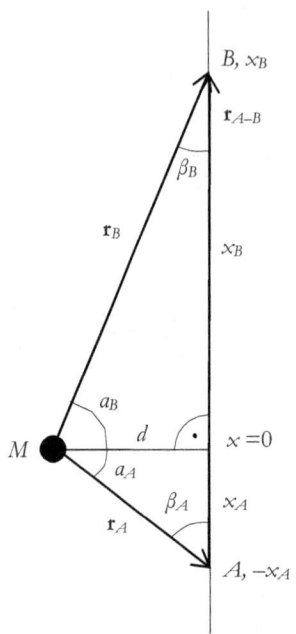

Figure 5.4.1-3. Calculation of ΔT for light propagation from point A to point B when the light path passes the shortest distance at $x = 0$.

Equations (5.4.1:19–24) give the Shapiro delay in the DU framework. In the GR framework, Shapiro delay is expressed without the last term in parentheses, which comes from the fact that the tangential component in the light path is not subject to lengthening.

When $x_A = 0$, then $r_A = d$. If, further, $r_B \gg d$ then $x_B \approx r_B$, and equation (5.4.1:19) obtains the form

$$\Delta T_{A,B} = \frac{GM}{c^3}\left\{2\ln\left[\frac{2r_B}{d}\right] - 1\right\} - \tag{5.4.1:25}$$

When $D_A \gg d$ and $D_B \gg d$ are at different directions from M, the total signal delay given by equation (5.4.1:21) reduces to

$$\Delta T_{D1,D2} = \frac{2GM}{c^3}\left\{\ln\left[\frac{4D_A D_B}{d^2}\right] - 1\right\} \tag{5.4.1:26}$$

(see Figure 5.4.1-4).

Equation (5.4.1:26) is applicable in cases like the experiments with Mariner 6 and 7 spacecrafts (see Section 7.3.4).

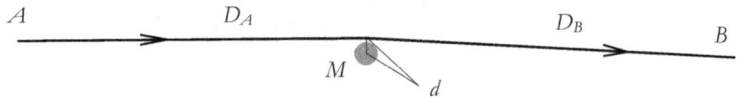

Figure 5.4.1-4. Distances D_A and D_B in equation (5.4.1:25) for the calculation of the delay of a signal traveling from A to B. The signal passes mass center M at distance d.

5.4.2 Shapiro delay in general relativity and in the DU

The prediction for Shapiro delay according to the general theory of relativity is given in the form

GR:
$$\Delta T_{A,B} = \frac{2GM}{c^3} \ln\left[\frac{x_B + r_B}{x_A + r_A}\right]$$
$$= \frac{GM}{c^3} \ln\left[\frac{x_B + r_B}{x_A + r_A}\right]_{path} + \frac{GM}{c^3} \ln\left[\frac{x_B + r_B}{x_A + r_A}\right]_{time},$$
(5.4.2:1)

which is equal to the first term of the Shapiro delay in equation (5.4.1:19). The GR prediction comes from two equal effects: the lengthening of the path, and the gravitational time dilation as demonstrated on the second line of equation (5.4.2:1). In the DU framework, the two effects on the Shapiro delay are different; the effect of the velocity of light (corresponding to gravitational time dilation in GR) affects equally the radial and tangential components of the propagation path. The lengthening of the path due to the curvature of space, however, has an effect only in the radial direction. As the counterpart of the second line in the GR equation (5.4.2:1), the two factors of the Shapiro delay in the DU framework are

DU:
$$\Delta T_{A,B(path)} = \frac{GM}{c_{0\delta}^3}\left\{\ln\left[\frac{x_B + r_B}{x_A + r_A}\right] - \left[\frac{x_B}{r_B} - \frac{x_A}{r_A}\right]\right\}$$
$$\Delta T_{A,B(velocity)} = \frac{GM}{c_{0\delta}^3} \ln\left[\frac{x_B + r_B}{x_A + r_A}\right].$$
(5.4.2:2)

The velocity term in the DU prediction is identical to the time dilation term in the GR prediction, but the lengthening of the path in the DU prediction takes into account the fact that the curvature of space occurs only in the radial component of the path, which results in the second term in the $\Delta t_{A,B(path)}$ in (5.4.2:2).

The difference between the GR prediction and DU prediction disappears when there is no tangential component in the propagation path, i.e., propagation occurs in the radial direction — in a direction towards or from the local mass center, resulting in the curvature ($x_A = r_A$ and $x_B = r_B$). Accordingly, the DU prediction in (5.4.1:20) is equal to the corresponding prediction in the GR.

The GR equation corresponding to the DU prediction (5.4.1:26) is

$$\Delta T_{D1,D2} = \frac{2GM}{c^3} \ln\left[\frac{4D_1 D_2}{d^2}\right].$$
(5.4.2:3)

5.4.3 Bending of light

Derivation of equation (5.4.1:26) results in

$$\frac{\partial(\Delta t)}{\partial d} = -\frac{4GM}{c^3 d},$$
(5.4.3:1)

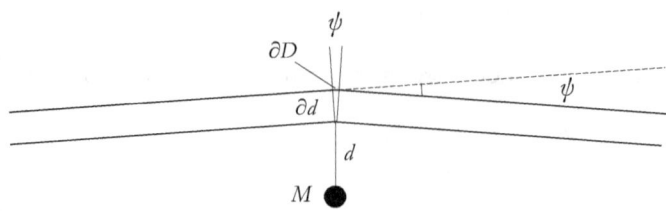

Figure 5.4.3-1. A light ray passing a mass center in space is bent due to a reduced velocity and increased distance close to a mass center.

which gives the difference in the propagation delay in a difference in the shortest distance d to the mass center M, a light ray is passing, Figure 5.4.3-1. The extra distance the outer side of the ray travels in Δt is $\partial D = c \cdot \partial(\Delta t)$, which can be expressed as arc ∂D

$$\partial D = \psi \cdot \partial d = c \cdot \partial(\Delta t) \quad \Rightarrow \quad \psi = \frac{\partial(\Delta t)}{\partial d} \cdot c. \tag{5.4.3:2}$$

Substitution of (5.4.3:1) into (5.4.3:2) gives the bending angle ψ towards the mass center

$$\psi = \frac{4GM}{c^2 d}. \tag{5.4.3:3}$$

The result is the same as the corresponding prediction derived from the general theory of relativity.

Both the Shapiro delay in (5.4.1:26) and the bending of light in (5.4.3:3) assume that the closest distance from the light path to the mass center is much larger than the critical radius.

5.4.4 Measurement of the Shapiro delay

Equation (5.4.1:2) for the total propagation time is written as the sum of the "flat space propagation time" $T_{(A,B)0\delta}$, and the Shapiro correction $\Delta T_{A,B}$ due to the curvature of space

$$T_{A,B} = T_{(A,B)0\delta} + \Delta T_{A,B}. \tag{5.4.4:1}$$

In principle, the flat space propagation time $T_{(A,B)0\delta}$ in (5.4.4:1) is based on the velocity of light in the apparent homogeneous space of the local gravitational frame, $c_{0\delta}$, which relates to the velocity of light at the receiver's location, c_B, as

$$c_{0\delta} = \frac{c_B}{(1-\delta_B)}. \tag{5.4.4:2}$$

In present practice, the velocity of light is fixed to the value measured on the Earth, which is lower than the velocity of light in the apparent homogeneous space of the Earth's gravitational frame or the solar gravitational frame. Also, the unit of time is fixed to the characteristic frequency of a Ce-clock on the Earth's geoid. The frequency of a clock in the apparent homogeneous space of the Earth's gravitational frame is

Mass, mass objects and electromagnetic radiation

$$f_{0\delta} = \frac{f_{Earth}}{(1-\delta_{Poles})} \approx \frac{f_{Earth}}{(1-\delta_{Equator})\sqrt{1-\beta_{Equator}}}. \qquad (5.4.4:3)$$

The two expressions in (5.4.4:3) are essentially equal due to the properties of the Earth's geoid (See Section 7.5). The frequency of a clock is essentially the same at different locations on the Earth. The velocity of light, however, is slightly higher at low latitudes (closer to the equator) because the radius of the Earth is higher at the equator than at the poles. At the poles, the frequency of an Earth clock and the local velocity of light can be related to the frequency of a hypothetical clock moving at velocity β_{Earth} in apparent homogeneous space, and the velocity of light in apparent homogeneous space as

$$\frac{f_{0\delta}}{c_{0\delta}} = \frac{f_B}{(1-\delta_{Poles})} \frac{1-\delta_B}{c_B} = \frac{f_B}{c_B}. \qquad (5.4.4:4)$$

The reading of a clock in apparent homogeneous space for the propagation time of a fixed distance X_{AB} in apparent homogeneous space is

$$N_{AB(0)} = \left(\frac{X_{AB}}{c_{0\delta}} + \Delta T_{AB}\right) f_{0\delta}, \qquad (5.4.4:5)$$

Assuming the local velocity of light for the light propagation, the reading of a clock at the North or South Pole for the same distance X_{AB} becomes

$$N_{AB(B)} = \frac{X_{AB}}{c_B} f_B + \Delta T_X f_B \approx N_{AB(0)}, \qquad (5.4.4:6)$$

which means that for a local observer, the reading of the clock used for the measurement of the propagation is essentially independent of the observer's gravitational state in the local frame.

A famous experiment of the Shapiro delay was performed with the Mariner 6 and 7 spacecrafts on their mission to the planet Mars. In the Mariner 6 and 7 experiments, the signal delay was studied by comparing the delays at different passing distances d between the signal path and the Sun. The observed quantity is the difference in delays at different passing distances d in (5.4.1:26). As a result, the effect of the constant −1 in (5.4.1:26) is ignored, and the experiment does not differentiate between the GR and DU predictions (see Section 7.3.4).

In the illustration of Figure 5.4.4-1, transmitter A is a satellite in its orbit, and receiver B is on the Earth. The velocity of light at the altitude of the satellite is higher than the velocity of light on the Earth.

In satellite communication, the signal path follows the shape of actual space, which means that in calculating the Shapiro effect we only need to take into account the change in the velocity of light between the satellite and the Earth station – i.e. the second term in equation (5.4.2:2). At a satellite's altitude, the velocity of light is higher than the velocity of light on the Earth. Accordingly, compared to the transmission time based on the velocity of light on Earth, the measured transmission time is shortened by a factor

$$\Delta T_{A,B(velocity)} \approx \frac{GM}{c^3} \ln\left[\frac{2h_{satellite}}{r_{Earth}}\right], \qquad (5.4.4:7)$$

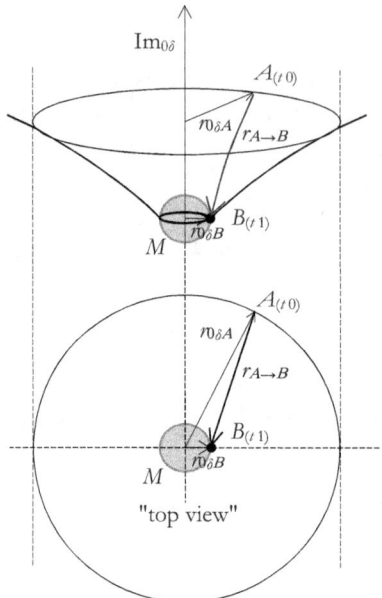

Figure 5.4.4-1. The light path $r_{A\rightarrow B}$ from location A at distance r_A from mass center M to location B on the surface of the mass center M. The upper figure illustrates the light path following the shape of the dent in space in the gravitational frame of mass center M. The path is subject to lengthening due to the "extra" distance in the direction of the $\mathrm{Im}_{0\delta}$ axis. The lower figure is "top view" giving the projection of the path on the apparent homogeneous space plane.

For GPS satellites, the altitude h is about 20,000 km. Applying the Earth mass $M \approx 6 \cdot 10^{24}$, and Earth radius $r_{Earth} \approx 6{,}400$ km, the Shapiro effect is about 0.03 ns, which corresponds to about 9 mm in distance, i.e., the actual distance to the satellite is 9 mm longer than that calculated from the velocity of light on the Earth.

In the case of Earth satellites, the effect of the curvature of space is too small to be detected — about 0.06 ns (max) corresponding to about 17 mm in the case of GPS satellite signals.

The reduced wavelength of radiation in the local gravitational frame is observed as a "gravitational blueshift". The deformation of the wave front is also observed as gravitational lensing as a consequence of the bent propagation direction.

5.4.5 Effects of a moving receiver and moving source

The propagation path of light or a radio signal from source A to receiver B moving in the same frame can be expressed as

$$\mathbf{r}_{AB} = \mathbf{r}_{B(t1)} - \mathbf{r}_{A(t0)}, \qquad (5.4.5{:}1)$$

where $\mathbf{r}_{A(t0)}$ is the location vector of the source in the propagation frame at the time the radiation is sent, and $\mathbf{r}_{B(t1)}$ is the location vector of the receiver at the time of the arrival of the signal. The propagation time of a signal from A to B is

$$T_{AB} = \int_{A(t0)}^{B(t1)} \frac{1}{c_\delta} d\mathbf{r}_p, \qquad (5.4.5{:}2)$$

where dr_p is the differential of the signal path $\mathbf{r}_{AB}$. Ignoring the gravitational variation of the velocity of light along the propagation path, equation (5.4.5:2) can be simplified into the form

$$T_{AB} = \frac{|\mathbf{r}_{AB}|}{c} = \frac{\mathbf{r}_{AB(t0)} \cdot \hat{\mathbf{r}}_{AB} + T \cdot (\mathbf{v}_B \cdot \hat{\mathbf{r}}_{AB})}{c} = \frac{\mathbf{r}_{AB(t0)} \cdot \hat{\mathbf{r}}_{AB}}{c} + T_{AB} \cdot \beta_{B(\mathbf{r})}, \qquad (5.4.5:3)$$

where c is the average local velocity of light along the propagation path and $\beta_{B(\mathbf{r})}$ is the velocity of the receiver in the direction of the signal received.

The first term in equation (5.4.5:3) gives the propagation time related to the path as it is when the signal is emitted, and the second term is a correction due to the motion of the receiver during the propagation. Equation (5.4.5:3) can be solved in the form

$$T_{AB} = \frac{\mathbf{r}_{AB(t0)} \cdot \hat{\mathbf{r}}_{AB}}{c\left(1 - \beta_{B(\mathbf{r})}\right)}. \qquad (5.4.5:4)$$

As shown by equation (5.4.5:4), the motion of the receiver has a first-order effect on the propagation time between a source and a receiver in the local gravitational frame.

The length of the propagation path from $\mathbf{r}_{A(t0)}$ to $\mathbf{r}_{B(t1)}$ is

$$L_{AB} = \mathbf{r}_{AB} = \frac{\mathbf{r}_{AB(t0)} \cdot \hat{\mathbf{r}}_{AB}}{1 - \beta_{B(\mathbf{r})}}, \qquad (5.4.5:5)$$

which is referred to as the transmission distance from A to B.

The effect of the rotation of the Earth on the transmission time and distance of signals sent from Earth satellites is often referred to as the Sagnac effect (see Section 7.3.2).

The frequency and wavelength of the radiation received by a moving receiver are subject to Doppler shift as discussed in Section 5.2.3. In the case of a source A at rest, with reference to equation (5.2.3:15), the wavelength observed at B moving at velocity $\beta_{B(\mathbf{r})}$ in the direction of $\mathbf{r}_{AB}$ is

$$\lambda_{A(B)} = \frac{\lambda_{A(0)}}{1 - \beta_{B(\mathbf{r})}}, \qquad (5.4.5:6)$$

and the corresponding cycle time $T = 1/f$ is

$$T_{A(B)} = \frac{T_{A(0)}}{1 - \beta_{B(\mathbf{r})}} = \frac{\lambda_{A(0)}/c}{1 - \beta_{B(\mathbf{r})}} = \frac{\lambda_{A(B)}}{c}, \qquad (5.4.5:7)$$

which means that the observed phase velocity, c, of the radiation received by the moving receiver is independent of the velocity of the receiver, and equal to the phase velocity of the radiation observed by an observer at rest

$$c = \frac{\lambda_{A(B)}}{T_{A(B)}} = \lambda_{A(B)} f_{A(B)} = \lambda_{A(0)} f_{A(0)}. \qquad (5.4.5:8)$$

In the analysis above, the radiation source was assumed to stay at rest in the propagation frame. In such a situation, the distance $\mathbf{r}_{AB(t0)}$ changes with time as

$$\mathbf{r}_{AB(0)}(t) = \mathbf{r}_{AB(t0)}\left[1 + \int_0^t \beta_{B(\mathbf{r},t)} c\, dt\right], \tag{5.4.5:9}$$

and, accordingly, the optical image of the radiation source is seen at the transmission distance

$$L_{AB} = \mathbf{r}_{AB} = \frac{\mathbf{r}_{AB(t0)} \cdot \hat{\mathbf{r}}_{AB}}{1 - \beta_{B(\mathbf{r})}}\left[1 + \int_0^t \beta_{B(\mathbf{r},t)} c\, dt\right]. \tag{5.4.5:10}$$

When measured in observed wavelengths, the transmission distance is

$$N_{L(AB)} = \frac{L_{AB}}{\lambda_{AB}} = \frac{\mathbf{r}_{AB(t0)}}{\lambda_A} \cdot \hat{\mathbf{r}}_{AB}\left[1 + \int_0^t \beta_{B(\mathbf{r},t)} c\, dt\right]. \tag{5.4.5:11}$$

When the velocity $\beta_{B(\mathbf{r})} = \pm\beta$ occurs in the direction of $\hat{\mathbf{r}}_{AB}$, equation (5.4.5:11) obtains the form

$$N_{L(AB)} = \frac{L_{AB}}{\lambda_{AB}} = \frac{L_{AB(0)}}{\lambda_A}\left[1 \pm \beta \cdot c(t - t_0)\right], \tag{5.4.5:12}$$

where $L_{AB(0)}$ is the physical distance from the source to the receiver at $t = 0$

$$L_{AB(0)} = \mathbf{r}_{AB(t0)} \cdot \hat{\mathbf{r}}_{AB} = |\mathbf{r}_{AB(t0)}|. \tag{5.4.5:13}$$

5.4.6 The effect of a dielectric propagation medium

The velocity of light propagating in a non-magnetic medium with relative permittivity $\varepsilon_r > 1$, and relative permeability $\mu_r \approx 1$ is expressed

$$c_n = \frac{1}{\sqrt{\mu_0 \varepsilon_0 \varepsilon_r}} = \frac{c}{\sqrt{\varepsilon_r}} = \frac{c}{n}, \tag{5.4.6:1}$$

where $n = \sqrt{\varepsilon_r}$ is referred to as the refractive index. The effect of relative permittivity is understood as polarization of the transmitting medium due to the electric field of the propagating electromagnetic wave. In the DU framework, the effect of polarization can be expressed as the buildup of the imaginary component of the momentum of the propagating wave. In a vacuum or in air, the momentum of a cycle of electromagnetic radiation is expressed

$$\mathbf{p}_0 = N\frac{\hbar_0}{\lambda}\mathbf{c}, \tag{5.4.6:2}$$

where N is the intensity factor. When entering into a medium with a refractive index $n > 1$, the momentum is reduced due to reflection, resulting in radiation pressure and a reduction in the intensity of radiation. The transmitted share of the momentum is

$$p_T = |\mathbf{p}_T| = (1 - R)p_0 = Tp_0 = \frac{4n}{(1+n)^2} p_0 = N_T p_0. \tag{5.4.6:3}$$

The effect of polarization in the transmitting medium is described as a buildup of the imaginary component of the momentum. The real component of the momentum is characterized by the reduced propagation velocity, $c' = c/n$,

$$\text{Re}\{p_T\} = N_T \frac{\hbar_0}{\lambda} \frac{c}{n}. \tag{5.4.6:4}$$

Conservation of the absolute value of the momentum requires the imaginary component

$$\text{Im}\{p_T\} = i N_T \frac{\hbar_0}{\lambda} c \sqrt{1 - \frac{1}{n^2}}, \tag{5.4.6:5}$$

i.e., the complex momentum is expressed as

$$\mathbf{p}_T = \text{Re}\{p_T\} + \text{Im}\{p_T\} = N_T \frac{\hbar_0}{\lambda} c \left(\frac{1}{n} + i \sqrt{1 - \frac{1}{n^2}} \right), \tag{5.4.6:6}$$

see Figure 5.4.6-1(a).

When the transmitting medium is in motion at velocity $\beta = v/c$ in the direction of the propagation of the radiation, the real component of the momentum of the radiation in the medium is increased as

$$\Delta \text{Re}\{p_T\} = N_T \frac{\hbar_0}{\lambda} \frac{\beta c}{\sqrt{1 - \beta^2}}, \tag{5.4.6:7}$$

resulting in total momentum

$$\mathbf{p}_{T(\beta)} = N_T \frac{\hbar_0}{\lambda} c \left[\left(\frac{1}{n} + \frac{\beta}{\sqrt{1 - \beta^2}} \right) + i \sqrt{1 - \frac{1}{n^2}} \right]. \tag{5.4.6:8}$$

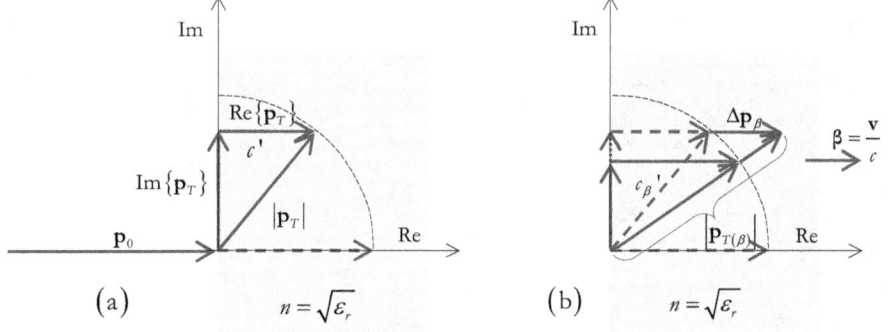

Figure 5.4.6-1. (a) Buildup of the imaginary component in the momentum of electromagnetic radiation in a transmitting medium with refractive index $n > 1$. The velocity related to the real component of the radiation is $c' = c/n$. (b) The effect of the velocity of the transmitting medium is an addition, $\Delta\mathbf{p}_\beta$, to the real component of the radiation. The effect of the increased momentum on the velocity of the radiation in the local frame, the frame where the medium is moving, is denoted as c_β' in the picture.

The modulus of the total momentum is (see Figure 5.4.6-1(b))

$$p_{T(\beta)} = \left|\mathbf{P}_{T(\beta)}\right| = N_T \frac{h_0}{\lambda} c \sqrt{\left(\frac{1}{n} + \frac{\beta}{\sqrt{1-\beta^2}}\right)^2 + \left(1 - \frac{1}{n^2}\right)}. \tag{5.4.6:9}$$

Relative to the local frame where the polarizing medium is moving, the velocity of the radiation, c_β', due to the motion of the medium, is increased as

$$c_\beta' = c \bigg/ \sqrt{1 + \left(1 - \frac{1}{n^2}\right) \bigg/ \left(\frac{1}{n} + \frac{\beta}{\sqrt{1-\beta^2}}\right)^2}. \tag{5.4.6:10}$$

Equation (5.4.6:10) is the DU replacement for the classical Fresnel equation

$$c_\beta'{}_{(Fresnel)} = c \left[\frac{1}{n} + \beta\left(1 - \frac{1}{n^2}\right)\right], \tag{5.4.6:11}$$

and its replacement based on the special theory of relativity

$$c_\beta'{}_{(SR)} = \frac{c/n + v}{1 + \frac{vc/n}{c^2}} = c\frac{1/n + \beta}{1 + \beta/n}. \tag{5.4.6:12}$$

A comparison of equations (5.4.6:10), (5.4.6:11), and (5.4.6:12) is presented in Figure 5.4.6-2. As shown by the comparison, for $\beta \ll 1$, the predictions given by all three equations are essentially equal. When the velocity of the transmitting medium approaches the velocity of light, the Fresnel equation (5.4.6:12) predicts radiation velocities exceeding c. The formula for adding velocities in special relativity in equation (5.4.6:13) corrects the Fresnel

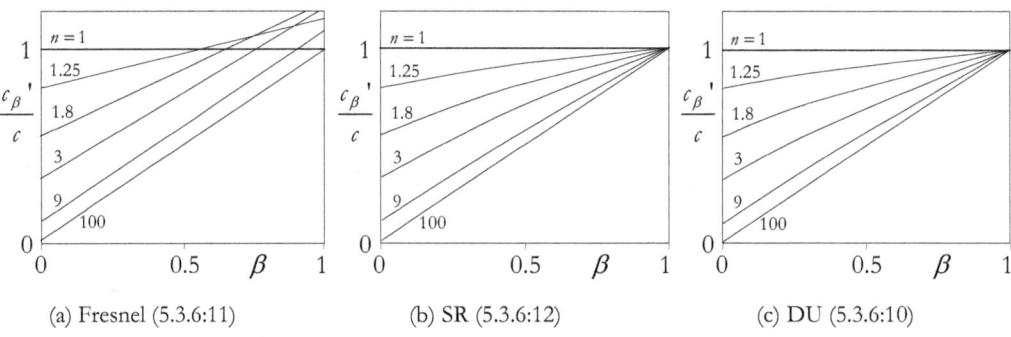

(a) Fresnel (5.3.6:11) (b) SR (5.3.6:12) (c) DU (5.3.6:10)

Figure 5.4.6-2. The effect of the refractive index n of the propagating medium on the velocity of light observed in a local frame where the velocity of the propagating medium is $v = \beta c$. The values of n used in the comparison are $n = 1$, $n = 1.25$, $n = 1.8$, $n = 3$, $n = 9$, and $n = 100$. Curves in (a) are based on equation (5.4.6:11), which is the Fresnel formula, curves in (b) are based on the prediction special relativity (5.4.6:12) obtained using the SR formula for the addition of velocities for $c_\beta' = v+c$. Curves (c) show the DU prediction in equation (5.4.6:10). At low velocities β, the Fresnel equation serves as an approximation equally for the SR and DU predictions.

predictions by forcing all curves, independent of the refractive index n, to meet at c when the velocity of the transmitting medium is equal to the velocity of light. The DU prediction is essentially equal to the SR prediction, although the physical basis, derivation, and mathematical form of the prediction are different.

According to the DU analysis, the effect of the refractive index on the velocity of light relies on the energy stored by polarization in a dielectric medium with $\varepsilon_r > 1$. The polarization energy can be expressed as a buildup of an imaginary component to the energy and momentum of the electromagnetic radiation propagating in the medium. Like Coulomb energy, the energy stored by polarization is reduced by the velocity of the polarizing medium, which means that the effective refractive index is reduced with an increasing velocity of the medium.

5.5 Propagation of light from stellar objects

5.5.1 Frame-to-frame transmission

The transmission time of electromagnetic radiation within a gravitational frame was given in equation (5.4.5:4). As in the derivation of the Doppler effect, the effect of nested frames on the transmission time can be calculated by regarding sub-frames as objects in the root parent frame. Accordingly, equation (5.4.5:4) can be written in the form

$$T_{A(t0) \to B(t1)} = T_{AB} = \frac{\mathbf{r}_{AB(t0)} \cdot \hat{\mathbf{r}}_{AB}}{c_k \prod_{j=k+1}^{m}\left(1-\beta_{[j]B(r)}\right)}, \qquad (5.5.1:1)$$

where frame k is the root parent frame common to the source and the receiver frames. With reference to equation (5.2.3:18), the denominator in equation (5.5.1:1) can be interpreted as the kinematic velocity of light in the receiver frame

$$c'_B = c_k \prod_{j=k+1}^{m}\left(1-\beta_{[j]B(r)}\right). \qquad (5.5.1:2)$$

Applying the kinematic velocity of light, the signal time can be expressed in the receiver frame as

$$T_{A(t0) \to B(t1)} = T_{AB} = \frac{\mathbf{r}_{AB(t0)} \cdot \hat{\mathbf{r}}_{AB}}{c'_B}. \qquad (5.5.1:3)$$

The kinematic velocity of the signal in the source frame can be related to the velocity of light in the root frame as

$$c'_A = c_k \prod_{i=k+1}^{n}\left(1-\beta_{[i]A(r)}\right). \qquad (5.5.1:4)$$

Dividing equation (5.5.1:2) by equation (5.5.1:4) relates the kinematic velocity of light in the receiver to the kinematic velocity of light in the source as

$$c'_B = c'_A \frac{\prod_{j=k+1}^{m}\left(1-\beta_{[j]B(r)}\right)}{\prod_{i=k+1}^{n}\left(1-\beta_{[i]A(r)}\right)} \approx c'_A\left[1-\left(\sum_{j=k+1}^{m}\beta_{[j]B(r)} - \sum_{i=k+1}^{n}\beta_{[i]A(r)}\right)\right]. \qquad (5.5.1:5)$$

Substitution of equation (5.5.1:5) for c'_B in equation (5.5.1:3) gives the signal time in terms of the kinematic velocity of light in the source frame as

$$T_{A(t0)\to B(t1)} = T_{AB} = \frac{\mathbf{r}_{AB(t0)} \cdot \hat{\mathbf{r}}_{AB}}{c'_A \left[1 - \left(\sum_{j=k+1}^{m} \beta_{[j]B(\mathbf{r})} - \sum_{i=k+1}^{n} \beta_{[i]A(\mathbf{r})}\right)\right]}. \quad (5.5.1:6)$$

The propagation of electromagnetic radiation from one energy frame to another can be understood as propagation in the root parent frame common to both the source and the receiver.

5.5.2 Gravitational lensing and momentum of radiation

Due to the local tilting of space, the velocity of light is reduced, and the transmission distance is increased near mass centers. As a result, the wave front of a plane wave is modified due to a higher delay close to the mass center as given in equations (5.4.1:19–24), Figure 5.5.2-1.

The reduction of the propagation velocity near a mass center is associated with a reduction of the wavelength. This reduction, which conserves the frequency and the momentum of the propagating radiation, can be expressed as

$$f = \frac{c_{0\delta}}{\lambda_{0\delta}} = \frac{c_\delta}{\lambda_\delta} \quad \Rightarrow \quad \lambda_\delta = \frac{c_\delta}{c_{0\delta}} \lambda_{0\delta}, \quad (5.5.2:1)$$

and

$$\mathbf{p} = N^2 X_\lambda \frac{h_0}{\lambda_{0\delta}} \mathbf{c}_{0\delta} = N^2 X_\lambda \frac{h_0}{\lambda_\delta} \mathbf{c}_\delta \quad \Rightarrow \quad \lambda_\delta = \frac{c_\delta}{c_{0\delta}} \lambda_{0\delta}. \quad (5.5.2:2)$$

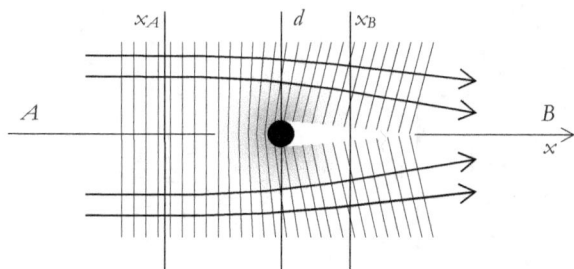

Figure 5.5.2-1. When radiation from a distant object passes a gravitational center, the wave front is distorted due to the lower velocity of light close to the center. As required by the conservation of momentum in the passing radiation, the wavelength is observed as reduced near the mass center.

5.5.3 Transversal velocity of the source and receiver

An electromagnetic wave is described as an energy object capable of carrying momentum in the direction of propagation (the direction of the Poynting vector). Motion of an emitter perpendicular to the emission direction of a plane wave does not contribute to the momentum of the wave emitted but results in "sliding the tail" of the transmitted beam in that direction. The velocity $\beta_\perp$ of the emitter results in a tilting angle in the apparent propagation direction, Figure 5.5.3-1(a)

$$\psi = \arctan \frac{v_\perp}{c} = \arctan \beta_\perp . \qquad (5.5.3:1)$$

At a fixed distance, L, in the actual propagation direction from the emitter, the phase of the wave is independent of the direction of the apparent propagation direction, Figure 5.5.3-1(b)

$$\Psi = \mathbf{k} \cdot \mathbf{r}' - \omega t = k r_\parallel - \omega t = 2\pi \left(\frac{L}{\lambda} - ft \right). \qquad (5.5.3:2)$$

The effect of a transversal motion of the receiver results in an apparent tilting of the direction of the radiation towards the motion of the receiver. The phenomenon is referred to as aberration. The aberration angle is formally identical to the apparent tilting of the emitted beam in equation (5.5.3:1)

$$\psi = \arctan \frac{v_\perp}{c} = \arctan \beta_\perp . \qquad (5.5.3:3)$$

The aberration angle does not contribute to the momentum of the incoming beam, Figure 5.5.3-2.

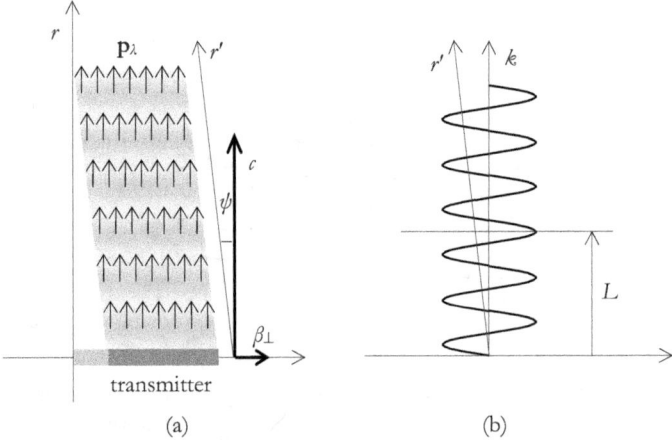

Figure 5.5.3-1. (a) Motion of a transmitter perpendicular to the emitted beam does not contribute to the momentum of the radiation but results in an apparent displacement of the propagation direction. (b) The phase of the wave at distance L in the actual propagation direction is independent of the apparent direction.

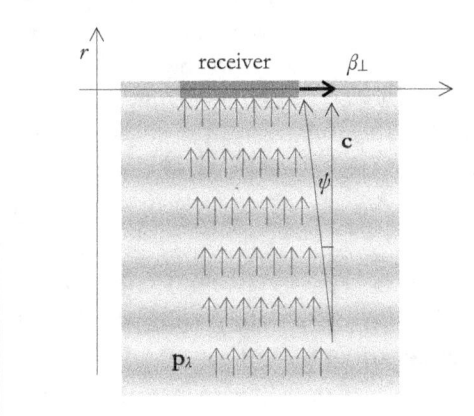

Figure 5.5.3-2. The aberration of light received by a receiver moving at velocity $\beta_\perp$ perpendicular to the momentum of the radiation received.

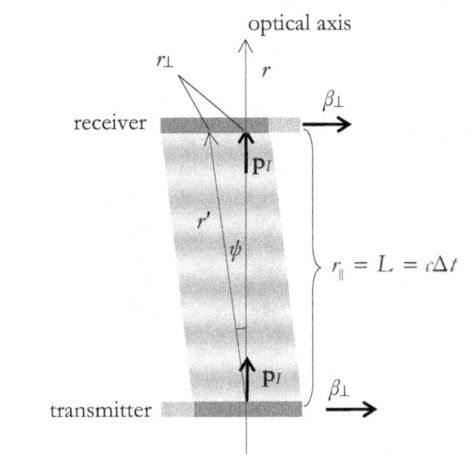

Figure 5.5.3-3. Propagation of a plane wave beam perpendicular to the velocity of the transmitter–receiver frame. In the Earth's gravitational frame, at about 40° latitude, for an optical axis of 1 meter in the south-north direction, the offset of the beam, $r_\perp$, is about 1 μm. This sets the lower limit for the width of a beam to be observed in a system where the beam is aligned in the direction of the optical axis perpendicular to the motion of the source–receiver frame.

When both source and receiver move at the same velocity perpendicular to the radiation transferred, a displacement of the receiver is needed in order for the full width of the transmitted beam to be received. The offset distance is

$$r_\perp = r_\parallel \tan\psi = \beta_\perp L, \tag{5.5.3:4}$$

where L is the physical distance from the source to the receiver in the frame moving at velocity $\beta_\perp = v_\perp/c$ perpendicular to the momentum of the beam in the propagation frame, Figure 5.5.3-3.

As is obvious from equation (5.5.3:2), the propagation time of a wave front is not affected by the motion of a frame perpendicular to the momentum of the radiation. The propagation time for distance L is

$$T_{\perp,prg} = \frac{L}{c}. \tag{5.5.3:5}$$

Radiation emitted by sources at cosmological distances can be regarded as plane wave fronts entering the solar gravitational frame and the Earth's gravitational frame. Each frame encountered by the radiation is in a position of a receiver relative to the incoming wave front. The total aberration due to the transverse velocity of each frame relative to the propagation frame of radiation (the root parent frame of the source and the receiver) can be expressed, Figure 5.5.3-4

$$\psi = \arctan \sum_i \beta_{i\perp} . \qquad (5.5.3{:}6)$$

On the Earth, the motion of the observatory with the rotation of the Earth results in a daily changing component (diurnal aberration) of the total aberration ($\psi_{EarthRot(max)} \approx 0.3"$). The orbital motion of the Earth gives a maximum contribution to the aberration of about $\psi_{EarthOrb(max)} \approx 20.4"$, which is referred to as the annual aberration. The maximum aberration due to the motion of the solar system in the Milky Way is about ten times the annual aberration. It is not, however, observable due to the very long orbital period of the solar system in the Milky Way. Aberration is a purely kinematic effect without any energy exchange between the observer and the radiation received.

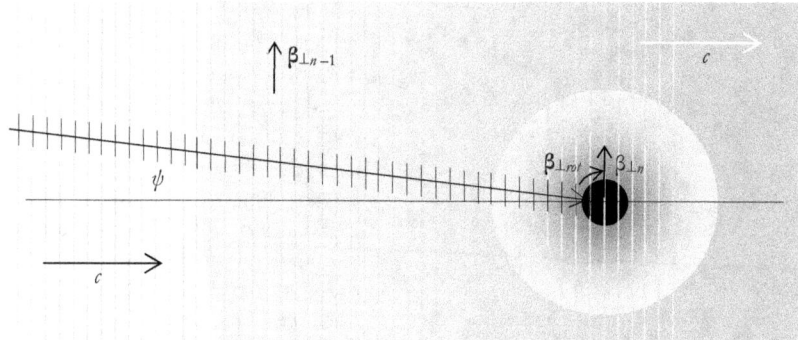

Figure 5.5.3-4. Radiation received in a local frame moving perpendicular to the radiation. Motions of the local frame and the parent frames result in aberration of the observation angle. Due to the kinematic nature of the aberration, the velocity of the observer in a local frame and the velocity of the local frame and the parent frames account for the total aberration.

5.6 The development of the lengths of a year, month, and day

5.6.1 Earth to Moon distance

Effect of the expansion of space on the Earth-Moon distance

As shown in Section 4.2.6, the orbital radii of stellar and planetary systems are subject to an increase in proportion to the increase of the R_4 radius due to the expansion of space

$$\frac{\Delta r_r}{r_r} = \frac{\Delta R_4}{R_4} = \frac{2}{3}\frac{\Delta t}{t}, \qquad (5.6.1{:}1)$$

where t is the time from the singularity of space, and Δt is the observation interval as given in equation (3.3.3:11). According to equations (3.3.3:10) and (5.1.4:15), the velocity of light and the ticking frequencies of atomic clocks decrease as

$$\frac{\Delta c}{c} = -\frac{1}{3}\frac{\Delta t}{t} = \frac{\Delta f}{f}. \qquad (5.6.1{:}2)$$

The signal transmission time from the Earth to the Moon increases as

$$\frac{\Delta T}{T} = \frac{\Delta r}{r} - \frac{\Delta c}{c} = \frac{2}{3}\frac{\Delta t}{t} + \frac{1}{3}\frac{\Delta t}{t} = \frac{\Delta t}{t}, \qquad (5.6.1{:}3)$$

and the time observed, the number of ticks of the atomic clock during the transmission time, is

$$\frac{\Delta n}{n} = \frac{\Delta f}{f} + \frac{\Delta T}{T} = -\frac{1}{3}\frac{\Delta t}{t} + \frac{\Delta t}{t} = \frac{2}{3}\frac{\Delta t}{t} = \frac{\Delta r}{r}. \qquad (5.6.1{:}4)$$

That is, the observed increase in transmission time gives, directly, the actual increase in the Earth-Moon distance. In one year, according to equation (5.6.1:4), the increase in the Earth to Moon distance due to the expansion of space is

$$\Delta r_{c0}/\Delta t = \frac{\Delta n}{n} r_{c0} \Big/ \Delta t = \frac{2}{3t} = \frac{2}{3 \cdot 9.3 \cdot 10^9} \approx 2.75 \; [\text{cm/year}], \qquad (5.6.1{:}5)$$

which is about 74% of the increase, $\Delta r_{obs} = 3.82 \pm 0.007$ cm/year, observed in the Lunar Laser Ranging program [72]. In the equation, the estimated time from the singularity, $t = 9.3 \cdot 10^9$ years, corresponds to the Hubble constant $H_0 = 70$ [(km/s)/Mpc]. The increase of the center-to-center distance given by equation (5.6.1:5) applies as such to the surface-to-surface distance because the radii of the Earth and the Moon are not subject to an increase due to the expansion of space. Another effect resulting in an increase of the Earth-Moon distance is the tidal system, which is assumed to be responsible for a difference of 1 cm/year between the observed value Δr_{obs} and the prediction of equation (5.6.1:5).

Annual perturbation of the Earth to Moon distance

In the Lunar Laser Ranging space program, the distance between the Earth and the Moon has been measured with high accuracy (up to a few centimeters) since 1970's [72]. The measurement is based on the two-way transmission time of a light pulse from the Earth to a reflector on the Moon and back to the Earth.

The distance between the Earth and the Moon is subject to many perturbations, such as tides, the radiation pressure of the Sun, and so on. A perturbation of special interest is related to the eccentricity of the Earth's orbit around the Sun. The orbital velocity of the Earth is highest at the perihelion point, resulting in a slower ticking frequency of the clock measuring the two-way time signal used for the distance measurement. Further, the gravitational potential of the Earth in the Sun's gravitational frame is at a minimum at the perihelion, which further slows down the ticking frequency of the clock. Likewise, the velocity of light used in the distance measurement is at a minimum in the minimum gravitational potential in the Sun's frame.

Owing to the energy balances in the gravitational frames of the Sun and the Earth, the Earth-to-Moon distance achieves its maximum when the Earth is at perihelion in its planetary orbit. At perihelion, due to the maximum orbital velocity of the Earth–Moon system, the radii of both the Earth and the Moon, as well as the dimensions of all material objects with them, are at their maximum.

According to equation (5.2.1:7), the frequencies of atomic clocks on the Earth depend on the orbital velocity and the gravitational state as

$$\frac{\Delta f}{f_{min}} \approx \frac{GM_{Sun}}{c^2}\left[\frac{1}{r_{min}} - \frac{1}{r_{max}}\right] - \frac{\beta^2_{r(max)} - \beta^2_{r(min)}}{2}. \tag{5.6.1:6}$$

The change in kinetic energy is equal to the change in gravitational energy because the energy of motion gained is equal to the energy of gravitation released. With reference to equation (5.2.1:8), we can express equation (5.6.1:6) as

$$\frac{\Delta f}{f_{min}} = \frac{2g_{Sun}\Delta r}{c^2}, \tag{5.6.1:7}$$

where Δr is the increase in the orbital radius of the Earth.

The change in the velocity of light, Δc, on the Earth due to the change in the distance to the Sun, Δc, can be determined from equation (4.1.10:2) as

$$\Delta c \approx \Delta\left[c\left(1 - \frac{GM_{Sun}}{rc^2}\right)\right] = \frac{g_{Sun}\Delta r}{c}, \tag{5.6.1:8}$$

corresponding to a relative change

$$\frac{\Delta c}{c} \approx \frac{g_{Sun}\Delta r}{c^2}. \tag{5.6.1:9}$$

As shown in Section 4.2.7, the radius of a local rotational system is increased when the system is taken closer to the central mass of its parent gravitational frame. According to equation (4.2.7:7), the decrease in the Earth to Moon (center to center) distance due to the increase in the Earth to Sun distance is

$$\frac{\Delta r_{cc}}{r_{cc}} \approx -\frac{g_{Sun}\Delta r}{c^2}. \tag{5.6.1:10}$$

The distance between the Earth and the Moon is measured from the surface of the Earth to the surface of the Moon. To obtain the center-to-center distance, the radii of the Earth and the Moon must be added. Because the radii of atoms change with the rest mass of electrons, the dimensions of all solid objects and, accordingly, the radii of the Earth and Moon change in accordance with equation (5.1.4:19) as

$$\frac{\Delta r_r}{r_r} \approx \frac{\beta_{r(\max)}^2 - \beta_{r(\min)}^2}{2} = -\frac{g_{Sun}\Delta r}{c^2}, \tag{5.6.1:11}$$

which means that the relative change in the radii of the Earth and Moon is equal to the relative change in the center-to-center distance given in equation (5.6.1:10). Combining equations (5.6.1:10) and (5.6.1:11), the prediction for the measured distance becomes

$$\frac{\Delta r_M}{r_M} = \frac{\Delta\left(r_{cc} - r_{r(E)} - r_{r(M)}\right)}{r_{cc} - r_{r(E)} - r_{r(M)}} = -\frac{g_{Sun}\Delta r}{c^2}. \tag{5.6.1:12}$$

The wavelength of light or electromagnetic radiation emitted by an object on the Earth is changed due to the change in the frequency emitted, as shown in equation (5.6.1:7), and to the change in the velocity of light given in equation (5.6.1:9),

$$\frac{\Delta\lambda}{\lambda} = \Delta\frac{(\Delta c/c)}{(\Delta f/f)} \approx \frac{\Delta c}{c} - \frac{\Delta f}{f} \approx -\frac{g_{Sun}\Delta r}{c^2}. \tag{5.6.1:13}$$

As shown by equations (5.6.1:12) and (5.6.1:13), the wavelength of the electromagnetic radiation emitted by an object on the Earth and the distances to the Moon and satellites change proportionally to the change in the distance from the Earth to the Sun. Accordingly, a hypothetical interferometric measurement of the distance from the Earth to the Moon gives an unchanging result in the course of the year.

The traveling time of light from the Earth to the Moon changes in the course of the year due to two factors: the change in the velocity of light and the change in the distance. The two factors change in the same direction and are equal in quantity. With reference to equations (5.6.1:9) and (5.6.1:12) we get

$$\frac{\Delta T}{T} = \frac{\Delta r_M}{r_M} - \frac{\Delta c}{c} \approx -\frac{2g_{Sun}\Delta r}{c^2}. \tag{5.6.1:14}$$

The relative change of the propagation time of light or a radio signal is opposite to the relative change of the ticking frequency of an atomic clock, given in equation (5.6.1:7).

The time interval observed, the count of cycles of an atomic clock during the traveling time of light or a radio signal from the Earth to the Moon, is

$$n = f_{clock}T. \tag{5.6.1:15}$$

Substituting equations (5.6.1:7) and (5.6.1:14) into equation (5.6.1:15), we obtain the change of the count with the change of distance from the Earth to the Sun,

$$\frac{\Delta n}{n} = \frac{\Delta f}{f} + \frac{\Delta T}{T} \approx \frac{2g_{Sun}\Delta r}{c^2} - \frac{2g_{Sun}\Delta r}{c^2} = 0 \qquad (5.6.1:16)$$

which means that the distance from the Earth to the Moon appears to be unaffected by the eccentricity of the planetary orbit when measured with the transmission time of a light signal, as done in the Apollo program.

The average radius of the Earth's planetary orbit is about $r \approx 150 \cdot 10^6$ km. The eccentricity of the orbit is $e \approx 0.0167$, resulting in a $\Delta r \approx 5 \cdot 10^6$ km difference between the orbital radii in the perihelion and aphelion points. The numerical value of the gravitational acceleration due to the Sun at the Earth's planetary orbit is about $g_{Sun} = 5.9 \cdot 10^{-3}$ [m/s], which means that, according to equation (5.6.1:11), the annual variation of the Earth to Moon distance due to the eccentricity of the Earth's orbit is

$$\Delta r_r \approx m - \frac{g_{Sun}\Delta r}{c^2} r_r \approx -12.6 \ [\text{cm}], \qquad (5.6.1:17)$$

where the minus sign means that the Earth to Moon distance decreases when the Earth–Moon system advances from the perihelion to aphelion in the Earth's planetary orbit, i.e., the Earth to Moon distance is at a minimum when the Sun to Earth distance is at maximum. As shown by equation (5.6.1:16), due to the counterbalancing changes in the clock frequency and the velocity of light, the change in the Earth to Moon distance given in equation (5.6.1:17) is not detectable with the two-way signal time measurement used in the Lunar Laser Ranging program.

The decrease of the Earth to Moon distance at aphelion results in a perturbation in the angular momentum of the Earth–Moon system, which is counterbalanced by an increase in the angular velocity of the Earth and, accordingly, a reduction in the length of a day. The resulting annual perturbation in the length of the day is about 0.08 ms. The maximum length of the day, relative to an atomic clock on the Earth, occurs when the Earth-Moon distance is at maximum at the Earth's perihelion in January.

The observed annual perturbation of the length of the day, resulting from meteorological and other effects, is about 2 ms/year (in phase with the 0.08 ms/year due to the annual fluctuation in the Earth to Moon distance).

5.6.2 Development of rotational and orbital velocities

As a consequence of the balance between the energies of motion and gravitation (δ and β in local energy frames are conserved), the orbital radii of stellar systems increase in proportion to the expansion of space. The increase in the radii is associated with a decrease in the orbital velocities in proportion to the degradation of the velocity of light.

The Earth draws away from the Sun at the rate

$$\frac{\Delta r_{orb}}{r_{orb}} = \frac{\Delta R_4}{R_4} = \frac{2}{3t} \approx 7.17 \cdot 10^{-11} \ / \ \text{year}, \qquad (5.6.2:1)$$

which, at present, means about 11 meters per year [see equation (3.3.3:11].

The orbital velocity decreases at the rate

Mass, mass objects and electromagnetic radiation

$$\frac{\Delta v_{orb}}{v_{orb}} = \frac{\Delta c_0}{c_0} = -\frac{1}{3t} \approx -3.6 \cdot 10^{-11} / \text{year}, \qquad (5.6.2{:}2)$$

see equation (3.3.3:10). The orbiting time, i.e., the length of a year, increases as

$$\frac{\Delta P_{orb}}{P_{orb}} = \frac{2\pi \Delta r}{2\pi r} - \frac{\Delta v_{orb}}{v_{orb}} = \frac{1}{t} = 1.1 \cdot 10^{-10} / \text{year}, \qquad (5.6.2{:}3)$$

which means that the present increase in the length of a year (in absolute time) is

$$\Delta P_{orb} = 1.1 \cdot 10^{-10} \cdot 86400 \cdot 365 = 3.4 \cdot 10^{-3} \text{ s / year}. \qquad (5.6.2{:}4)$$

Likewise, the ticking frequencies of clocks decrease in proportion to the degradation of the velocity of light. Accordingly, the increase in the year observable with clocks on the Earth is

$$\Delta P_{orb} = \left(\frac{1}{t} - \frac{1}{3t}\right) \cdot P_{orb} = \frac{2}{3t} \cdot P_{orb} = 2.3 \cdot 10^{-3} \text{ s / year}. \qquad (5.6.2{:}5)$$

Because the Bohr radius is conserved in the expansion of space, the radius of the Earth is conserved. However, the rotational velocity (and the angular momentum) of the Earth decreases in proportion to the decrease of c_0. Accordingly, with reference to equation (5.6.2:2), the rotation time, i.e., the length of a day, increases as

$$\begin{aligned}\Delta P_{rot} &= \frac{1}{3t} \cdot P_{rot} = 3.6 \cdot 10^{-11} \cdot 86400 = 3.1 \cdot 10^{-6} \ (\text{s/day})/\text{year} \\ &= 0.31 \ (\text{ms/day})/\text{century}.\end{aligned} \qquad (5.6.2{:}6)$$

The lengthening of a day expressed in equation (5.6.2:6) results from the expansion of space and does not include the lengthening of a day due to the tidal interactions with the Moon and the Sun. The effect of the expansion of space on the length of a day is fully compensated by the reduction of the ticking frequency of atomic clocks, which is reduced in direct proportion to the velocity of light. Accordingly, changes in the length of a day observed with an atomic clock come from effects other than the direct effect of the expansion of space on the rotation of the Earth.

5.6.3 Days in a year based on coral fossil data

An interesting indication of the development of the length of a day comes from coral fossil data dating back 100–800 million years. Fossil layers preserve both the daily and annual variations, thus giving the number of days in a year. At least partly, tidal variations can also be detected, which allows an estimate of the development of the number of days in a lunar month.

Equations (5.6.2:1–3) relate differential changes in the radii, orbiting velocities, and orbiting times to the changes in the 4-radius and the velocity of light. For the interpretation of observations over long periods of time, as in the case of coral fossil data, it is necessary to peg the changes to the absolute times of the observations.

With reference to equations (3.3.3:7) and (4.2.6:6) and by denoting the present semi-major axis of an orbiting system as a_0 and the present time from the singularity as t_0, the semi-major axis a at time t can be expressed as

$$\frac{a}{a_0} = \frac{R_4}{R_{4(0)}} = \frac{At^{2/3}}{At_0^{2/3}} = \left(\frac{t}{t_0}\right)^{2/3}, \qquad (5.6.3:1)$$

which gives the change in the semi-major axis

$$\frac{\Delta a}{a} = \frac{a_0 - a}{a_0} = 1 - \frac{a}{a_0} = 1 - \left(\frac{t}{t_0}\right)^{2/3}. \qquad (5.6.3:2)$$

With reference to equation (5.6.2:3), the change in the orbital period obtains the form

$$\frac{\Delta P_{orb}}{P_{orb}} = \frac{P_{orb(0)} - P_{orb}}{P_{orb(0)}} = 1 - \frac{P_{orb}}{P_{orb(0)}} = 1 - \frac{t}{t_0}, \qquad (5.6.3:3)$$

and with reference to equation (5.6.2:6), the change in the rotational period of solid objects obtains the form

$$\frac{\Delta P_{rot}}{P_{rot}} = 1 - \left(\frac{t}{t_0}\right)^{1/3}. \qquad (5.6.3:4)$$

As shown by equations (5.6.3:3) and (5.6.3:4), the increase of an orbital period is greater than the increase of a rotational period. This is because the orbiting radii are subject to an increase with the expansion of space, whereas the radii of spinning solid objects are not. If the expansion of space were the only factor relevant to the length of a day and the length of a year, the number of days in a year would increase with the expansion of space. The increase in the Earth's distance from the Moon, however, results in an extra decrease in the rotational velocity of the Earth, which makes the number of days decrease with the expansion of space. In other words, in the past, the number of days in a year was greater.

With reference to equation (5.6.3:3), the length of the year, Y_t, for Δt years ago, at the time $t = t_0 - \Delta t$, can be calculated as

$$\Delta Y = Y_0 - Y_t = Y_0 - Y_0 \frac{t_0 - \Delta t}{t_0} \quad ; \quad Y_t = Y_0 \left(1 - \frac{\Delta t}{t_0}\right), \qquad (5.6.3:5)$$

and, with reference to equation (5.6.3:4), the length of a day, D_t, due to the expansion of space as

$$D_{t(\exp)} = D_0 \left(1 - \frac{\Delta t}{t_0}\right)^{1/3}. \qquad (5.6.3:6)$$

When the lengthening of a day due to the lengthening of the Earth to Moon distance and the tidal interaction in the Earth–Sun system are included, the length of a day as it was for Δt years ago can be expressed as

$$D_t = D_{t(\exp)}\left(1 - \Delta t \frac{dD/D}{100Y}\right) = D_0\left(1 - \frac{\Delta t}{t_0}\right)^{1/3}\left(1 - \Delta t \frac{dD_0/D}{100Y}\right), \qquad (5.6.3{:}7)$$

where the relative shortening of a day due to the interactions of the Earth with the Moon and the tidal effect of the Sun in the last term is assumed to increase in direct proportion to the time from the singularity (as does the period of the Moon). To match observations in coral fossils, the last term, $(dD_0/D)/100Y$, should have a value of 2.5 [(ms/day)/century] = $2.9 \cdot 10^{-10}$ [1/y], giving equation (5.6.3:7) in form

$$D_t = D_0\left(1 - \frac{\Delta t}{t_0}\right)^{1/3}\left(1 - 2.9 \cdot 10^{-10}\,\Delta t\right), \qquad (5.6.3{:}8)$$

corresponding to the number of days in a year

$$N_D = \frac{Y_t}{D_t} = \frac{Y_0}{D_0}\frac{Y_t/Y_0}{D_t/D_0} = N_0\left(1 - \frac{\Delta t}{t_0}\right)^{2/3}\bigg/\left(1 - 2.9 \cdot 10^{-10}\,\Delta t\right). \qquad (5.6.3{:}9)$$

In the Introduction, Figure 1.3.4-1 illustrates the development of the length of the year (in current days) and the number of days in a year during the last 1000 million years. The number of days given by equation (5.6.3:9) follows well the development of the number of days in a year counted in fossil samples since almost one billion years back. Experimental values, shown as squares in the figure, have been collected from papers comprising coral fossil data [56,57,58] and stromatolite data from the Bitter Springs Formation [62] (the data from the samples going back to more than 800 million years). In all data points in Figure 1.3.4-1, the DU correction in the age estimate is made according to equation (6.4.3:10).

5.7 Timekeeping in the Dynamic Universe

5.7.1 Periodic phenomena and timescales

Characteristic wavelength and frequency of atomic objects

When expressed in terms of the Bohr radius, Balmer's formula for the characteristic or characteristic emission wavelength obtains the form [equation (5.1.4:21)]

$$\lambda_{(\beta_i)(n1,n2)} = \frac{4\pi a_{0(\beta_i)}}{\alpha Z^2 \left[1/n_1^2 - 1/n_2^2\right]} = \frac{4\pi a_{0(0)}}{\alpha Z^2 \left[1/n_1^2 - 1/n_2^2\right] \prod_{i=0}^{n} \sqrt{1-\beta_i^2}}, \qquad (5.7.1:1)$$

which shows the increase of the Bohr radius and the emission wavelength as functions of the motion of the emitting atom.

The linkage of the emission wavelength to the Bohr radius is not a specific property of the Dynamic Universe only. The linkage is also present in the standard quantum mechanical solution of the hydrogen atom.

The additional feature given by the Dynamic Universe model is the linkage of the Bohr radius and the emission wavelength to the velocity of the atom in the local frame and the parent frames. An important consequence of the linkage is the conservation of the resonance condition in moving resonators [see Section 5.5.2].

Because the definition of a second is based on the frequency of a Ce-clock, Balmer's formula for the characteristic oscillation frequencies of atomic oscillators is of special importance [see equations (5.1.4:14) and (5.1.4:15)]

$$f_{n1,n2} = \frac{\alpha^2}{2b_0} Z^2 \left[\frac{1}{n_1} - \frac{1}{n_2}\right] m_{e(0)} c_0 \prod_{i=0}^{n}\left[(1-\delta_i)\sqrt{1-\beta_i^2}\right], \qquad (5.7.1:2)$$

or

$$f_{n1,n2} = \frac{\alpha^2}{2b_0} Z^2 \left[\frac{1}{n_1} - \frac{1}{n_2}\right] m_{e(nucleus)} c, \qquad (5.7.1:3)$$

where $m_{e(Nucleus)}$ is the rest mass of the electron in the nucleus frame as given in equation (5.1.4:9) and c is the local velocity of light related to the velocity of light in hypothetical homogeneous space, c_0, by equation (5.1.4:10).

Substitution of equation (3.3.3:8) for c_0 in equation (5.7.1:2) Balmer's formula for frequencies yields the form [see equation (5.1.4:15)]

$$f_{n1,n2} = A_{n1,n2} B_{(\delta,\beta)} \cdot t^{-1/3}, \qquad (5.7.1:4)$$

where the constant $A_{n1,n2}$ is

Mass, mass objects and electromagnetic radiation

$$A_{n1,n2} = \frac{a^2 m_{e(0)}}{2 h_0} \left(\frac{2GI_g M_\Sigma}{3}\right)^{1/3} Z^2 \left[\frac{1}{n_1} - \frac{1}{n_2}\right], \qquad (5.7.1:5)$$

and $B_{(\delta,\beta)}$ is

$$B_{(\delta,\beta)} = \prod_{i=0}^{n}\left[(1-\delta_i)\sqrt{1-\beta_i^2}\right]. \qquad (5.7.1:6)$$

Equations (5.7.1:4–6) give the characteristic oscillation frequencies of hydrogen-like atoms in terms of the time since the singularity of space, the state of motion and gravitation of the oscillating atom in space, and the main quantum numbers defining the transition in the atom.

In a defined state of motion and gravitation in space, the frequency of an atomic oscillator is a function of time t only

$$f_{n1,n2(\delta,\beta)} = A_{n1,n2} B_{(\delta,\beta)} \cdot t^{-1/3}, \qquad (5.7.1:7)$$

or in differential form

$$\frac{\Delta f}{f} = -\frac{1}{3}\frac{\Delta t}{t} \quad \left(= -\frac{\Delta c_0}{c_0}\right). \qquad (5.7.1:8)$$

Natural periodic phenomena

Timekeeping is based on selected periodic phenomena like the orbital motion of the Earth, the rotation of the Earth, or the frequency of characteristic oscillations of atomic objects. In the Dynamic Universe, none of the periodic phenomena mentioned is constant but slows down with the expansion of space and the related dilution of the energy excitation of all matter and processes in space. The lengthening of a day due to the expansion of space corresponds to the increase of a second based on the period of the Ce-clock. However, the increase in lunar distance and tidal interactions results in extra lengthening of a day, which makes the length of a day increase by about 2 ms/century when measured with atomic time, Table 5.7.1-I.

Periodic phenomenon	Proportionality to the time from the singularity	Annual increase of the period (due to expansion) $\Delta P/P$	Change due to other identified mechanisms $\Delta P/P$
Atomic clocks	$\sim t^{1/3}$	$\approx 3.6 \cdot 10^{-11}$	
Rotational period (a day on the Earth)	$\sim t^{1/3}$	$\approx 3.6 \cdot 10^{-11}$	$\approx 2.7 \cdot 10^{-10}$
Orbit time of planets (a year for the Earth)	$\sim t$	$\approx 1.1 \cdot 10^{-10}$	

Table 5.7.1-I. The direct effect of the expansion of space on the period of atomic clocks, the length of a day, and the length of a year. The length of a day is also affected by the increase in the lunar distance related to tidal interactions.

In the DU framework, most natural phenomena are related to the slowing velocity of light, determined by the expansion velocity of space in the fourth dimension. It means that, like the ticking frequency of atomic clocks, the radioactive decay rate also slows down with the expansion of space. Accordingly, radiometric dating results based on constant decay rates of radioactive isotopes must be corrected for faster decay rates in the past. The decay rates approach infinity in the singularity, which excludes the possibility of radiometric ages beyond the age of the expanding space.

Radioactive decay rates, like the frequencies of atomic clocks, are subject to slowing like the frequencies of atomic clocks. In radiometric dating, linear extrapolation of ages based on constant decay rates gives a major error in ages approaching the age of the expanding space.

For terrestrial observations, it is, however, convenient to fix the timescale and the unit of time, a second, to the frequency of local atomic clocks, which are the most accurate known instruments available for measuring time. In present timekeeping, the unit of time, the **SI Second**, is based on the frequency of radiation from the transition between two hyperfine levels of the ground state of the cesium-133 atom on the Earth's geoid.

Coordinated Universal Time

The early definition of a second was based on the average rotation time of the Earth (1/86400 of the average day). Currently, the average day is about 2 milliseconds longer than 86400 seconds. To match the *Coordinated Universal Time* (UTC) to UT1, the timescale following the rotation of the Earth, a leap second is added almost annually to the UTC time, Figure 5.7.1-1.

The DU framework does not require different timescales (coordinate times) in different gravitational frames. For example, the effect of local gravitational environment required in precise ephemeris calculations is included in the equations of motions (see equation (4.2.2:3)).

The frequencies of atomic clocks in different gravitational environments are expressed in terms of the energy state of the clock based on the nested energy frames. The system of nested energy frames links the frequency of a clock in an arbitrary energy environment to the frequency of a Ce-clock in a fixed position on the Earth's geoid, thus linking together the Terrestrial Time (TT), Geocentric Coordinate Time (TCG), and Barycentric Coordinate Time (TCB) and allowing a completion of the list with Milky Way Coordinate Time and Extragalactic Coordinate Time.

Due to the eccentricity of the Earth's orbit, the frequencies of atomic clocks on the Earth and in the Earth's gravitational frame are subject to annual variations. At perihelion, the orbital velocity of the Earth is at its maximum, and the distance to the Sun is at its minimum. Both of these factors result in the slowing of clock frequencies. Such variation is observable only in comparison to a reference clock outside the solar gravitational frame.

The rotation of the Earth results in a local daily variation in the distance to the Sun. With reference to the analysis in Section 4.1.5, such variation, as internal variation in the Earth's gravitational frame, does not affect the frequency of the clocks. The same is true for the frequency of Earth satellite clocks; the frequency of an Earth satellite clock is independent of the momentary distance from the satellite to the Sun.

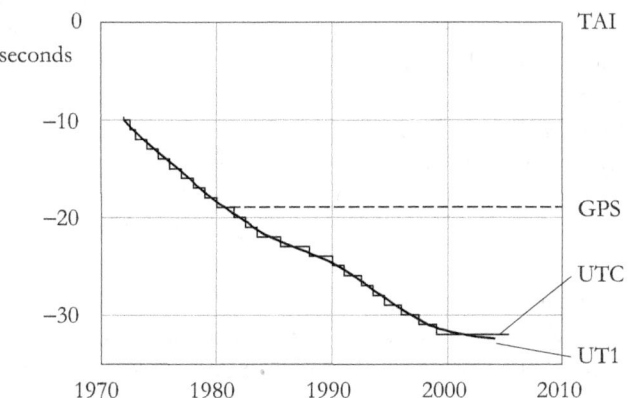

Figure 5.7.1-1. A leap second has been added since 1972 to the Coordinated Universal Time (UTC) in order to keep it synchronized to the rotational velocity of the Earth and the related UT1-time. GPS time was fixed to UTC in 1972 which means that in 2004 GPS is 13 seconds ahead of UTC. The reference is the International Atomic Time (TAI) based on atomic clocks on Earth's geoid.

5.7.2 Units of time and distance, the frames of reference

The Earth second

Equation (5.7.1:4) gives a general expression for the frequency of an atomic oscillator, the primary standard for the **SI Second**. Relative to the frequency of the oscillator at a selected moment t_0, the frequency at t can be expressed as

$$f_{n1,n2(\delta,\beta)(t)} = f_{n1,n2(\delta,\beta)(t_0)} \cdot \left(\frac{t}{t_0}\right)^{-1/3}, \tag{5.7.2:1}$$

where frequency $f_{n1,n2(\delta,\beta)(t_0)}$ is

$$f_{n1,n2(\delta,\beta)(t_0)} = A_{n1,n2} \prod_{i=0}^{n} \left[(1-\delta_i)\sqrt{1-\beta_i^2}\right] t_0^{-1/3}, \tag{5.7.2:2}$$

where the factor $A_{n1,n2}$ [see equation (5.7.1:5)] defines the effect of the quantum states the oscillation is related to, and t is the time since the singularity, which now is about 9.3 billion years.

The unit of time, the second, was originally defined as the fraction 1/86400 of the mean solar day. Now, according to the definition accepted in 1967, the length of a second, the **SI unit of time**, is equal to the duration of 9192631770 periods of the radiation corresponding to the transition between two hyperfine levels of the ground state of the cesium-133 atom.

The timescale based on the SI second is referred to as **International Atomic Time (TAI)**, which is a statistical atomic time scale based on a large number of clocks operating in laboratories around the world. **TAI** time is maintained by the Bureau International des Poids et Mesures; its unit interval is exactly one SI second on the Earth's geoid, "*the equipotential surface of the Earth's gravity field which best fits global mean sea level*".

With reference to equation (3.3.3:10), in absolute time, the SI-second lengthens as

$$\frac{\Delta T}{T} = -\frac{\Delta f}{f} = -\frac{\Delta c}{c} \approx 3.6 \cdot 10^{-11} \,[1/\text{year}] \quad \Rightarrow \quad \approx 1.13\,[\text{ms}/\text{year}], \qquad (5.7.2:3)$$

and is subject to the gravitational state and motion of the standard.

Because important quantities observed are proportional to the velocity of light on the Earth, it is useful to apply the presently defined **SI Second** as the **Terrestrial SI second**, binding the **TAI** time to **the Earth's geoid and the slowing velocity of light on the Earth**. By definition, the SI second is

$$\Delta t[1\ \textbf{SI Second}] = \Delta N \big/ \overline{f_{Ce,G(\Phi),t}} \qquad (5.7.2:4)$$

or

$$\Delta t[1\ \textbf{SI Second}] = \Delta N \Big/ \Big[f_{E,0}\big(1-\delta_{G(\Phi)}\big)\sqrt{1-\beta_{G(\Phi)}^2}\,\Big] = \Delta N/f_{E,G}, \qquad (5.7.2:5)$$

where $\Delta N = 9192631770$ is the number of cycles defining the length of a second, and frequency $\overline{f_{Ce,G(\Phi),t}}$ is the average frequency of the SI-second standard on the Earth's geoid

$$\overline{f_{Ce,G(\Phi),t}} = f_{E,0}\big(1-\delta_{G(\Phi)}\big)\sqrt{1-\beta_{G(\Phi)}^2} = f_{E,G}, \qquad (5.7.2:6)$$

where $\delta_{G(\Phi)}$ and $\beta_{G(\Phi)}$ are the gravitational factor and velocity of rotation at latitude Φ on the geoid. The geoid is defined so that the total effect of the gravitational factor and motion is constant at any latitude

$$G_G = \big(1-\delta_{G(\Phi)}\big)\sqrt{1-\beta_{G(\Phi)}^2} = \left(1-\frac{GM_{G(\Phi)}}{c^2 R_{G(\Phi)}}\right)\sqrt{1-\frac{v_{G(\Phi)}^2}{c^2}}$$

$$\approx 1 - \frac{GM_{G(\Phi)}}{c^2 R_{G(\Phi)}} - \frac{1}{2}\frac{v_{G(\Phi)}^2}{c^2} = 1 - \Delta G_{G\delta} - \Delta G_{G\beta} = \text{constant}. \qquad (5.7.2:7)$$

In equation (5.7.2:6), the frequency $\overline{f_{Ce,G(\Phi),t}}$ is the average frequency of the SI-second standard on the Earth's geoid

$$\overline{f_{Ce,G(\Phi),t}} = f_{E,0}\big(1-\delta_{G(\Phi)}\big)\sqrt{1-\beta_{G(\Phi)}^2}, \qquad (5.7.2:8)$$

where $f_{E,0}$ is the frequency of the standard at rest in apparent homogeneous space of the Earth's gravitational frame

$$f_{E,0} = f_{Ce(0,t)} \prod_X \big(1-\delta_X\big)\sqrt{1-\beta_X^2} = f_{Ce(0,t)} \prod_X G_X, \qquad (5.7.2:9)$$

combining the effects of motion and gravitation in the Extragalactic frame ($X=XG$), Milky Way ($X=MW$) frame, and the solar frame ($X=S$).

The reference frequency $f_{Ce(0,t)}$ in equation (5.7.2:9)

$$f_{Ce(0,t)} = A_{n1,n2} \cdot t^{-1/3}, \tag{5.7.2:10}$$

is the frequency of the Ce_{133} standard at rest in hypothetical homogeneous space slowing down with the deceleration of the expansion of space in the direction of the 4-radius.

In a general form, the factors of gravitation and motion on the frequency can be expressed in the form of factors G_X is

$$G_X = (1-\delta_X)\sqrt{1-\beta_X^2} = \left(1-\frac{GM_X}{c^2 R_X}\right)\sqrt{1-\frac{v_X^2}{c^2}} = G_{X\delta} \cdot G_{X\beta}, \tag{5.7.2:11}$$

or by approximating

$$G_{X\delta} \approx 1 - \frac{GM_X}{c^2 R_X} = 1 - \Delta G_{X\delta}, \tag{5.7.2:12}$$

and

$$G_{X\beta} = \sqrt{1-\beta_X^2} \approx 1 - \tfrac{1}{2}\beta_X^2 = 1 - \Delta G_{X\beta}. \tag{5.7.2:13}$$

Using these approximations, the total effect of the state factors, G_X, can be expressed as

$$G_{\Pi(X)} \approx \prod_X \left(1 - \sum_i \Delta G_{X(i)}\right), \tag{5.7.2:14}$$

or in terms of frequency

$$f_{Ce,G(\theta),t} \approx f_{Ce(0,t)} \prod_X \left(1 - \sum_i \Delta G_{X(i)}\right), \tag{5.7.2:15}$$

where $f_{Ce(0,t)}$ is the frequency of the Ce-standard at rest in hypothetical homogeneous space.

For an arbitrary terrestrial clock, index X has values $X = XG, MW, S, E$, and L, where L refers to a local energy system to which the clock is bound in the Earth's gravitational frame. State factor G_E gives the effects of gravitation and motion in the Earth gravitational frame, G_S gives the gravitation and motion of the local frame in the solar barycenter frame, G_{MW} gives the gravitation and motion of the solar frame in the Milky Way frame, and G_{XG} gives the effects of gravitation and motion of the Milky Way frame in extragalactic space, which is regarded as the frame next to hypothetical homogeneous space.

The frequency of a clock in an arbitrary state of gravitation and motion can be related to the SI standard clock as

$$f_{Ce,\delta_i,\beta_i,t} = f_{Ce,SI,t} \frac{\prod_X \left(1 - \sum_i \Delta G_{X(i)}\right)}{\prod_{X,SI} \left(1 - \sum_i \Delta G_{X(i),SI}\right)}, \tag{5.7.2:16}$$

where the state factors in the denominator refer to the state of the SI-second standards on the Earth's geoid.

For objects in Kepler's orbit in a gravitational frame, the orbiting radius R can be expressed as

$$\frac{1}{R} = \frac{1+e\cos\varphi}{a(1-e^2)} \approx \frac{1}{a}(1+e\cos\varphi), \qquad (5.7.2:17)$$

where a is the semimajor axis, e is the eccentricity of the orbit, and φ is the angle from perihelion in the direction of the motion. The last form of equation (5.7.2:17) applies for orbits with small eccentricity.

The energy relation for Kepler's orbit can be expressed as

$$\frac{v_{E(S)}^2}{2} = GM_S\left[\frac{1}{R_{E(S)}} - \frac{1}{2a_{E(S)}}\right]. \qquad (5.7.2:18)$$

Substitution of equations (5.7.2:17) and (5.7.2:18) into equation (5.7.2:9) gives the factor G_{Xi} of an object in Kepler's orbit, combining the effects of motion and gravitation in form

$$G_X \approx 1 - \frac{3}{2}\frac{GM}{ac^2} - \frac{2GM}{ac^2}e\cdot\cos\varphi = 1 - \frac{GM}{ac^2}\left(\frac{3}{2} + 2e\cdot\cos\varphi\right). \qquad (5.7.2:19)$$

The meter

In 1980, the iodine-stabilized Helium-Neon laser wavelength was accepted as a length standard. It had a wavelength uncertainty of 1 part in 10^{10} at the time. In 1983, the meter was redefined again. The definition states that one meter is the length traveled by light in a vacuum during 1/299792458 of a second, based on the velocity of light equal to c = 299,792,458 m/s. The General Conference on Weights and Measures, however, accepted the iodine-stabilized Helium-Neon laser[73] as a recommended radiation for realizing the meter. The wavelength of the HeNe laser is λ_{HeNe} = 632.99139822 nm with an estimated relative standard uncertainty of ± 2.5 x 10^{-11}.

By applying equation (5.7.2:5), the 1983 definition of the SI-meter can be expressed as

$$\begin{aligned}\Delta L1\,[\text{SI Meter}] &= c_{E0}\left(1-\delta_{G(\Phi)}\right)\cdot\Delta N \Big/ \left[f_{E,0}\left(1-\delta_{G(\Phi)}\right)\sqrt{1-\beta_{G(\Phi)}^2}\right] \\ &= c_{E0}\cdot\Delta N \Big/ \left[f_{E,0}\sqrt{1-\beta_{G(\Phi)}^2}\right],\end{aligned} \qquad (5.7.2:20)$$

where $c_{E0}(1-\delta_{G(\Phi)})$ is the local velocity of light at latitude Φ. Comparison of equations (5.7.2:5) and (5.7.2:20) shows that the SI-second is constant on the Earth geoid, but the SI-meter is increased at low latitudes (close to the equator) due to the rotational motion of the Earth.

By applying equation (5.1.4:16), the length of a meter based on the characteristic wavelength of the iodine stabilized Helium-Neon laser on the Earth geoid can be expressed in the form

$$\Delta L\,[\text{SI Meter}] = N_\lambda \cdot \lambda_{\text{HeNe}[G(\Phi)]} = \frac{N_\lambda \cdot \lambda_{\text{HeNe}[E,0]}}{\sqrt{1-\beta_{G(\Phi)}^2}}, \qquad (5.7.2{:}21)$$

where $N_\lambda = 632\,991\,398\,220$ is the number of waves in a SI-meter, and wavelength $\lambda_{\text{HeNe}[E,0]}$ is the wavelength of the laser at rest in the Earth gravitational frame (like at the poles). In absolute meters, equation (5.7.2:21) makes the [SI]-wavelength of the $\lambda_{\text{HeNe(SI)}}$ laser a function of the latitude through the effect of the rotational velocity $\beta_{G(\Phi)}$. Comparison of equations (5.7.2:20) and (5.7.2:21) shows that the SI-meter based on the light time in 1/299792458 of an SI-second and on the wavelength of a Helium-Neon laser are equally functions of latitude. At the equator on the Earth geoid the SI-meter is about $1.2 \cdot 10^{-12}$ meters longer than it is at the poles.

Because the Bohr radius is directly proportional to the characteristic wavelengths, the classical meter standard based on a platinum rod is also a function of latitude. Lengthening of the one-meter platinum rod by $1.2 \cdot 10^{-12}$ is about 1% of the radius of a platinum atom in the rod. In the case of the HeNe-laser, the lengthening is about $2 \cdot 10^{-6}$ of the wavelength in a meter.

The Earth geoid

The effects of gravitation and motion in the Earth gravitational frame [or in Earth Centered Inertial frame (ECI-frame)] are expressed in terms of factor G_E in equation (5.7.2:6). With reference to equation (5.7.2:7), the factor G_E consists of the term $\Delta G_{E\delta}$ related to the gravitational state in the Earth gravitational frame and the term $\Delta G_{E\beta}$ related to the velocity in the Earth gravitational frame.

On the Earth geoid, $\Delta G_{E\delta}$ is determined by the local radius of the geoid and the effect of the quadrupole moment of gravitation due to the ellipsoidal shape of the Earth

$$\Delta G_{E\delta} = \delta_{\Phi(E)} = -\frac{GM}{c^2 r_{\Phi(E)}}\left\{1 - J_2\left[\frac{r_{0(E)}}{r_{\Phi(E)}}\right]^2 \frac{1}{2}\left(3\sin^2\Phi - 1\right)\right\}, \qquad (5.7.2{:}22)$$

where J_2 is the quadrupole moment coefficient $J_2 = 1.0826300 \cdot 10^{-3}$ and $r_{0(E)}$ is the equatorial radius (latitude $\Phi = 0$) of the Earth. Radius $r_{\Phi(E)}$ at latitude Φ is approximated as

$$\begin{aligned}r_{\Phi(E)} =\ & 6356742.025 \\ & + 21353.642\,\cos^2\Phi + 39.832\cos^4\Phi \\ & + 0.798\cos^6\Phi + 0.003\cos^8\Phi \quad [\text{m}].\end{aligned} \qquad (5.7.2{:}23)$$

On the geoid, the velocity term $\Delta G_{E\beta}$ can be expressed as

$$\Delta G_{E\beta} \approx \tfrac{1}{2}\beta_{\Phi(E)}^2 = \frac{1}{2}\frac{v_{\Phi(rot)}^2}{c^2} = \frac{\omega_E^2 r_{\Phi(E)}^2 \cos^2\Phi}{c^2}, \qquad (5.7.2{:}24)$$

where $v_{\Phi(rot)}$ is the rotational velocity, $r_{\Phi(E)}$ the radius on the geoid at latitude Φ, and ω_E the sidereal angular velocity of the rotation of the Earth

$$\omega_E = \frac{2\pi}{86164.0989} \left[\frac{1}{s}\right]. \quad (5.7.2:25)$$

The effect of latitude on the gravitational and rotational terms $\Delta G_{E\delta}$ and $\Delta G_{E\beta}$ is opposite. Close to the poles, the velocity due to rotation is lower, reducing the rotational velocity term while the gravitational term is increased due to the smaller radius of the Earth. The average value of factor G_E on the Earth geoid (the value applicable to an SI-second standard) is

$$G_E \simeq 1 - \Delta G_{E\delta}(\Phi) - \Delta G_{E\beta}(\Phi) \simeq 1 - 6.969291 \cdot 10^{-10}. \quad (5.7.2:26)$$

The variation of $\Delta G_E(\Phi) = [\Delta G_{E\delta}(\Phi) + \Delta G_{E\beta}(\Phi)] - \Delta G_{E\delta}(\Phi=\pi/2)$ as a function of latitude Φ is presented in Figure 5.7.2-1. The reference level in the curve in Figure 5.7.2-1 has been fixed to the value of ΔG_E at the poles where only the gravitational term $\Delta G_{E\delta}$ applies. The calculation is based on the CODATA 2006 value of the gravitational constant $G = 6.67428 \cdot 10^{-11}$ [Nm²/kg²] and the Earth mass $M_E = 5.974 \cdot 10^{24}$ [kg], Figure 5.7.2-1.

As illustrated by Figure 5.7.2-1, at the geoid defined by equation (5.7.2:22), the effects of gravitation and the rotational velocity of the Earth are compensated at an accuracy better than $5 \cdot 10^{-16}$. However, for a precise definition of the **Terrestrial SI Second**, the location of the primary standard should be defined at a fixed latitude. A preferred location would be at the North or South Pole, where the effect of motion in the Earth frame is zero. This is also a preferred condition for the definition of one SI meter.

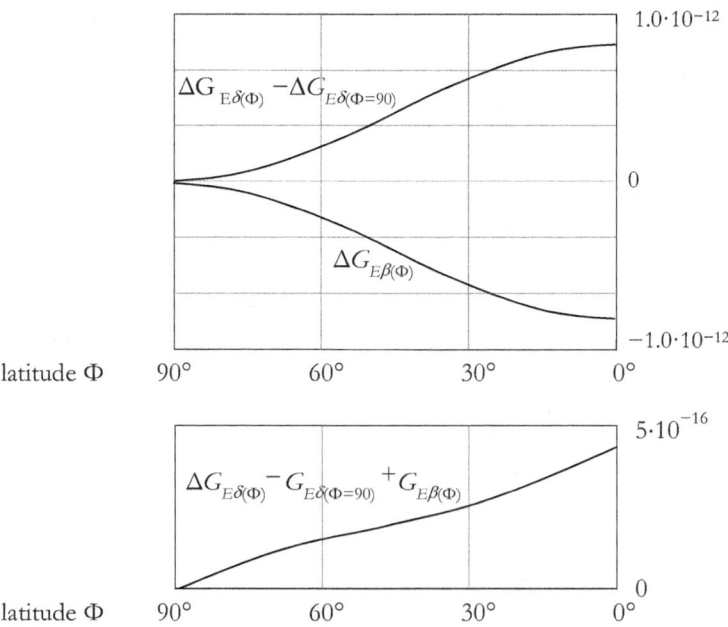

Figure 5.7.2-1. Dependence of the ticking frequency of atomic clocks on latitude on the Earth geoid according to equations (5.7.2:22) and (5.7.2:24). Upper curves show the opposite effects of gravitation and motion and the lower curve the sum of the two effects. The frequency of a clock at either pole, where the angular velocity is zero, is used as the reference.

5.7.3 Periodic fluctuations in Earth clocks

The effect of the eccentricity of the Earth-Moon barycenter orbit

The motion of the Earth in the solar barycenter frame can be described as Kepler's orbit. Due to the eccentricity of Earth's orbit, the factor $G_{S(B)}$ for Earth in the solar gravitational frame is not a constant but varies during the year, reaching a minimum on January 2 when the Earth is at the perihelion of its orbit.

The factor $G_{S(B)}$ in equation (5.7.2:19) gives the correction due to the gravitational state and orbital velocity of the Earth-Moon barycenter frame, which lies within the solar barycenter frame

$$G_{S(B)} \approx 1 - \frac{3}{2}\frac{GM_S}{a_{E(S)}c^2} - \frac{2GM_S}{a_{E(S)}c^2} e_{E(S)} \cdot \cos\varphi, \qquad (5.7.3:1)$$

where $a_{E(S)}$ is the semimajor axis of the Earth-moon barycenter orbit, and angle φ is the true anomaly.

Substitution of numerical values

$a_{E(S)} = 1.4959789 \cdot 10^{11}$ [m],

$G = 6.674 \cdot 10^{-11}$ [Nm²/kg²],

$M_S = 1.9891 \cdot 10^{30}$ [kg], and

$e = 0.0167$

gives the numerical value for $G_{S(B)}$ as function of φ

$$G_{S(B)} \approx 1 - 1.4808 \cdot 10^{-8} - 3.3037 \cdot 10^{-10} \cdot \cos\varphi. \qquad (5.7.3:2)$$

The anomaly angle φ can be expressed in terms of the day in a year as

$$\varphi \approx \frac{D - D_{PH}}{365} \cdot 2\pi, \qquad (5.7.3:3)$$

where D is the number of the day counted from the beginning of the year, and $D_{PH} = 2$ is the number of the perihelion day (January 2). Substitution of equation (7.3.4:3) for φ in equation (5.7.3:2) gives $\Delta G_{S(\varphi)}$ in the form

$$\Delta G_{S(B)} \approx 1.4808 \cdot 10^{-8} + 3.3037 \cdot 10^{-10} \cdot \cos\left(\frac{D - D_{PH}}{365} \cdot 2\pi\right). \qquad (5.7.3:4)$$

Applying (5.7.3:4), the frequency of Earth clocks in the solar gravitational frame can be expressed

$$f_{(Earth)} \approx f_{0(Sun)}\left[1 - 1.4808 \cdot 10^{-8} - 3.3037 \cdot 10^{-10} \cdot \cos\left(\frac{D - D_{PH}}{365} \cdot 2\pi\right)\right], \qquad (5.7.3:5)$$

where $f_{0(Sun)}$ is the frequency of a hypothetical reference clock at rest in the apparent homogeneous space of the solar gravitational frame.

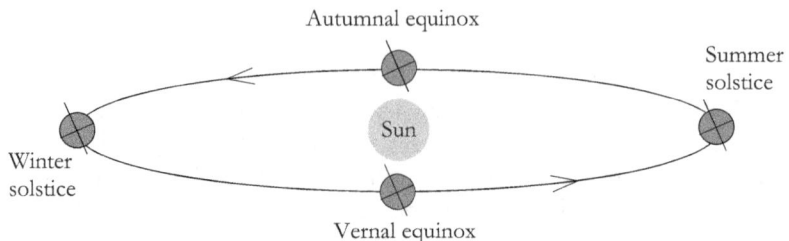

Figure 5.7.3-1. The rotation of the Earth and the inclination angle between the equatorial plane and the ecliptic result in local variations in the distance to the Sun.

The frequency observed from a pulsar is subject to the Roemer effect, resulting from the Doppler effect due to the orbital velocity of the Earth towards or from the pulsar observed.

The Doppler effect due to the orbital velocity of the Earth may be of the order of

$$\frac{\Delta f_{max}}{f} \approx \frac{v_{orb}}{c} \approx 10^{-4}, \qquad (5.7.3:6)$$

corresponding to a situation when the pulsating object is at the orbital plane of the Earth. Accordingly, the effect of the eccentricity of the Earth's orbit is only of the order of ppm compared to the Roemer effect in the observed pulsar frequency.

Rotation and the inclination angle of the Earth

The rotation of the Earth results in a daily periodic variation in the distance from a particular location on the Earth to the Sun. The position of a location in the Earth's gravitational frame relative to the solar gravitational frame can be illustrated by the projection of the location in the Sun frame, which is the apparent homogeneous space of the Earth frame (see Figure 4.1.5-1). Due to the inclination angle between the equatorial plane and the ecliptic plane, there is a small annual variation in the daily distance variation. The inclination also results in a small annual periodic displacement in the solar distance, Figure 5.7.3-1.

With reference to equation (4.1.5:4), within the Earth frame, a difference Δr, in the distance to the Sun ($\Delta r \ll r_s$) does not affect the rest energy of mass. Accordingly, there is no daily period in the frequency of clocks on the Earth due to the variation in the distance to the Sun. This is because the effects of gravitation and motion of the locations in the solar gravitational frame cancel each other (see Section 4.1.5).

The velocity of light, however, is subject to change due to the variation in the distance to the Sun. With reference to equation (4.1.10:3), such variation is

$$\frac{dc}{c} \approx \delta \frac{dr_s}{r_s}, \qquad (5.7.3:7)$$

where r_S is the distance from the center of the Earth to the Sun, and dr_S is the difference in the distance to the Sun. The difference $dr_S(t,D,\Phi,H)$ is a function of the inclination angle, the day of the year, the latitude, and the hour of the day in local time. The maximum daily variation occurs at the equator at the vernal and autumnal equinox, then $\Delta r_S = +r_E$ at midnight and $\Delta r_S = -r_E$ at noon. The maximum of the annual variation occurs at the North Pole and South Pole; at the winter solstice, the distance is at maximum at the North Pole and at the minimum at the South Pole.

5.7.4 Galactic and extragalactic effects

Solar system in the Milky Way frame

The speed of the solar system in the Milky Way is about 220 km/sec. Applying the estimated mass of the Milky Way $M_{MW} \approx 2 \cdot 10^{11}$, the solar mass $\approx 4 \cdot 10^{41}$ [kg], and the distance $r_{MW} \approx 25\,000$ l.y. $\approx 2.4 \cdot 10^{20}$ [m], the factors of $\Delta G_{MW(\delta)}$ and $\Delta G_{MW(\beta)}$ due to gravitation and motion in equation (5.7.2:7) are

$$\Delta G_{MW(\delta)} \approx 10^{-6}$$
$$\Delta G_{MW(\beta)} \approx 2.6 \cdot 10^{-7}. \tag{5.7.4:1}$$

According to the values in equation (5.7.4:1), the velocity factor $\Delta G_{MW(\beta)}$ of the solar system in the Milky Way is only about half that required by circular Kepler's motion. In terms of velocity, this means that the velocity is about 0.7 of the velocity on a circular orbit.

Milky Way galaxy in Extragalactic space

The Milky Way is part of the Local Group of galaxies, of which the Andromeda Nebula is the other large member. The Milky Way is estimated to move towards the common center of the Local Group at a speed of about 40 km/s. The Local Group is part of the Local Supercluster, a much larger collection of galaxies including the Virgo Cluster at about 45 million light-years from the Milky Way. The Local Group is estimated to move at a speed of about 600 km/s towards the Great Attractor in the direction of the constellation Centaurus. When estimated from the dipole pattern of the microwave background radiation, the solar system is moving at 350-400 km/sec relative to the background.

The gravitational structure at the extragalactic level is not known in detail. Assuming that a velocity of 600 km/s has been obtained in free fall from gravitation, we end up with factors of $\Delta G_{XG(\delta)}$ and $\Delta G_{EG(\beta)}$ as high as

$$\Delta G_{XG} = \Delta G_{XG(\delta)} + \Delta G_{XG(\beta)} \approx 2 \cdot \tfrac{1}{2}\left(2 \cdot 10^{-3}\right)^2 = 4 \cdot 10^{-6}. \tag{5.7.4:2}$$

5.7.5 Summary of timekeeping

Average frequency of the SI-second standard

In universal time, the frequency of an SI-second standard, as well as that of any clock based on characteristic oscillation frequencies of atomic objects, can be expressed in the form

$$f_{\delta,\beta} = f_{0,0} \prod_{i=0}^{n}\left[(1-\delta_i)\sqrt{1-\beta_i^2}\right], \tag{5.7.5:1}$$

where $f_{0,0}$ is the frequency of the clock at rest in hypothetical homogeneous space, and δ_i and β_i characterize the states of gravitation and motion of the clock in the local energy frame and the parent frames. The frequency $f_{0,0}$ is subject to slowing due to the gradual release of the energy excitation of matter with the expansion of space

$$f_{0,0} = A \cdot t^{-1/3}. \tag{5.7.5:2}$$

The slowdown of the frequency $f_{0,0}$ occurs in parallel with the slowdown of the velocity of light with the expansion of space.

Clock frequencies on the Earth and in the Earth's gravitational frame are subject to annual periodic variation due to the eccentricity of the Earth's orbit. The effect of local periodic variations in solar distance within the Earth's gravitational frame on the clock frequencies is cancelled due to the opposite effects of gravitation and velocity in the solar frame (Section 4.1.5).

By substituting all non-periodic factors of gravitation and motion (G&M) of a location on the Earth geoid in the Earth gravitational frame (E-frame) in equation (5.7.2:26), in the Sun frame (S-frame) in equation (5.7.3:4), in the Milky Way frame (MW-frame) in equations (5.7.4:1), and in Extragalactic space (XG-frame) in equation (5.7.4:2) into equation (5.7.2:11), the average frequency of the SI-second standard, a Ce-clock on the Earth geoid, can be related to the frequency of a Ce-clock at rest in hypothetical homogeneous space as

$$\begin{aligned}f_{Ce(\delta,\beta)} &\approx f_{Ce(0,0)} \cdot \left(1 - 4\cdot 10^{-6}\right) \quad \text{G\&M of MW-frame in the XG-frame} \\ &\quad \cdot \left(1 - 1.26 \cdot 10^{-6}\right) \quad \text{G\&M of S-frame in the MW-frame} \\ &\quad \cdot \left(1 - 1.4808 \cdot 10^{-8}\right) \quad \text{G\&M of E-frame in the S-frame} \\ &\quad \cdot \left(1 - 6.969291 \cdot 10^{-10}\right) \quad \text{G\&M of Earth geoid the E-frame}.\end{aligned} \tag{5.7.5:3}$$

The average frequency of a clock moving at velocity β_L in a closed energy system L in a laboratory at gravitational state δ_G and velocity β_G in the Earth's gravitational frame is

Mass, mass objects and electromagnetic radiation

$$f_{Ce,E(\delta,\beta),t} \approx f_{Ce(0,0)} \quad \cdot (1-4\cdot 10^{-6}) \quad \text{G\&M of MW-frame in the XG-frame}$$

$$\cdot (1-1.26\cdot 10^{-6}) \quad \text{G\&M of S-frame in the MW-frame}$$

$$\cdot (1-1.4808\cdot 10^{-8}) \quad \text{G\&M of E-frame in the S-frame} \quad (5.7.5:4)$$

$$\cdot (1-\delta_G)\sqrt{1-\beta_G^2} \quad \text{G\&M of L-frame in the E-frame}$$

$$\cdot \sqrt{1-\beta_L^2} \quad \text{velocity factor in the } L\text{-frame}.$$

A closed energy system in the Earth's frame may be a particle accelerator, a centrifuge, or any other system in which local potential energy is converted into motion. For a clock stationary on the Earth in the Earth's gravitational frame, in an airplane, or in a satellite, $\beta_L = 0$. By dividing equation (5.7.5:4) by equation (5.7.5:3), we can relate the average frequency of a Ce-standard in δ_E, β_E state in the Earth frame to the average frequency of the Ce-standard on the Earth geoid

$$f_{Ce,t} = f_{Ce,SI,t} \frac{(1-\delta_E)\sqrt{1-\beta_E^2}}{1-6.969291\cdot 10^{-10}} \approx f_{Ce,SI,t} \frac{1-\delta_E-\tfrac{1}{2}\beta_E^2}{1-6.969291\cdot 10^{-10}}. \quad (5.7.5:5)$$

For two objects in the Earth frame, all the G&M factors in the parent frames are equal.

In equation (5.7.5:5), the frequency $f_{Ce,SI,t}/(1-6.969291\cdot 10^{-10})$ is the reference frequency of a Ce-standard for the Earth gravitational frame

$$\frac{f_{Ce,SI,t}}{1-6.969291\cdot 10^{-10}} = f_{Ce,0E} \approx f_{Ce,SI,t}(1+6.969291\cdot 10^{-10}). \quad (5.7.5:6)$$

Physically, the frequency $f_{Ce,0E}$ has the meaning of a reference Ce-clock at rest in the apparent homogeneous space of the Earth frame. Substitution of $f_{Ce,0E}$ back into equation (5.7.5:5) gives the frequency of a Ce-clock in a δ_E, β_E state in the Earth frame in the form

$$f_{Ce,t} = f_{Ce,0E}(1-\delta_E)\sqrt{1-\beta_E^2} \approx f_{Ce,0E}(1-\delta_E-\tfrac{1}{2}\beta_E^2). \quad (5.7.5:7)$$

Equations (5.7.5:7) and (5.7.5:6) apply for clocks in the Earth's gravitational frame, on the Earth, and in Earth satellites, in full agreement with the Geocentric Coordinate time definition.

> **Geocentric Coordinate Time (TCG)** is a *coordinate time* having its spatial origin at the center of mass of the Earth. TCG differs from TT as: TCG − TT = Lg x (JD − 2443144.5) x 86400 seconds, with Lg = 6.969291·10⁻¹⁰.

Following the same procedure as done for an arbitrary clock on Earth, the frequency of a clock outside the Earth frame, at gravitational state δ_S and velocity β_S in the solar gravitational frame, can be expressed as

$$f_{Ce,t} \approx f_{Ce,SI,t} \frac{1-\delta_S - \frac{1}{2}\beta_S^2}{(1-6.969291\cdot 10^{-10})(1-1.4808\cdot 10^{-8})} \quad (5.7.5:8)$$

$$\approx f_{Ce,SI,t} \frac{1-\delta_S - \frac{1}{2}\beta_S^2}{1-1.5505\cdot 10^{-8}},$$

where all the G&M factors outside the Sun frame are equal and, accordingly, reduced. The frequency $f_{Ce,SI,t}/[(1-6.969291\cdot 10^{-10})(1-1.4808\cdot 10^{-8})]$ is the reference frequency of a Ce-standard in the solar gravitational frame

$$f_{Ce,0S} = \frac{f_{Ce,SI,t}}{(1-6.969291\cdot 10^{-10})(1-1.4808\cdot 10^{-8})} \quad (5.7.5:9)$$

$$\approx f_{Ce,SI,t}(1+1.5505\cdot 10^{-8}).$$

Equation (5.7.5:9) is closely related to the definition of the Barycentric Coordinate time. Substituting $f_{Ce,0S}$ back into equation (5.7.5:8), the frequency of a Ce-clock in a δ_S, β_S state in the Sun frame becomes

$$\begin{aligned}f_{Ce,t} &= f_{Ce,0S}(1-\delta_S)\sqrt{1-\beta_S^2}\,(1-\delta_E)\sqrt{1-\beta_E^2} \\ &\approx f_{Ce,0S}(1-\delta_S - \tfrac{1}{2}\beta_S^2)(1-\delta_E - \tfrac{1}{2}\beta_E^2).\end{aligned} \quad (5.7.5:10)$$

Equation (5.7.5:10) is needed for spacecraft clocks in the solar gravitational frame. Far from the Earth, the frame factors δ_E and β_E approach zero. When applied to clocks within the Earth frame, equation (5.7.5:10) combines the effects of the Earth frame and the Sun frame. Physically, frequency $f_{Ce,0S}$ has the meaning of a reference clock at rest in the apparent homogeneous space of the solar gravitational frame.

> **Barycentric Coordinate Time (TCB)** is a *coordinate time* having its spatial origin at the solar system barycenter. TCB differs from TDB in rate. The two are related by:
> TCB−TDB = Lb x (JD−2443144.5) x 86400 seconds, with Lb = 1.550505·10⁻⁸.

For the Milky Way frame, the reference frequency obtains the form

$$f_{Ce,0MW} = \frac{f_{Ce,SI,t}}{(1-6.969291\cdot 10^{-10})(1-1.4808\cdot 10^{-8})(1-1.26\cdot 10^{-6})} \quad (5.7.5:11)$$

$$\approx f_{Ce,SI,t}(1+1.275\cdot 10^{-6}).$$

For extragalactic frequency comparisons, the hypothetical homogeneous space reference $f_{Ce,0,t}$ is needed. It can be solved from equation (5.7.5:1), which results in

$$f_{Ce,0,t} \approx f_{Ce,SI,t}(1+5.27\cdot 10^{-6}). \quad (5.7.5:12)$$

In principle, equation (5.7.5:12) applies in any energy frame in a nested energy system [see equation (5.7.2:9)]. Specifically, it is needed in cosmological observations at extragalactic distances.

6. The dynamic cosmology

The DU model means a major reorientation in our picture of the Universe. We live in a much more structured world than suggested by the standard cosmology model based on the theory of relativity. Energy structures in space are created via diversification of mass and the expressions of energy in space, maintaining a link to the whole through a system of nested energy frames. As a consequence, the dimensions of galaxies and the radii of orbiting stellar systems expand in direct proportion to the expansion of the whole space.

The precise geometry, dynamics, and energy development of space allow the derivation of precise mathematical expressions for redshifts, optical distances, angular sizes, and apparent magnitudes of cosmological objects without additional parameters.

The expansion of space will continue until infinity at a decelerating rate until the energy of motion gained from the gravitational energy in the pre-singularity contraction is paid back to the gravitational energy of the structure.

6.1 Redshift and the Hubble law

Redshift observations by Edwin Hubble in the 1920's provided the observational underpinning for the conclusion that space is undergoing homogeneous expansion. The recession velocity of galaxies appeared to be proportional to their distance. The relationship between the recession velocity and the distance of objects is known as Hubble's law.

6.1.1 Expanding and non-expanding objects

As shown by the analysis of the Bohr radius, material objects built of atoms and molecules are not subject to expansion with space, Figure 6.1.1-1. As shown by equations (5.1.4:16) and (5.1.4:19), like the Bohr radius, the characteristic emission wavelengths of atomic objects are likewise unchanged in the course of the expansion of space. When propagating in space, the wavelength of electromagnetic radiation is increased in direct proportion to the expansion. Accordingly, when detected after propagation in space, characteristic radiation is observed redshifted relative to the wavelength emitted by the corresponding transition in situ at the time of observation.

The theory of relativity and the standard cosmology model assume that the rest energy of matter has remained constant since the formation of matter after the "Big Bang". Likewise, the velocity of light is assumed to have remained constant. In the DU framework, the energy excitation of motion and gravitation is diluted during the expansion, which is directly reflected in the rest energy of matter

$$E_{rest} = c_0 mc = \frac{GmM''}{R''}. \tag{6.1.1:1}$$

The Dynamic Cosmology

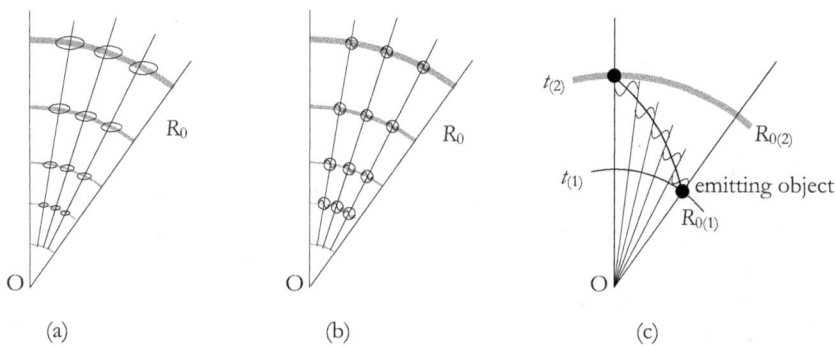

Figure 6.1.1-1. (a) Dimensions of galaxies and other gravitationally bound systems expand in direct proportion to the expansion of space. (b) Localized objects bound by electromagnetic forces conserve their size. The characteristic wavelength emitted by atomic objects is conserved. (c) The wavelength of electromagnetic radiation propagating in space increases in direct proportion to the expansion of space. As a consequence, the observed wavelength is redshifted.

The velocity of the expansion of space, c_0, changes with the expansion of space in accordance with equation (3.3.3:8) as

$$c_0 = \frac{2}{3}\frac{R_4}{t}, \qquad (6.1.1:2)$$

where t refers to the time from the singularity in the primary energy buildup process. According to equation (5.1.1:15), the unit energy of a cycle of electromagnetic energy emitted is expressed as

$$E_\lambda = \frac{h_0}{\lambda} c_0 c = m_\lambda c_0 c, \qquad (6.1.1:3)$$

where m_λ is the mass equivalence carried by a radiation quantum. As a requirement of the conservation of total mass in the course of the expansion of space, the mass equivalence of radiation propagating in space is conserved.

Comparison of equations (6.1.1:1) and (6.1.1:3) shows that the rest energy of an emitting object and the energy of electromagnetic radiation emitted change in the same way in the course of the development of the universe.

As shown by equation (5.1.4:16), the wavelength of the characteristic radiation emitted by an object is a function of the emitting object but independent of the expansion of space

$$\lambda_{(n1,n2)} = \frac{\lambda_{0(n1,n2)}}{\prod_{i=1}^{n}\sqrt{1-\beta_i^2}}. \qquad (6.1.1:4)$$

A hydrogen atom, a billion years ago, emitted the same wavelength for a specific energy transition as it does today in the same local environment. The wavelength of radiation propagating in expanding space is assumed to increase in direct proportion to the expansion, but the mass equivalence of radiation is conserved.

6.1.2 Redshift and Hubble law

Optical distance and redshift in DU space

As a consequence of the spherical symmetry and the zero-energy balance in space, the velocity of light in DU space is determined by the velocity of space in the fourth dimension. The momentum of electromagnetic radiation has the direction of propagation in space. Although the actual path of light comprising the expansion of space in the direction of the 4-radius and the propagation of light in space is a spiral in four dimensions, the length of the optical path in the direction of the momentum of radiation in space is the tangential component of the spiral.

Due to the equal velocities of space in the fourth dimension and the propagation velocity of light in space, the length of the optical path in space is equal to the increase of the 4-radius, Figure 6.1.2-1

$$D = R_4 - R_{4(0)}. \qquad (6.1.2:1)$$

The differential of optical distance can be expressed in terms of R_0 and the distance angle θ as

$$dD = R_4 d\theta = c_0 dt = dR_4. \qquad (6.1.2:2)$$

By first solving for the distance angle θ,

$$\theta = \int_{R_{4(0)}}^{R_4} \frac{dR_4}{R_4} = \ln \frac{R_4}{R_{4(0)}} = \ln \frac{R_4}{R_4 - D}, \qquad (6.1.2:3)$$

the optical distance D obtains the form

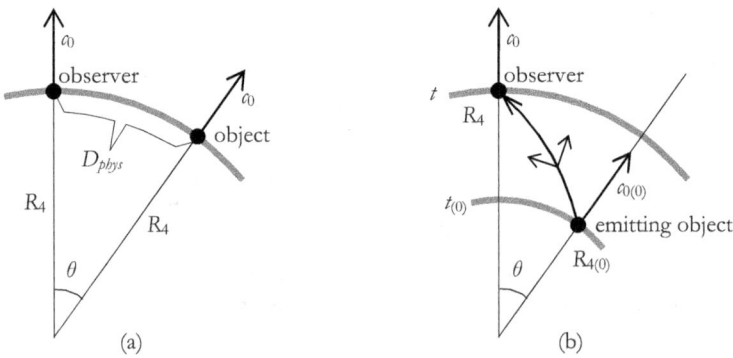

Figure 6.1.2-1. (a) The classical Hubble law corresponds to Euclidean space, where the distance of the object is equal to the physical distance, the arc D_{phys}, at the time of the observation. (b) When the propagation time of light from the object is taken into account, the optical distance is the length of the integrated path over which light propagates in space in the tangential direction on the 4-sphere, so that $D_{opt} = D = \int dD_\perp$. Because the velocity of light in space is equal to the expansion of space in the direction of R_4, the optical distance is $D = R_4 - R_{4(0)}$, the lengthening of the 4-radius during the propagation time.

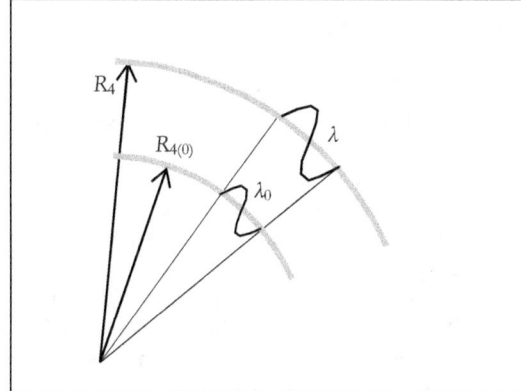

Figure 6.1.2-2. The increase in the wavelength of electromagnetic radiation propagating in expanding space.

$$D = R_4\left(1 - e^{-\theta}\right), \qquad (6.1.2{:}4)$$

where R_0 means the value of the 4-radius at the time of the observation.

The wavelength of radiation propagating in expanding space is assumed to be subject to an increase in direct proportion to the expansion of space, Figure 6.1.2-2. Combining equations (6.1.2:1) and (6.1.2:4), the redshift, the increase in the wavelength, becomes

$$z = \frac{\lambda - \lambda_0}{\lambda_0} = \frac{R_4 - R_{4(0)}}{R_{4(0)}} = \frac{D/R_4}{1 - D/R_4} = e^{\theta} - 1. \qquad (6.1.2{:}5)$$

Solved from (6.1.2:6), the optical distance D can be expressed as

$$D = R_0 \frac{z}{1+z}. \qquad (6.1.2{:}6)$$

The observed recession velocity, the velocity at which the optical distance increases, obtains the form

$$v_{rec(optical)} = \frac{dD}{dt} = c_0\left(1 - e^{-\theta}\right) = \frac{D}{R_4} c_0. \qquad (6.1.2{:}7)$$

As demonstrated by equation (6.1.2:7), the maximum value of the observed (optical) recession velocity never exceeds the velocity of light, c, at the time of the observation, but approaches it asymptotically when distance D approaches the length of 4-radius R_0.

Figure 6.1.2-3 illustrates the development of the size, the optical path, and the redshift in spherically closed space. The optical distance D is the tangential length of the optical path. D is equal to the increase of the 4-radius during the propagation of light from the object to the observer.

Classical Hubble law

The classical Hubble law is written in terms of recession velocity as

$$v_H = H_0 D, \qquad (6.1.2{:}8)$$

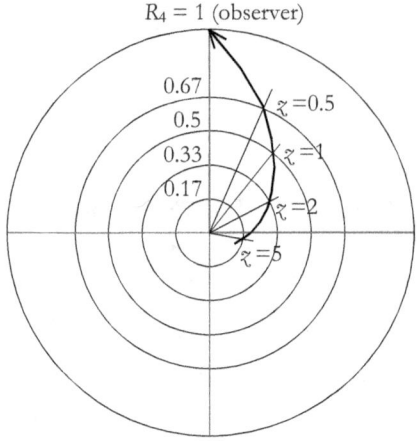

Figure 6.2.1-3. The length of the R_0 radius and the location of objects for redshifts $z = 0$ to 5. Location $R_4 = 1$ is the observer's location. The optical distance to the object is the tangential length of the path, which is equal to the difference between the present R_4 radius and the $R_{4(0)}$ radius as it was when the light was emitted. For example, the light emitted at $R_4 = 0.17$ in the drawing is redshifted by $z = 5$ after traveling the distance $D = (1-0.17)R_0$ in expanding space.

where H_0 is the Hubble constant. The classical Doppler effect expresses a linear connection between the classical redshift and the recession velocity as

$$z_H = \frac{v_H}{c}. \quad (6.1.2:9)$$

In terms of the redshift, which is the observed quantity, the classical Hubble law can be written as

$$z_H = \frac{H_0}{c} D = \frac{D}{R_H}, \quad (6.1.2:10)$$

where

$$R_H \equiv \frac{c}{H_0} = R_4 \quad (6.1.2:11)$$

is referred to as the "Hubble radius" or the radius of the present curvature of space. In the Big Bang model, the Hubble radius can be interpreted as the distance traveled by light at velocity c since the Big Bang ($t = 1/H_0$).

Redshift in the standard cosmology model

In the standard cosmology model, the general expression for the redshift is

$$z \equiv \frac{\Delta\lambda}{\lambda} = \frac{a(t_r)}{a(t_e)} - 1, \quad (6.1.2:12)$$

where $a(t_r)$ and $a(t_e)$, following the formalism of the general theory of relativity, are distances proportional to the dimensions of space when a light signal is received and emitted, respectively [49]. In the DU geometry, the natural measure of expanding distances in space is the length of the 4-radius, which at the time of observation is $a(t_r) = a \cdot R_0$. Because light in space propagates at the same velocity as space expands in the direction of the 4-radius, the length of the 4-radius at the time of the emission of the light is $a(t_e) = a \cdot (R_0 - D)$, where D is the optical distance of the emitting object. Substituting the DU values of $a(t_r)$ and $a(t_e)$ into equation (6.1.2:12) gives

$$z = \frac{a(t_r)}{a(t_e)} - 1 = \frac{R_4}{R_4 - D} - 1 = \frac{D/R_4}{1 - D/R_4}, \qquad (6.1.2:13)$$

which is the same result as that given in equation (6.1.2:5) as the result of a more elaborate derivation.

A present estimate [74] for the value of the Hubble constant is $H_0 = 70.5 \pm 1.3$ [(km/s)/Mpc], which corresponds to a Hubble radius (equal to R_0 in the DU) of about 14 billion light years and corresponds to about 9.3 billion years from the singularity.

Recession velocity of cosmological objects

The physical recession velocity of an object as it is at the time of the observation can be expressed in terms of the distance angle θ and the present velocity of light in accordance with equation (3.3.4:2) as

$$v_{rec} = \theta c_0. \qquad (6.1.2:14)$$

The value of θ can be determined from the redshift z and the optical distance D in accordance with equations (6.1.2:1) and (6.1.2:3) as

$$\theta = \ln \frac{R_4}{R_{4(0)}} = \ln \frac{zR_4}{D}, \qquad (6.1.2:15)$$

which gives the actual recession velocity at the time of the observation as

$$v_{rec(physical)} = c_0 \ln \frac{zR_4}{D}. \qquad (6.1.2:16)$$

The recession velocity we observe is the optical recession velocity, which is the velocity at which the optical distance increases

$$v_{rec(optical)} = \frac{dD}{dt} = \left(1 - 1/e^\theta\right)\frac{dR_4}{dt} = \frac{D}{R_4} c_0, \qquad (6.1.2:17)$$

where equation (6.1.2:4) has been substituted for D.

The curves in Figure 6.1.2-4 summarize the different definitions of the recession velocity as a function of the redshift. The discussion of the redshift above excluded the effects of the local gravitational environment and any local motions of the object and the observer.

Effects of local motion and gravitation on redshift

The effect of the gravitational state and the local motion of source A and the receiver B on the wavelength of source A observed at B can be derived from equation (5.2.3:23) into the form

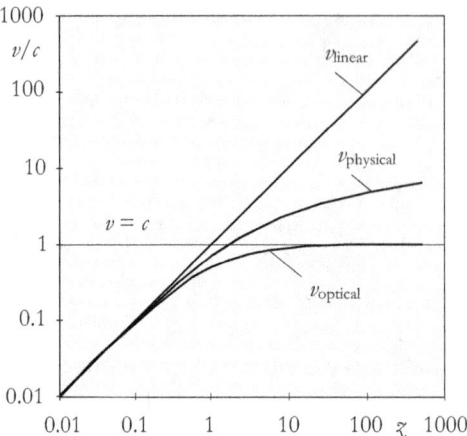

Figure 6.2.1-4. The curves show the development of recession velocities as function of the redshift. The optical velocity, the velocity of the lengthening of the optical distance, never exceeds the velocity of light at the time of the observation but approaches it asymptotically with the increasing redshift.

$$\frac{\lambda_{A(B)}}{\lambda_B} = \frac{\prod_{j=k+1}^{n}\left(1-\delta_{Bj}\right)\sqrt{1-\beta_{Bj}^2}\ \prod_{i=k+1}^{m}\left(1-\beta_{iA(\mathbf{r})}\right)}{\prod_{i=k+1}^{m}\left(1-\delta_{Ai}\right)\sqrt{1-\beta_{Ai}^2}\ \prod_{j=k+1}^{n}\left(1-\beta_{Bj(\mathbf{r})}\right)}. \qquad (6.1.2{:}18)$$

Equation (6.1.2:18) does not take into account the redshift due to the expansion of space. The redshift of wavelength from source A at distance D from observer B due to the expansion of space (6.1.2:5) is

$$\frac{\lambda_{A(B)}}{\lambda_B} = 1+z = \frac{1}{1-D/R_4}. \qquad (6.1.2{:}19)$$

Substitution of D/R_4 solved from (6.1.2:17) into (6.1.2:5) gives the redshift as a function of the optical recession velocity as

$$z = \frac{\lambda - \lambda_0}{\lambda_0} = \frac{\lambda_{A(B)} - \lambda_A}{\lambda_A} = \frac{D/R_4}{1-D/R_4} = \frac{\beta_{opt}}{1-\beta_{opt}}, \qquad (6.1.2{:}20)$$

where $\beta_{opt} = v_{opt}/c$. The wavelength of equal sources is conserved in the course of the expansion, i.e., $\lambda_B = \lambda_A$. Accordingly, the effect of expansion on the wavelength can be expressed in terms of the optical recession velocity as

$$\frac{\lambda_{A(B)}}{\lambda_B} = 1+z = \frac{1}{1-\beta_{opt}}. \qquad (6.1.2{:}21)$$

Combining the effects due to local gravitational states and motions of the source and the receiver in equations (6.1.2:18) and (6.1.2:21), we get

$$\frac{\lambda_{A(B)}}{\lambda_B} = \frac{1}{1-\beta_{opt}}\frac{\prod_{j=k+1}^{n}\left(1-\delta_{Bj}\right)\sqrt{1-\beta_{Bj}^2}\ \prod_{i=k+1}^{m}\left(1-\beta_{iA(\mathbf{r})}\right)}{\prod_{i=k+1}^{m}\left(1-\delta_{Ai}\right)\sqrt{1-\beta_{Ai}^2}\ \prod_{j=k+1}^{n}\left(1-\beta_{Bj(\mathbf{r})}\right)}. \qquad (6.1.2{:}22)$$

The Dynamic Cosmology

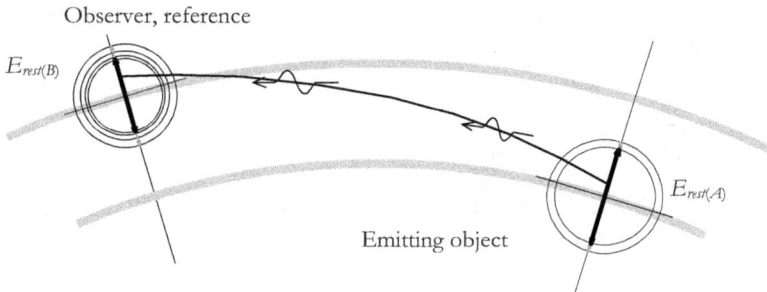

Figure 6.1.2-5. When the rest energy of an emitting object is higher than that of the reference in the observer's energy environment, a blueshift component is added to the observation. For objects at high distances the effect of the energy state of the source is generally orders of magnitude smaller than the redshift due to the expansion of space.

Equation (6.1.2:22) shows that the effect of the optical recession velocity is equal to the effect of the Doppler shift due to the recession velocity of the observer.

The Doppler effect due to local orbital motion results in both recessive and approaching velocity components, which appear as a broadening of spectral lines in the signal received. Typically, the value of the Doppler term is of the form $1\pm\Delta$, where $\Delta \ll 1$, whereas the redshift $1+z$ due to the expansion is always larger than 1. The Doppler term may be determining for objects at low distances, which allows blueshift instead of redshift, Figure 6.1.2-5.

6.1.3 Light propagation time in expanding space

In the Dynamic Universe, the velocity of light slows down with the expansion of space. Light observed from distant objects has propagated at a velocity higher than the velocity of light at the time of arrival. Because the velocity of light follows the velocity of space in the direction of the 4-radius, the optical distance (the distance the light travels) of an object is equal to the increase of the 4-radius during the light travel time [see equation (6.1.2:1)]. The length of the present 4-radius of space, R_4, as a function of time from the singularity is given in equation (3.3.3:7) which can be solved for t as

$$t = \frac{2}{3}\frac{1}{\sqrt{GM''}} R_4^{3/2}. \qquad (6.1.3:1)$$

The time required for R_4 to grow from value $R_{4(0)}$ to its present value R_4 is

$$\Delta T = \frac{2}{3}\frac{1}{\sqrt{GM''}}\left(R_4^{3/2} - R_{4(0)}^{3/2}\right) = \frac{2}{3}\frac{R_4^{3/2}}{\sqrt{GM''}}\left[1-\left(\frac{R_{4(0)}}{R_4}\right)^{3/2}\right]$$

$$= \frac{2}{3}\frac{R_4}{c_0}\left[1-\left(\frac{R_{4(0)}}{R_4}\right)^{3/2}\right], \qquad (6.1.3:2)$$

where c_0 is the velocity of light at radius R_0. Substitution of equation (6.1.2:5) for the ratio $R_{4(0)}/R_4$ in equation (6.1.3:2) gives the light propagation time from an object at $R_{4(0)}$ to the observer at R_4 in terms of the redshift as

$$\Delta T = \frac{2}{3}\frac{R_4}{c_0}\left[1 - \frac{1}{(1+z)^{3/2}}\right] = \frac{2}{3H_0}\left[1 - \frac{1}{(1+z)^{3/2}}\right], \qquad (6.1.3:3)$$

where H_0 is the Hubble constant (6.1.2:11). For low redshifts ($z \ll 1$), equation (6.1.3:3) can be developed into the form

$$\Delta T = \frac{2}{3H_0}\left[\frac{(1+z)^{3/2}-1}{(1+z)^{3/2}}\right] \approx \frac{2}{3H_0}\left[\frac{1+\frac{3}{2}z-1}{(1+z)^{3/2}}\right] = \frac{1}{H_0}\frac{z}{(1+z)^{3/2}}, \qquad (6.1.3:4)$$

where the term $(1+z)^{3/2}$ in the numerator is approximated with the first-order term of a series expansion.

Application of the velocity of light at the time of observation to the optical distance D of the object [see equation (6.1.3:7)] produces an apparent propagation time

$$\begin{aligned}\Delta T' &= \frac{D}{c_0} = \frac{R_4 - R_{4(0)}}{c_0} = \frac{R_4}{c_0}\left(1 - \frac{R_{4(0)}}{R_4}\right) \\ &= \frac{R_4}{c_0}\left(1 - \frac{1}{1+z}\right) = \frac{1}{H_0}\frac{z}{(1+z)}.\end{aligned} \qquad (6.1.3:5)$$

By combining equations (6.1.3:4) and (6.1.3:5), the real propagation time ΔT can be expressed in terms of the apparent propagation time as

$$\Delta T \approx \frac{\Delta T'}{\sqrt{(1+z)}} = \frac{D}{c_0\sqrt{(1+z)}} = \frac{D}{c_0}\sqrt{1-\frac{D}{R_4}} = \Delta T'\sqrt{1-\frac{D}{R_4}}, \qquad (6.1.3:6)$$

which applies for $z \ll 1$. In the last two forms of equation (6.1.3:6), the redshift is expressed in terms of the optical distance and the 4-radius R_0. Based on equation (6.1.3:6), we can define an effective propagation velocity of light (the average velocity of light) based on the local velocity of light at the time of the observation

$$c_{e\!f\!f} \approx c\sqrt{(1+z)}. \qquad (6.1.3:7)$$

The effective velocity of light $c_{e\!f\!f}$ includes the effect of the slowing velocity of light during propagation from an object with redshift z ($z \ll 1$).

The effect of the local structure of space

The local structure of space is known at a certain level within the Milky Way gravitational frame. The propagation time of light from a stellar object in the Milky Way gravitational system can be calculated from equation (5.5.1:1), which, by including the correction in equation (6.1.3:7), takes into account the change in the velocity of light during propagation

$$T_{A(t0) \to B(t1)} = T_{AB} = \frac{\mathbf{r}_{AB(t0)} \cdot \hat{\mathbf{r}}_{AB}}{c_k \sqrt{1+\tilde{z}} \prod_{j=k}^{m}\left(1-\beta_{j,B(\mathbf{r})}\right)}. \tag{6.1.3:8}$$

In equation (6.1.3:8), the velocity of propagation c_k is the velocity of light in the Milky Way gravitational frame at the time of the observation. Velocities $\beta_{j,B(\mathbf{r})}$ are the velocities of the solar frame in the Milky Way frame, the velocity of the Earth in the solar frame, and the rotational velocity of the observer in the Earth's gravitational frame.

In the case of stellar objects in the Milky Way or extragalactic space, the distance $\mathbf{r}_{AB(t0)}$ is known only through the magnitude or the redshift information, which is mixed with the Doppler shift and the gravitational shift of the frequency resulting from the gravitational environment of the object observed. In the case of continuous monitoring of the propagation times, such as in the observation of pulsar frequencies, the propagation time is subject to the effects of the daily and annual changes in observation distance due to the orbital motion and rotation of the Earth.

The distance variation due to the observer's motion relative to the propagation frame is known as the Roemer delay, known since the 18th century from observations of the moons of Jupiter. The Roemer delay for pulsars far from the solar system can be calculated from the difference in the length of the signal path relative to the solar barycenter.

$$\Delta t_{P-SB} = \frac{\mathbf{r}_{SB-O} \cdot \hat{\mathbf{r}}_{P-O}}{c_{E(S)}} = \frac{1}{c_{E(S)}}\left(\mathbf{r}_{SB-S} + \mathbf{r}_{S-E} + \mathbf{r}_{E-O}\right) \cdot \hat{\mathbf{r}}_{P-O}, \tag{6.1.3:9}$$

where distance $\mathbf{r}_{SB-O}$ is the distance from the solar barycenter to the observatory given as the sum of its components, $\mathbf{r}_{SB-S}$ as the distance from the solar system barycenter to the center of the Sun, $\mathbf{r}_{S-E}$ from the center of the Sun to the center of the Earth, and $\mathbf{r}_{E-O}$ from the center of the Earth to the observatory. The vector $\hat{\mathbf{r}}_{P-O}$ in equation (6.1.3:9) is the unit vector in the direction of the signal path from the pulsar to the observatory. In principle, the signal time in equation (6.1.3:9) should also include the effect of the change in distance due to the motion of the solar gravitational frame.

In a precise calculation, the velocity $c_{E(S)}$ must be chosen as the velocity of light at Earth's location in the solar gravitational frame in order to include the Shapiro delay [equation 5.4.1:23)] in the solar gravitational frame, which becomes observable when the signal path passes the Sun at a small distance

$$\Delta T_{P-E} = \frac{GM}{c_{0\delta}^3}\left\{2\ln\left[\cot\frac{\beta_{E-S}}{2}\cot\frac{\beta_{P-S}}{2}\right] - \left(\cos\beta_{E-S} + \cos\beta_{P-S}\right)\right\}, \tag{6.1.3:10}$$

where

$$\begin{aligned}\beta_{E-S} &= \text{Earth} \angle \text{Sun} \\ \beta_{P-S} &= \text{pulsar} \angle \text{Sun}\end{aligned} \tag{6.1.3:11}$$

There is a minor difference in the Shapiro delay between the DU and GR (see Section 5.3). The classical Roemer delay is a natural part of the DU framework; in prevailing practice, it is added as a separate correction to the predictions of general relativity.

6.2 Angular sizes of a standard rod and expanding objects

6.2.1 Angular size of a standard rod in FLRW space

In FLRW space, both solid objects and gravitationally bound local systems like galaxies and quasars are assumed to conserve their dimensions in the course of the expansion of space. Accordingly, in FLRW space, practically all observed objects can be regarded as standard rods. The prediction for the angular size of the standard rod d_s in FLRW space is [75]

$$\psi = \frac{d_s}{D_A} = \frac{d_s}{R_H} \Bigg/ \left[\frac{1}{(1+z)} \int_0^z \frac{1}{\sqrt{(1+z)^2(1+\Omega_m z) - z(2+z)\Omega_\Lambda}} dz \right], \quad (6.2.1:1)$$

which is based on the angular diameter distance D_A. Angular diameter distance is related to the comoving distance D_C in FLRW space

$$D_A = \frac{D_C}{1+z} = \frac{R_H}{1+z} \int_0^z \frac{1}{\sqrt{(1+z)^2(1+\Omega_m z) - z(2+z)\Omega_\Lambda}} dz, \quad (6.2.1:2)$$

where $R_H = c/H_0$ is referred to as the Hubble radius, which corresponds to R_0 in the DU.

As shown by (6.2.1:2), the angular diameter distance D_A turns to a decreasing trend at redshifts above $z > 3$, resulting in an increasing angular size.

6.2.2 Angular size of a standard rod in DU space

The standard rod is a hypothetical celestial object that conserves its dimensions in the course of the expansion of space. In DU space, solid objects may be regarded as standard rods.

Radiation from an object $A(z)$ at a distance angle θ from the observer is seen at its apparent location $A'(z)$, at distance D, redshifted by

$$z = e^\theta - 1 = \frac{D/R_4}{1 - D/R_4}, \quad (6.2.2:1)$$

where R_0 is the 4-radius of space at the time of observation, Figure 6.2.2-1. As given in equation (6.1.2:6), the optical distance D is

$$D = \frac{z}{1+z} R_4. \quad (6.2.2:2)$$

The optical angle or angular size $\psi_{r(s)}$ subtended by a standard rod can be expressed as the ratio of the length of the rod d_s and the optical distance D

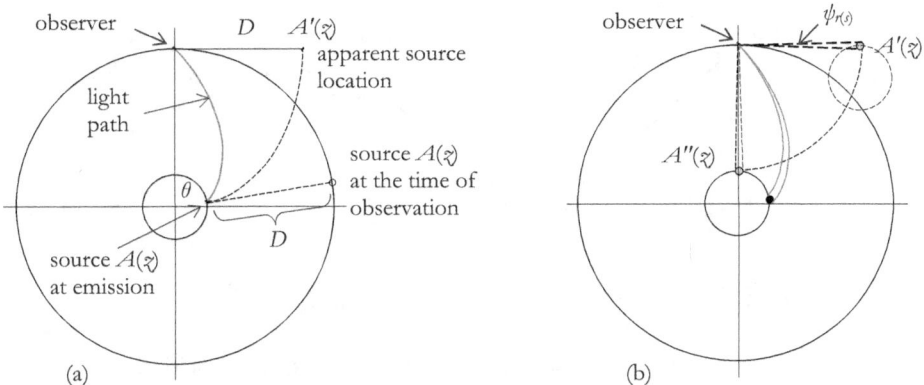

Figure 6.2.2-1. (a) Propagation of light in an expanding spherically closed space. The apparent line of sight is the straight tangential line. The distance to the apparent source location $A'(z)$ is at the optical distance $D = R_{(observation)} - R_{(emission)}$ along the apparent line of sight. (b) The symmetry of expansion in the three space dimensions and in the fourth dimension makes the observed optical angle $\psi_{r(s)}$ of the apparent source $A'(z)$ equal to the optical angle of a hypothetical image $A''(z)$ at distance D in the direction of the R_0 radius.

$$\psi_{r(s)} = \frac{d_s}{D} = \frac{d_s}{R_4} \frac{1+z}{z}, \qquad (6.2.2:3)$$

or, when normalized to (r_s/R_0), as

$$\frac{\psi_{r(s)}}{d_s/R_4} = \frac{1+z}{z}. \qquad (6.2.2:4)$$

Expression of the optical angle $\psi_{r(s)}$ as the ratio d_s/D assumes symmetry of expansion for light front elements dV_λ propagating in the optical path

$$dV_\lambda = \Delta\lambda \cdot \Delta A. \qquad (6.2.2:5)$$

At redshifts $z < 0.1$, the observation angle of the standard rod ψ_{DU} follows the Euclidean $1/z$ dependence. At high redshifts, the normalized observation angle approaches $\psi_{DU}/(d_s/R_4) = \psi_{DU}/\theta_{r(s)} \Rightarrow 1$.

6.2.3 Angular size of expanding objects in DU space

As a major difference to FLRW cosmology, all gravitationally bound systems like galaxies and quasars, are expanding objects in the DU framework. The angular diameter of an *expanding object* in DU space can be expressed

$$d(z) = \frac{d_R}{(1+z)}, \qquad (6.2.3:1)$$

where d_R is the diameter of the object at the time of observation. Substitution of $d(z)$ in (6.2.3:1) for d_s in (6.2.2:3) gives the angular size of an expanding objects

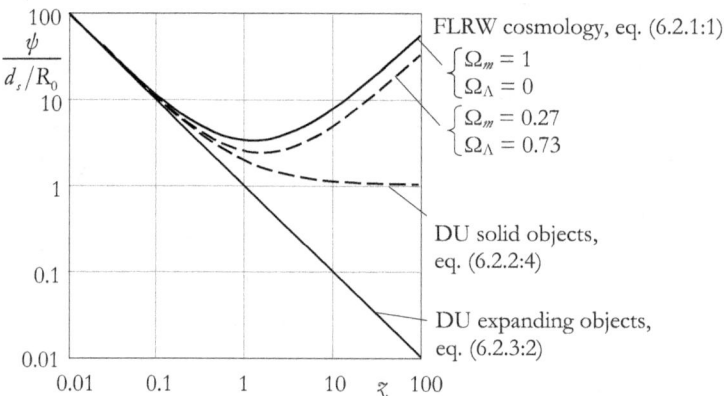

Figure 6.2.3-1. Angular diameter of objects as a function of redshift in FLRW space and in DU space.

$$\psi = \frac{d(z)}{D} = \frac{d_R}{(1+z)}\frac{(1+z)}{R_4 z} = \frac{d_R}{R_4}\frac{1}{z} = \frac{\theta_d}{z} \quad ; \quad \frac{\psi}{d_R/R_4} = \frac{\psi}{\theta_d} = \frac{1}{z}, \quad (6.2.3:2)$$

where the ratio $d_R/R_0 = \theta_d$ means the angular size of the expanding object as seen from the barycenter of space. Equation (6.2.3:2) implies a Euclidean appearance of expanding objects in space. A comparison of equations (6.2.1:2), (6.2.2:4), and (6.2.3:2) is given in Figure 6.2.3-1.

The DU prediction for solid objects (standard rod) approaches asymptotically to the angular size of the object as it would appear from the barycenter of space (i.e., from M'').

It can be concluded that an essential factor in the Euclidean appearance of galaxy space in the DU is the linkage of the gravitational energies of local systems to the gravitational energy in the whole space. Such a linkage is missing in the GR-based FLRW cosmology due to the local nature of general relativity.

In Figure 6.2.3-2, the DU prediction (6.2.3:2) and the FLRW prediction (6.2.1:2) are compared to observations of the Largest Angular Size (LAS) of galaxies and quasars in the redshift range $0.001 < z < 3$ [54].

In Figure 6.2.3-2 (a), the observation data is set between two Euclidean lines of the DU prediction in equation (6.2.3:2). The FLRW prediction is calculated for the conventional Einstein de Sitter case ($\Omega_m=1$ and $\Omega_\Lambda=0$) shown by the solid curve, and for the recently preferred case with a share of dark energy included as $\Omega_m=0.27$ and $\Omega_\Lambda=0.73$ (dashed curves). Both FLRW predictions deviate significantly from the Euclidean lines in (a) that enclose the set of data uniformly in the whole redshift range. As shown in Figure 6.2.3-2 (b), the effect of the dark energy contribution on the FLRW prediction of the angular size is quite marginal.

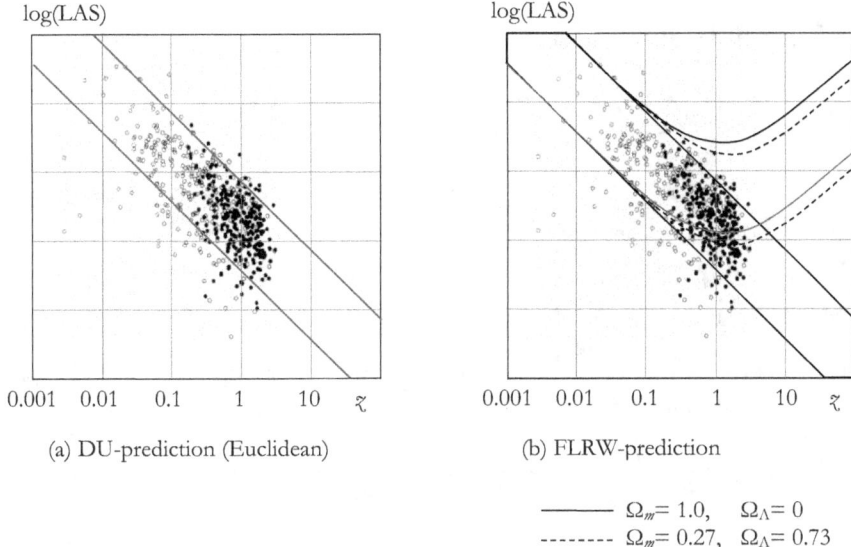

Figure 6.2.3-2. Dataset of the observed Largest Angular Size (LAS) of quasars and galaxies in the redshift range $0.001 < z < 3$. Open circles are galaxies, filled circles are quasars. In (a), observations are compared with the DU prediction (6.2.3:2). In (b), observations are compared with the FLRW prediction (6.2.1:2) with $\Omega_m = 0$ and $\Omega_\Lambda = 0$ (solid curves), and $\Omega_m = 0.27$ and $\Omega_\Lambda = 0.73$ (dashed curves).

6.3 Magnitude and surface brightness

6.3.1 Luminosity distance and magnitude in FLRW space

In the classical Euclidean space, radiation flux F from a spherically symmetric source is assumed to be diluted in proportion to the square of distance

$$F_{classical} = \frac{L}{A} = \frac{L}{4\pi d^2} \quad \left[\frac{W}{m^2}\right], \qquad (6.3.1:1)$$

where L [W] is the luminosity of the radiation source. Applying (6.3.1:1), the classical definition of apparent magnitude becomes

$$m = -2.5\log F + C = -2.5\log\frac{L}{4\pi D^2} + 2.5\log\frac{L}{4\pi D^2_{10pc}} + M, \qquad (6.3.1:2)$$

where the constant M is the absolute magnitude, which is the observed magnitude of the object as it would be at 10 parsec distance from the observer. Using the reference distance $D_0 = 10$ pc, equation (6.3.1:2) can be rewritten into the form

$$m = M - 5\log\frac{D_0}{R_H} + 5\log\frac{D_L}{R_H}, \qquad (6.3.1:3)$$

where $D = D_L$ is the luminosity distance, and R_H is the Hubble radius.

The luminosity distance D_L in FLRW cosmology is

$$D_L = (1+z)^2 D_A = R_H (1+z)\int_0^z \frac{1}{\sqrt{(1+z)^2(1+\Omega_m z) - z(2+z)\Omega_\Lambda}} dz, \qquad (6.3.1:4)$$

where D_A is the angular size distance given in (6.2.1:2). The ratio $(1+z)^2$ between D_L and D_A means that in FLRW space, the classical distance dilution of radiation that is proportional to D_A^2 is further diluted by a factor $(1+z)^4$ due to expansion

$$F_{FLRW} = \frac{F_{classical}}{(1+z)^4} = \frac{L}{4\pi D_A^2}\frac{1}{(1+z)^4}. \qquad (6.3.1:5)$$

As first proposed by Tolman [47] and later concluded by Hubble and Humason [52], de Sitter [50], and Robertson [51], the energy of a quantum is reduced by $(1+z)$ as a consequence of the effect of Planck's equation $E = hf$ as an "energy effect", a reduction of the "intensity of the radiation" due to reduced frequency. When receiving the redshifted radiation at a lowered frequency, a second $(1+z)$ factor was assumed as a "number effect". Hubble considered that the latter is relevant only in the case that the redshift is due to recession velocity [49]. The double dilution $(1+z)^2$ due to redshift has stayed in the FLRW cosmology since the early work in the 1930s [53].

In the power density in (6.3.1:5), the other $(1+z)^2$ factor is referred to as "aberration factor" or reciprocity effect based on the analysis of distances in general relativity by

Tolman in 1930 [47], and Etherington in 1933 [76]. The magnitude prediction based on luminosity distance D_L in FLRW cosmology assumes reduction of the observed power densities to power densities in "emitter's rest frame" — the prediction is compared to observations corrected with the *K-correction*, which in addition to correction of instrumental factors, cancels the reciprocity factor by adding a $(1+z)^2$ attenuation factor to the power densities observed in bolometric multi bandpass photometry (see Section 6.3.3).

The prediction for *K*-corrected magnitudes in FLRW cosmology is given by the equation

$$m = M + 5\log\frac{R_H}{D_0} + K_{instr} + 5\log\left[(1+z)\int_0^z \frac{1}{\sqrt{(1+z)^2(1+\Omega_m z) - z(2+z)\Omega_\Lambda}} dz\right],$$

(6.3.1:6)

where $D_0 = 10$ pc is the distance of the reference object.

6.3.2 Magnitude of standard candle in DU space

Conservation of the mass equivalence of radiation in DU space negates the basis for an "energy effect" as a violation of the conservation of energy. An analysis of the linkage between Planck's equation and Maxwell's equations shows that Planck's equation describes the energy conversion at the *emission* of electromagnetic radiation. Redshift should be understood as a *dilution of the energy density due to an increase in the wavelength* in the direction of propagation, not as a *loss of energy*. Accordingly, the observed energy flux $F = E_\lambda f$ is subject only to a single $(1+z)$ dilution factor, the "number effect" in the historical terms

$$F_{rec(z)} = E_\lambda f = \frac{h_0}{\lambda_e} cc_0 \cdot f = \frac{h_0}{\lambda_e} cc_0 \frac{c}{\lambda_r} = \frac{h_0}{\lambda_e} \frac{c_0 c^2}{\lambda_e(1+z)} = \frac{h_0 c_0 c^2}{\lambda_e^2(1+z)},$$

(6.3.2:1)

where λ_e is the wavelength at the emission of the radiation. The emission wavelength of characteristic radiation from atomic emitters is constant in the course of the expansion of space. Accordingly, the reference flux of characteristic radiation from a reference source at the time and location the redshifted radiation is received is ($\lambda_{e(ref)} = \lambda_e$)

$$F_{emit(ref)} = E_{\lambda(ref)} f_{ref} = \frac{h_0}{\lambda_{e(ref)}} cc_0 f_{ref} = \frac{h_0}{\lambda_{e(ref)}} cc_0 \frac{c}{\lambda_{e(ref)}} = \frac{h_0 c_0 c^2}{\lambda_e^2}.$$

(6.3.2:2)

Relative to the reference flux with zero redshift, the power density in the redshifted flux is

$$F_{rec(z)} = \frac{F_{e(ref)}}{(1+z)}.$$

(6.3.2:3)

Equation (6.3.2:3) gives the dilution due to redshift but ignores the areal dilution related to the optical distance of the source. When the redshifted radiation is received from a source at distance D from the reference source, the observed energy flux is

$$F_{(D,z)} = \frac{L}{4\pi D^2} \frac{h_0 c_0 c^2}{\lambda_e^2 (1+z)}, \qquad (6.3.2:4)$$

where L is the luminosity of the source. Related to the flux density $F_{e(ref)}$ from a reference source with the same luminosity at distance d_0 ($z \approx 0$), the energy flux is

$$F_{(D,z)} = F_{e(ref)} \cdot \frac{\dfrac{L}{4\pi D^2} \dfrac{h_0 c_0 c^2}{\lambda_e^2 (1+z)}}{\dfrac{L}{4\pi D_0^2} \dfrac{h_0 c_0 c^2}{\lambda_e^2}} = F_{e(ref)} \cdot \frac{d_0^2}{D^2} \frac{1}{(1+z)}. \qquad (6.3.2:5)$$

Substitution of equation (6.2.2:2) for D in (6.3.2:5) gives

$$F_{(D,z)} = F_{e(ref)} \cdot \frac{d_0^2}{R_0^2} \frac{(1+z)^2}{z^2} \frac{1}{(1+z)} = F_{e(ref)} \cdot \frac{d_0^2}{R_0^2} \frac{(1+z)}{z^2}, \qquad (6.3.2:6)$$

which corresponds to the apparent magnitude

$$m = M + 5\log\frac{R_0}{d_0} + 5\log z - 2.5\log(1+z) + K_{instr}. \qquad (6.3.2:7)$$

Equation (6.3.2:7) applies for the *bolometric energy flux* observed for radiation from a source at optical distance $D = R_0 \cdot z /(1+z)$ from the observer in DU space. In equation (6.3.2:7), possible effects of galactic extinction, spectral distortion in the Earth's atmosphere, or effects due to the local motion and gravitational environment of the source and the receiver are included in K_{instr}.

6.3.3 Bolometric magnitudes in multi-bandpass detection

For analyzing the detection of bolometric flux densities and magnitudes by multi-bandpass photometry, the source radiation is assumed to have the spectrum of blackbody radiation. The bandpass system applied consists of a set of UBVIZYJHK filters approximated with transmission curves of the form of a normal distribution

$$f_X(\lambda) = e^{-\left[(\lambda - \lambda_{C(X)})^2 / 2(\Delta\lambda_X/\sigma_{1/2})^2\right]} = e^{-\frac{\sigma_{1/2}^2}{2}\left(\frac{\lambda_{C(X)}}{\Delta\lambda_X}\right)^2 \left(\frac{\lambda}{\lambda_{C(X)}} - 1\right)^2} = e^{-\frac{2.773}{W_X^2}\left(\frac{\lambda}{\lambda_{C(X)}} - 1\right)^2}, \qquad (6.3.3:1)$$

where $\lambda_{C(X)}$ is the peak wavelength of filter X, $\Delta\lambda_X$ the half-width of the filter, $W_X = \Delta\lambda_X/\lambda_{C(X)}$ the relative width, and $\sigma_{1/2} = 2.35481$ is the half-width deviation of the normal distribution, Figure 6.3.3-1.

For the numerical calculation of the energy flux from a blackbody source, equation (A.2:10) in Appendix 1 is rewritten for a relative wavelength differential $d\lambda_z/\lambda_z = d\lambda/\lambda \ll W_X$:

The Dynamic Cosmology

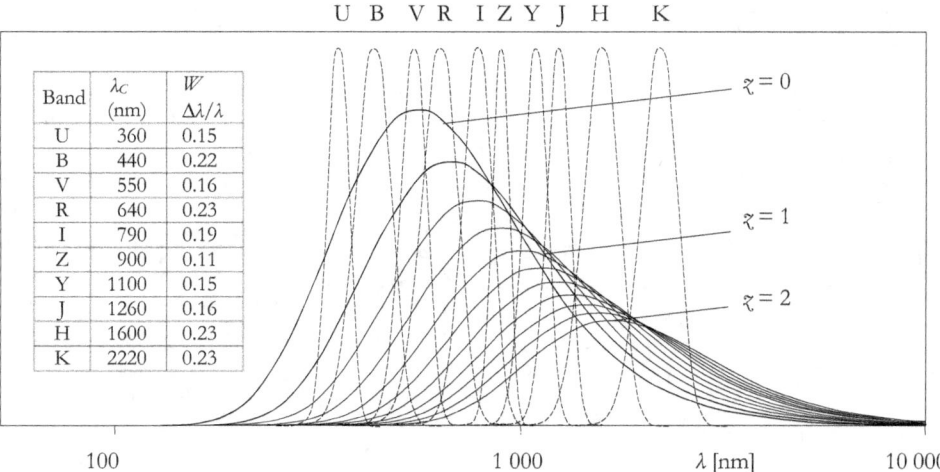

Band	λ_C (nm)	W $\Delta\lambda/\lambda$
U	360	0.15
B	440	0.22
V	550	0.16
R	640	0.23
I	790	0.19
Z	900	0.11
Y	1100	0.15
J	1260	0.16
H	1600	0.23
K	2220	0.23

Figure 6.3.3-1. The effect of redshift $z = 0\ldots2$ (shown in steps of 0.2) on the energy flux density per relative bandwidth of the blackbody radiation spectrum from a $T = 6600$ °K blackbody source corresponding to $\lambda_T = 440$ nm and $\lambda_W = 557$ nm (solid curves). Transmission curves of UBVRIZYJHK filters listed in the table are shown with dashed lines. The half-widths of the filters follow the widths of standard filters in the Johnson system. All transmission curves are approximated with a normal distribution. The horizontal axis shows the wavelength in nanometers on a logarithmic scale.

$$F\left(\frac{d\lambda_z}{\lambda_z}\right) = \frac{15}{\pi^4} \frac{F_{bol(z=0)}}{(1+z)} \left[\left(\lambda_0 \bigg/ \frac{\lambda}{1+z}\right)^4 \bigg/ \left(e^{\left(\lambda_0 / \frac{\lambda}{1+z}\right)} - 1\right)\right] \frac{d\lambda}{\lambda}. \quad (6.3.3:2)$$

Equation (6.3.3:2) excludes the areal dilution due to the distance from the source to the observer. Integration of (6.3.3:2) gives the bolometric radiation

$$F_{bol} = \int_0^\infty F\left(\frac{d\lambda_z}{\lambda_z}\right) = \frac{F_{bol(z=0)}}{1+z}. \quad (6.3.3:3)$$

The transmission through filter X, normalized to the bolometric flux by applying equation (A2:12), can now be calculated by applying the transmission function of equation (6.3.3:1) to the flux in (6.3.3:2)

$$dF_{X(z)}\left(\frac{d\lambda_z}{\lambda_z}\right) = \frac{15}{\pi^4} \frac{F_{bol(z=0)}}{(1+z)} \int_0^\infty \left[\left(\lambda_0 \bigg/ \frac{\lambda}{1+z}\right)^4 \bigg/ \left(e^{\left(\lambda_0 / \frac{\lambda}{1+z}\right)} - 1\right)\right] e^{-\frac{2.773}{W_X^2}\left(\frac{\lambda/(1+z)}{\lambda_{C(X)}} - 1\right)^2} \frac{d\lambda}{\lambda}, \quad (6.3.3:4)$$

which gives the flux observed through filter X as a function of the redshift of the radiation, Figure 6.3.3-2.

The energy flux of equation (6.3.3:4) from sources at a small distance d_0 ($z_{d0} \approx 0$) and at distance D ($z_D > 0$) are related by

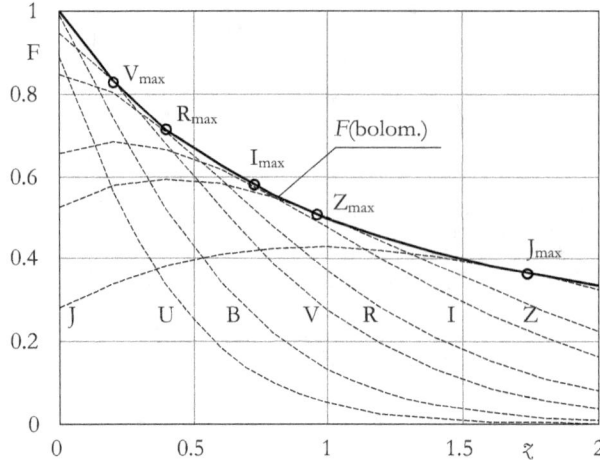

Figure 6.3.3-2. Transmission curves obtained by numerical integration of (6.3.3:4) for filters UBVRIZJ for radiation in the redshift range $z = 0...2$ from a blackbody with $\lambda_T = 350$ nm ($\lambda_W = 440$ nm, $T = 8300$ °K). Each curve touches the bolometric curve of equation (6.3.3:3) at the redshift matching maximum of the radiation flux to the nominal wavelength λ_W of the filter (small circles in the figure).

$$\frac{F_{X(D)}}{F_{X0(d_0)}} = \frac{d_0^2}{D^2} \frac{\int_0^\infty dF_{X(z)}}{\int_0^\infty dF_{X0(0)}}. \qquad (6.3.3{:}5)$$

Substitution of equation (6.2.2:2) for D and equation (6.3.3:4) for $F_{X(D)}$ and $F_{X0(do)}$ in (6.3.3:5) gives the radiation power observed in filters X and $X0$ from standard sources at distances D and d_0, respectively

$$\frac{F_{X1(D)}}{F_{X2(d_0)}} = \frac{d_0^2}{R_4^2} \frac{(1+z)^2}{z^2} \frac{1}{(1+z)} \frac{\int_0^\infty \left[\left(\lambda_0 \big/ \frac{\lambda}{1+z}\right)^4 \Big/ \left(e^{\left(\lambda_0 / \frac{\lambda}{1+z}\right)} - 1\right)\right] e^{-\frac{2.773}{W_{X1}^2}\left(\frac{\lambda/1+z}{\lambda_{C(X1)}} - 1\right)^2} \frac{d\lambda}{\lambda}}{\int_0^\infty \left[\left(\lambda_0/\lambda\right)^4 \Big/ \left(e^{(\lambda_0/\lambda)} - 1\right)\right] e^{-\frac{2.773}{W_{X2}^2}\left(\frac{\lambda}{\lambda_{C(X2)}} - 1\right)^2} \frac{d\lambda}{\lambda}}. \qquad (6.3.3{:}6)$$

By denoting the integrals in the numerator and denominator in (6.3.3:6) by $I_{X(D)}$ and $I_{X0(do)}$, respectively, energy flux $F_{X(D)}$ can be expressed

$$F_{X(D)} = F_{X0(d_0)} \frac{d_0^2}{R_4^2} \frac{(1+z)}{z^2} \frac{I_{X(D)}}{I_{X0(d_0)}}. \qquad (6.3.3{:}7)$$

Choosing $d_0 = 10$ pc, the apparent magnitude for flux through filter X at distance D can be expressed as

$$m_{X1} = M + 5\log\left(\frac{R_4}{10\text{pc}}\right) + 5\log(z) - 2.5\log(1+z) + 2.5\log\left(\frac{I_{X2(d_0)}}{I_{X1(D)}}\right), \qquad (6.3.3{:}8)$$

where M is the absolute magnitude of the reference source at a distance of 10 pc.

For $R_4 = 14 \cdot 10^9$ ly., consistent with Hubble constant $H_0 = 70$ [(km/s)/Mpc], the numerical value of the second term in (6.3.3:8) is $5 \cdot \log(R_4/10\text{pc}) = 43.16$ magnitude units. For Ia supernovae, the numerical value for the absolute magnitude is about $M \approx 19.5$.

When filter X is chosen to match $\lambda_{C(X)} = \lambda_W(1+z)$ and $\lambda_{C(X0)} = \lambda_W$ [or $\lambda_{C(X)} = \lambda_T(1+z)$ and $\lambda_{C(X0)} = \lambda_T$, the integrals $I_{X(D)}$ and $I_{X0(do)}$ are related as the relative bandwidths

$$\frac{I_{X0(d_0)}}{I_{X(D)}} = \frac{W_{X0}}{W_X}, \qquad (6.3.3:9)$$

which means that for optimally chosen filters with equal relative widths, the last term in equation (6.3.3:8) is zero, and equation (6.3.3:8) obtains the form of equation (6.3.4:10) for bolometric energy flux

$$m_{X(opt)} = M + 5\log\left(\frac{R_4}{10\text{pc}}\right) + 5\log(z) - 2.5\log(1+z). \qquad (6.3.3:10)$$

Figures 6.3.3-3 (a,b) illustrate the magnitudes calculated for filters $X = B, V, R, I, Z, J$ from equation (6.3.3:8) in the redshift range $z = 0...2$. Each curve touches the solid curve of equation (6.3.3:10) corresponding to the bolometric magnitude obtainable with optimal filters at each redshift in the redshift range studied.

In Figure 6.3.3-3(c), the predictions are compared to magnitudes collected from Table 7 in Tonry et al. [55]. The magnitudes given by Tonry et.al. are values that a "normal" SN Ia might achieve at maximum, derived from the colors of SN 1995D at maximum and the spectral energy distribution of SN 1994S.

6.3.4 K-corrected magnitudes

In the observation praxis based on the Standard Cosmology Model, direct observations of magnitudes in the bandpass filters are treated with the *K-correction*, which corrects the filter mismatch and converts the observed magnitude to the "emitter's rest frame" presented by observations in a bandpass matched to a low redshift reference of the objects studied. The *K-correction* for observations in the X_j band relative to the rest frame reference in the X_i band is defined [77] as

$$K_{i,j}(z) = 2.5\log(1+z) + 2.5\log\left\{\frac{\int_0^\infty F(\lambda)S_i(\lambda)\,d\lambda \int_0^\infty Z(\lambda)S_j(\lambda)\,d\lambda}{\int_0^\infty F(\lambda/(1+z))S_j\,d\lambda \int_0^\infty Z(\lambda)S_i(\lambda)\,d\lambda}\right\}. \qquad (6.3.4:1)$$

In the case of a blackbody source and filters with transmission functions described by a normal distribution, equation (6.3.4:1) can be expressed by substituting equation (6.3.3:2) for the energy flux integrals, equation (6.3.3:1) for the transmission curves of the filters, and the relative bandwidths of filters i and j for the transmission integrals

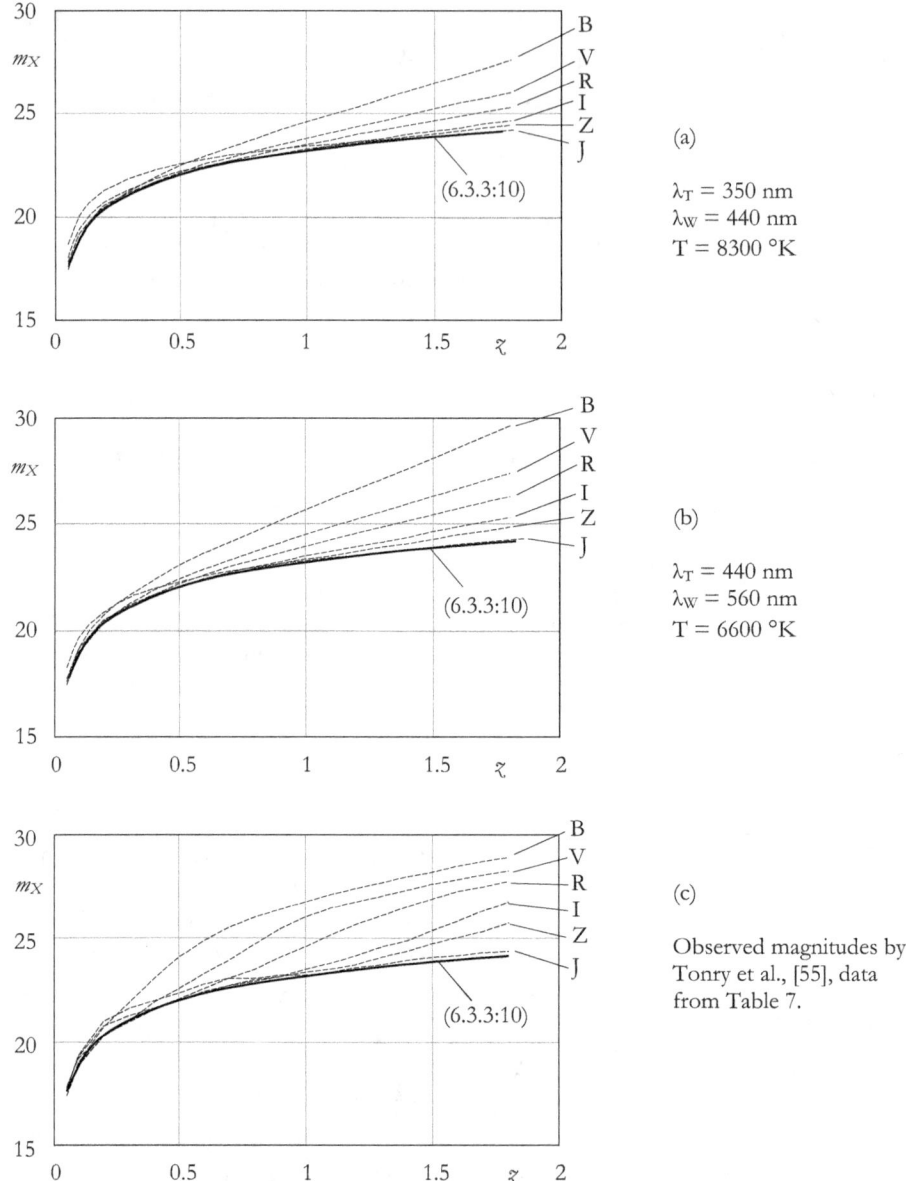

Figure 6.3.3-3 (a) The magnitudes predicted by (6.3.3:8) for filters BVRIZJ as functions of redshift are shown as the families of curves drawn with dashed lines (see Appendix 1 for the definitions of λ_T and λ_W characterizing blackbody radiation). The transmission functions of the filters used by Tonry et al. [55], Table 7, are slightly different from the transmission functions used in calculations for (a) and (b). The DU prediction (6.3.3:10) for the magnitudes in optimally chosen filters is shown by the solid DU curve in each figure.

The Dynamic Cosmology

$$K_{i,j(W)}(z) = 2.5\log(1+z)$$

$$+2.5\log\left\{\frac{\int_0^\infty \left[(\lambda_0/\lambda)^5 / \left(e^{(\lambda_0/\lambda)}-1\right)\right] e^{-\frac{2.773}{W_i^2}\left(\frac{\lambda}{\lambda_{C(i)}}-1\right)^2} d\lambda}{\frac{1}{1+z}\int_0^\infty \left[\left(\lambda_0 \bigg/ \frac{\lambda}{1+z}\right)^5 \bigg/ \left(e^{\left(\lambda_0 \big/ \frac{\lambda}{1+z}\right)}-1\right)\right] e^{-\frac{2.773}{W_j^2}\left(\frac{\lambda/(1+z)}{\lambda_{C(j)}}-1\right)^2} d\lambda} \cdot \frac{W_j}{W_i}\right\}, \quad (6.3.4:2)$$

where the relative differential $d\lambda/\lambda$ of (6.3.3:2) is replaced by differential $d\lambda$ to meet the definition of (6.3.4:1).

Figure 6.3.4-1 (a) illustrates the K_{BX} corrections calculated for radiation from a blackbody source with $\lambda_T = 440$ nm equivalent to 6600 °K blackbody temperature. An optimal choice of filters, matching the central wavelength of the filter to the wavelength of the maximum of redshifted radiation, leads to the K-correction

$$K(z) \approx 5\log(1+z). \quad (6.3.4:3)$$

The K-correction of (6.3.4:3) gives an accuracy of better than 0.1 magnitude units in the whole range of redshifts covered with the set of filters used. The difference between the K-corrections in equation (6.3.4:2) and (6.3.4:3) is presented in Figure 6.3.4-1(b).

Substitution of (6.3.4:3) for K in equation (6.3.2:7) gives the DU space prediction for K-corrected magnitudes

$$m_{K(DU)} = M + 5\log\frac{R_4}{D_0} + 5\log z + 2.5\log(1+z). \quad (6.3.4:4)$$

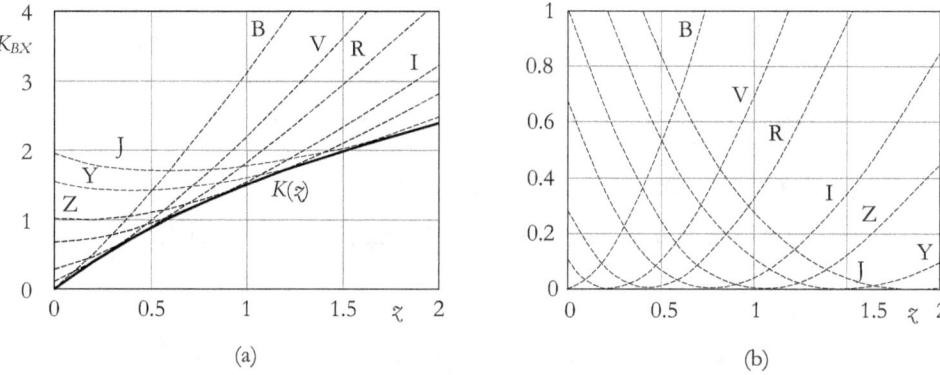

Figure 6.3.4-1. (a) K_{BX}-corrections (in magnitude units) according to (6.3.4:2) for the B band as the reference frame, calculated in the redshift range $z = 0\ldots2$ for radiation from a blackbody source with $\lambda_T = 440$ nm equivalent to 6600 °K blackbody temperature. All of the K_{BX}-correction curves touch the solid $K(z)$ curve, which shows the $K(z) = 5\cdot\log(1+z)$ function. (b) The difference $K_{BX}-K(z)$. With an optimal choice of filters, the difference $K_{BX}-K(z)$ is smaller than 0.05 magnitude units in the whole range of redshifts $z = 0\ldots2$ covered by the set of filters B…J, demonstrating the bolometric detection with optimally chosen filters.

The prediction for *K*-corrected magnitudes in the standard model is given by the equation

$$m = M + 5\log\left(\frac{R_H}{10\text{ pc}}\right) + 5\log\left(\frac{D_L}{R_H}\right) = M + 43.2$$

$$+ 5\log\left[(1+z)\int_0^z \frac{1}{\sqrt{(1+z)^2(1+\Omega_m z) - z(2+z)\Omega_\lambda}}dz\right],$$

(6.3.4:5)

where $R_H = c/H_0 \approx 14 \cdot 10^9$ l.y. is the Hubble radius, the standard model replacement of R_4 in DU space, and D_L is the luminosity distance defined in equation (6.3.1:4). Mass density parameters Ω_m and Ω_Λ give the density shares of mass and dark energy in space. For a flat space condition, the sum $\Omega_m + \Omega_\Lambda = 1$.

The best fit of equation (6.3.4:5) to the *K*-corrected magnitudes of Ia supernova observations has been obtained with $\Omega_m = 0.26\ldots0.31$ and $\Omega_\Lambda = 0.74\ldots0.69$ [55,78,79,80,81,82,83,84,85]. Figure 6.3.4-2 shows a comparison of the prediction given by equation (6.3.4:5) with $\Omega_m \approx 0.31$, $\Omega_\Lambda \approx 0.69$, and $H_0 = 64.3$ used by Riess et al. [79] and the DU space prediction for *K*-corrected magnitudes in equation (6.3.4:4).

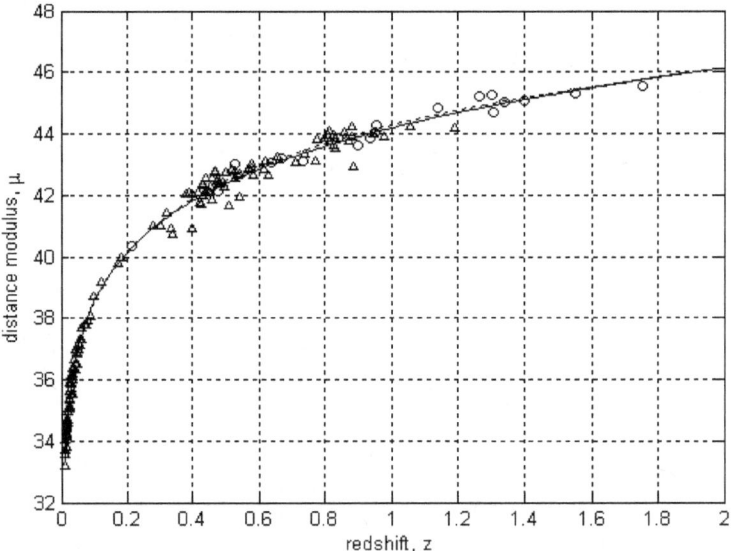

Figure 6.3.4-2. Distance modulus $\mu = m - M$, vs. redshift for Riess et al.'s gold dataset and the data from the HST. The triangles represent data obtained via ground-based observations, and the circles represent data obtained by the HST [79]. The optimum fit for the standard cosmology prediction (6.3.4:5) is shown by the dashed curve, and the fit for the DU prediction (6.3.4:4) is shown, slightly below, by the solid curve [86].

The Dynamic Cosmology

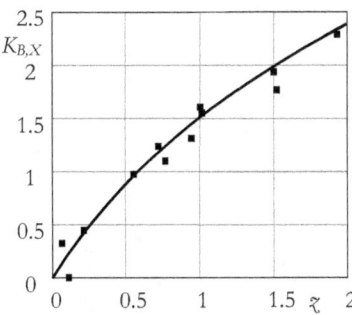

Figure 6.3.4-3. Average $K_{B,X}$-corrections (black squares) collected from the $K_{B,X}$ data in Table 2 used by Riess et al. [79] for the K-corrected distance modulus data shown in Figure 6.3.4-2. The solid curve gives the theoretical K-correction (6.3.4:3), $K = 5 \cdot \log(1+z)$, derived for filters matched to redshifted spectra (see Fig. 6.3.4-1) and applied in equation (6.3.4:4) for the DU prediction for K corrected apparent magnitude.

In the redshift range $z = 0...2$, the apparent magnitude of equation (6.3.4:5) coincides accurately with the magnitudes of equation (6.3.4:4). The K-corrections used by Riess et al. [79], Table 2, follow the $K(z) = 5 \cdot \log(1+z)$ prediction of equation (6.3.4:3) as illustrated in Figure 6.3.4-3.

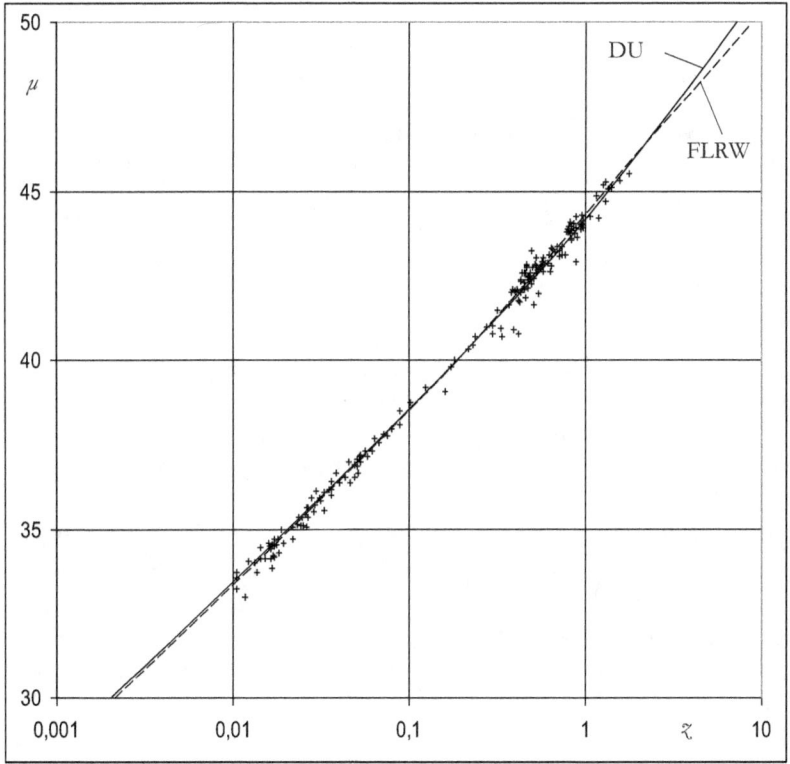

Figure 6.3.4-4. Distance modulus $\mu = m - M$, vs. redshift for Riess et al. "high-confidence" dataset and the data from the HST, presented on a logarithmic scale.

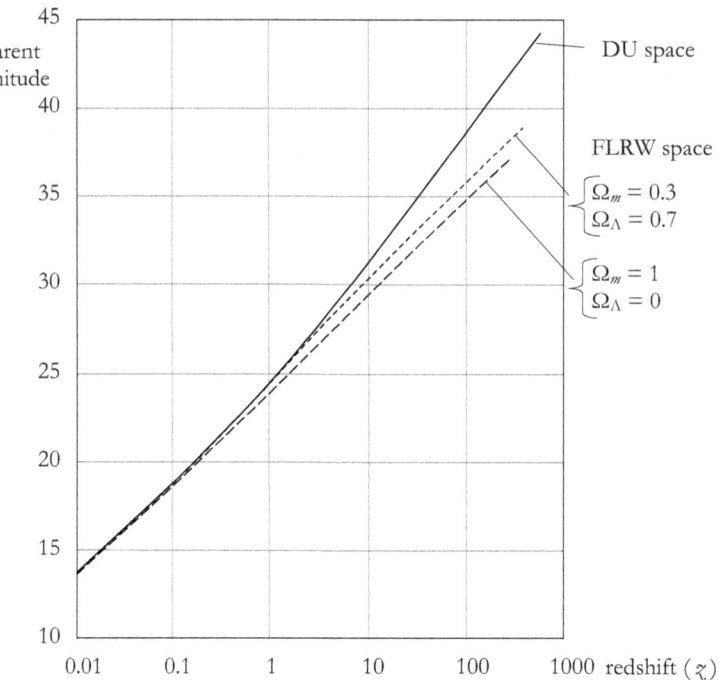

Figure 6.3.4-5. Comparison of predictions for the K-corrected apparent magnitude of standard sources in the redshift range 0.01…1000 given by the Standard Cosmology Model with $\Omega_m=0.3/\Omega_\Lambda=0.7$ and $\Omega_m=1/\Omega_\Lambda=0$ according to equation (6.3.4:5), and DU space given by equation (6.3.4:4). In each curve, the absolute magnitude used is $M = -19.5$. The $\Omega_m=0.3/\Omega_\Lambda=0.7$ prediction follows the DU prediction closely up to redshift $z \approx 2$, the $\Omega_m=1/\Omega_\Lambda=0$ prediction of the standard model shows remarkable deviation even at smaller redshifts.

Figure 6.3.4-4 converts the data and the predictions in Figure 6.3.4-2 to a logarithmic scale. At redshifts above $z > 2$, the difference between the two predictions, (6.3.4:4) and (6.3.4:5), becomes noticeable and grows up to several magnitude units at $z > 10$, Figure 6.3.4-5. For comparison, Figure 6.3.4-5 also shows the standard model prediction for $\Omega_m=1$ and $\Omega_\Lambda=0$.

As demonstrated by the FLRW curve calculated for $\Omega_m=0.3/\Omega_\Lambda=0.7$, the effect of the dark energy appears as a buildup of a certain S-shape in the magnitude/redshift curve in the redshift range $0.1 < z < 10$; at redshifts below 0.1, the effect of the assumed dark energy is negligible, and the curve is parallel to the curve without dark energy. At redshifts higher than 10, the dark energy curve again becomes parallel to the FLRW curve calculated without dark energy.

6.3.5 Time delay of bursts

Thermal transients, such as explosions related to a particular energy environment, can be assumed to emit the same number of quanta at any state of the cosmic expansion. As a conclusion, not only the number of quanta emitted in parallel at each phase of the transient but also the number of quanta emitted in a sequence is independent of the stage of the cosmic expansion.

The time required to receive a sequence of n cycles of radiation with a particular wavelength is

$$T_{(r)n\lambda} = \frac{n_\lambda \lambda_r}{c} = \frac{\chi n_\lambda \lambda_r}{c_0}. \qquad (6.3.5\!:\!1)$$

Applying equation (6.1.2:5), the wavelength received can be related to the redshift and the emitted wavelength (with zero redshift) as

$$\lambda_r = (z+1)\lambda_e. \qquad (6.3.5\!:\!2)$$

Substitution of equation (6.3.5:2) into equation (6.3.5:1) gives

$$T_{(r)n\lambda(z)} = \frac{\chi n_\lambda \lambda_e}{c_0}(z+1) = (z+1)T_{(r)n\lambda(e)}, \qquad (6.3.5\!:\!3)$$

where $T_{(r)n\lambda(e)}$ is the time required to receive a burst of n cycles of radiation at wavelength $\lambda_r = \lambda_e$ with zero redshift. A supernova explosion is regarded as a standard candle emitting a characteristic number of quanta. The stretching of supernova explosions and also gamma bursts according to equation (6.3.5:3) is supported by observations [87,88]. In the standard model, the lengthening of radiation bursts from distant objects is referred to as cosmological time dilation.

6.3.6 Surface brightness of expanding objects

The Tolman test [46,50] is considered a critical test for an expanding universe model. In expanding space, according to Tolman's prediction, the observed surface brightness of standard objects decreases by the factor $(1+z)^4$ with the redshift. Following the properties of FLRW space, Tolman's prediction assumes that galaxies and quasars are non-expanding objects. In the DU space, galaxies and quasars are expanding objects. With reference to equation (6.2.3:2) the angular area of spherical expanding objects with diameter $d_s = d_{s0}/(1+z)$ is

$$\Omega_D \approx \frac{\pi r^2}{4\pi D^2} = \left(\frac{r_0/(1+z)}{2R_0 z/(1+z)}\right)^2 = \frac{r_0^2}{4R_0^2 z^2}. \qquad (6.3.6\!:\!1)$$

Applying (6.3.2:6) for the power density, the surface brightness of an object at distance D relates to the surface brightness of a reference object at distance D_0 ($z_{d0} \ll 1$, $\Omega_{D0} = r_0/4D_0^2$) as

$$\frac{SB_{(D)}}{SB_{(D_0)}} = \frac{D_0^2}{R_0^2}\frac{(1+z)}{z^2}\frac{\Omega_{D_0}}{\Omega_D} = \frac{(1+z)z^2}{z^2} = (1+z), \qquad (6.3.6\!:\!2)$$

or

$$SB_{(D)} = SB_{(D_0)}(1+z). \tag{6.3.6:3}$$

When related to the *K*-corrected power densities in a multi-bandpass photometry with nominal filter wavelengths matched to the redshifted radiation (6.3.6:3) is converted into

$$SB_{(D,K)} = SB_{(D_0)}(1+z)^{-1}. \tag{6.3.6:4}$$

The predictions of equations (6.3.6:3) and (6.3.6:4) do not include the effects of possible evolutionary factors.

6.4 Observations in distant space

6.4.1 Microwave background radiation

The bolometric energy density of cosmic microwave background (CMB) radiation, $4.2 \cdot 10^{-14}$ [J/m³], corresponds, with high accuracy, to the energy density *in a blackbody cavity* at 2.725 °K. The rest energy calculated for the total mass in space is $E_{rest} = M_\Sigma \cdot c^2 \approx 2 \cdot 10^{70}$ [J] corresponding to energy density $E_{rest}/(2\pi^2 R_4^3) = 4.6 \cdot 10^{-10}$ [J/m³] in DU space. Accordingly, the share of the CMB energy density of the total energy density in space is about 10^{-4}.

According to the FLRW cosmology, the CMB is the afterglow of the hot early stage of the universe about 380,000 years after the Big Bang. In the standard cosmology model, the Planck constant is interpreted as an attribute of radiation, rather than an attribute of the emission/absorption process. As a consequence, standard cosmology concludes that the energy density of blackbody radiation, as any electromagnetic radiation dilutes in inverse proportion to a factor $(1+z)^4$, while the volume of space expands in proportion to the factor $(1+z)^3$. This means that, according to the standard interpretation, the radiation loses energy due to the expansion of space (without assuming absorption losses). In the case of CMB, with redshift $z \approx 1000$, the loss of energy due to the expansion of space would be about 1000 times the present CMB energy, which means the disappearance of about 10% of the total energy in space.

As discussed in Section 1.1.2 and analyzed in detail in Section 5.1.1, the Planck equation is interpreted as describing the emission/absorption process rather than an intrinsic property of radiation propagating in space. Such an interpretation conserves the mass equivalence of radiation but results in a reduction in the energy density due to the increase in the wavelength with the increasing volume. Accordingly, the energy density of blackbody radiation, like the energy density of any electromagnetic radiation, decreases in inverse proportion to the increase of volume in expanding space, which conserves the total mass and the overall energy balance in space. The energy density of radiation in a blackbody cavity is given in Appendix 1, equation (A1:7), as the energy density of a cycle of radiation at wavelength λ_T, corresponding to the temperature T as defined in equation (A1:1). The energy density in a blackbody cavity at the temperature T is

$$E_{bol,T} = \frac{8\pi^5}{15 \cdot \lambda_T^3} \cdot \frac{h_0}{\lambda_T} c_0 c, \tag{6.4.1:1}$$

where the quantity h_0/λ_T is the unit mass equivalence of a cycle of radiation at the wavelength λ_T. Conserving the mass equivalence $h_0/\lambda_{T(e)}$ at the emission, the energy density of blackbody radiation redshifted by the factor $(1+z)$ is

$$E_{bol,T}(z) = \frac{1}{(1+z)^3} \frac{8\pi^5}{15 \cdot \lambda_T^3} \cdot \frac{h_0}{\lambda_{T(e)}} c_0 c = \frac{8\pi^5}{15 \cdot \lambda_{T(z)}^3} \cdot \frac{h_0}{\lambda_{T(e)}} c_0 c, \tag{6.4.1:2}$$

where $\lambda_{T(z)}$ is

$$\lambda_{T(z)} = \lambda_T (1+z),\tag{6.4.1:3}$$

corresponding to temperature $T(z) = T/(1+z)$.

The energy density of CMB radiation is equal to the nominal energy density of radiation in a blackbody cavity at 2.725 °K.

6.4.2 Double image of an object

Because of spherical symmetry, uniform radiation from all space directions can also be received from objects at an angle $n \cdot 180°$ relative to the observer. Applying the 180° transmission to the background radiation, we get

$$R_{4(180)} = R_4 \, e^{-\pi} = \frac{R_4}{23.1} \approx 6 \cdot 10^6 \ (\text{l.y.}). \tag{6.4.2:1}$$

Radiation from an object at exactly the opposite side of spherical space has a redshift

$$z_\pi = e^\pi - 1 = 22.14, \tag{6.4.2:2}$$

which means an infrared peak in the background radiation. Considering the galaxy density in space, the probability that there is a luminous object exactly at 180° of angular distance, however, is small.

It can further be concluded that an object located close to the 180° angular distance may be visible in opposite directions with different redshifts, Figure 6.4.2-1.

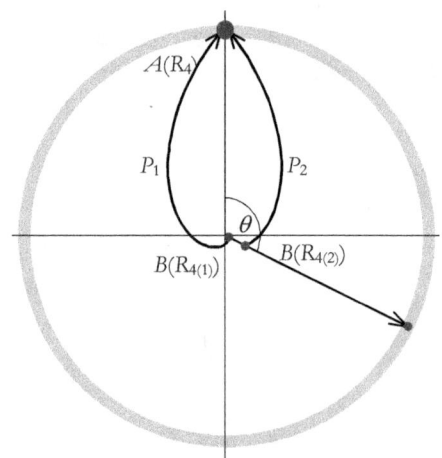

Figure 6.4.2-1. Light emitted by object B when the 4-radius of space $R_{4(1)}$ reaches the observer at $A(R_4)$ through path P_1. Light emitted by object B at $R_{4(2)}$ reaches the observer through path P_2 which appears in the direction opposite to P_1. The opposite images show the opposite faces of the object.

As can be seen from Figure 6.4.2-1, the light emitted from object B at states $B(R_{4(1)})$ and $B(R_{4(2)})$ may be observed at object $A(R_4)$ through light paths P_1 and P_2, respectively. The corresponding redshifts are

$$z_1(P_1) = e^{(2\pi - \theta)} - 1, \tag{6.4.2:3}$$

and

The Dynamic Cosmology

$$\zeta_2(P_2) = e^\theta - 1. \tag{6.4.2:4}$$

Note that the radiation observed through the longer path originates from an earlier stage of the object.

6.4.3 Radiometric dating

We can assume that the rate of nuclear decay processes is proportional to the rest momentum of the decaying object, which on a cosmological scale is a function of the time, T, since the singularity (3.3.3.8)

$$R_D = \frac{dN/N}{dt} \sim mc_0 \sim \frac{1}{T^{1/3}} = 1 \bigg/ \left[\tau_0 \left(\frac{T}{T_0} \right)^{1/3} \right] = 1/\tau(T), \tag{6.4.3:1}$$

where τ_0 is the time constant of the decay at T_0, when the decay was started. Beginning at T_0, the decay rate decreases in proportion to $(T/T_0)^{1/3}$, Figure 6.4.3-1, which means a corresponding increase in the decay time constant τ

$$\tau = \tau_0 \left(\frac{T}{T_0} \right)^{1/3}. \tag{6.4.3:2}$$

In radiometric dating, a decreasing decay rate means shortening of the estimated ages obtained with data based on a constant decay rate.

In the case of a constant decay rate, as assumed in radiometric dating, the decrease of decaying nuclei is expressed as

$$N(t) = N_0 e^{-t/\tau}, \tag{6.4.3:3}$$

where τ is the time constant of the decay.

The decay time t is

$$t = T - T_0 = \tau \ln \frac{N_0}{N(T)}, \tag{6.4.3:4}$$

where the decay process was started at time T_0, when the number of decaying nuclei is N_0.

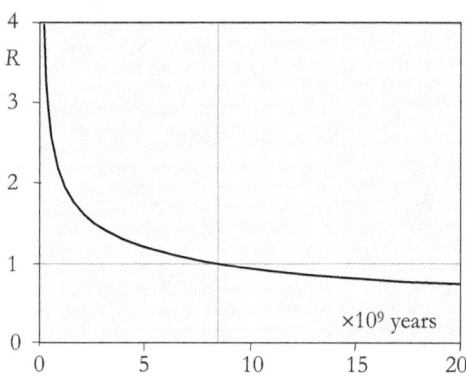

Figure 6.4.3-1. Development of the nuclear decay rate R with time from singularity. At present, $t \approx 9.3 \cdot 10^9$ years, the decay rate is about half of the decay rate at $t = 1 \cdot 10^9$ years. The reduction of the decay rate has a major effect on the determination of ages of objects existing billions of years ago.

In the case of decreasing decay rate (6.4.3:2), the decrease of decaying nuclei is expressed

$$\ln\frac{N_0}{N(T)} = \frac{T_0^{1/3}}{\tau_0}\int_{T_0}^{T}\frac{dT}{T^{1/3}} = \frac{T_0^{1/3}}{\tau_0}\frac{3}{2}\left(T^{2/3} - T_0^{2/3}\right). \qquad (6.4.3:5)$$

Substitution of (6.4.3:2) for τ_0 in (6.4.3:5) gives $N(T)$ in form

$$\ln\frac{N_0}{N(T)} = \frac{T}{\tau}\frac{3}{2}\left(1 - \left(\frac{T_0}{T}\right)^{2/3}\right), \qquad (6.4.3:6)$$

and

$$T = \frac{2}{3}\tau\cdot\ln\frac{N(T)}{N_0}\bigg/\left(1 - \left(\frac{T_0}{T}\right)^{2/3}\right). \qquad (6.4.3:7)$$

Time T in equations (6.4.3:5–7) means the time since the singularity to the time of observing the $N(T)/N_0$ ratio. Today, $T \approx 9.3$ billion years. For decay times $t = T - T_0 \ll T$, equation (6.4.3:7) can be written in the form

$$T_0 + t \approx \tau\cdot\ln\frac{N(T)}{N_0}\bigg/\frac{3}{2}\left(1 - \left(1 - \frac{2}{3}\frac{t}{T_0}\right)\right) \approx \tau\cdot\ln\frac{N(T)}{N_0}\bigg/\frac{t}{T_0}, \qquad (6.4.3:8)$$

and further as

$$t \approx \frac{\tau}{1 + t/T_0}\cdot\ln\frac{N(T)}{N_0}. \qquad (6.4.3:9)$$

For $t \ll T_0$, (6.4.3:9) is equal to (6.4.3:4)

$$t \approx \tau\cdot\ln\frac{N(T)}{N_0}, \qquad (6.4.3:10)$$

which is equal to equation (6.4.3:4) for constant decay rate in the standard model, Figure 6.4.3-2.

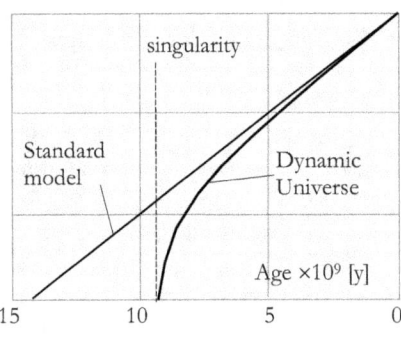

Figure 6.4.3-2. Accumulation of decay products according to the standard model with a constant decay rate and the DU model with a decreasing decay rate.

7. Summary

7.1 The picture of reality behind theory and experiments

The relativistic reality

We may say that the need for the rethinking of Newtonian space in the late 19th century arose primarily from philosophical problems in combining the local and the global. Maxwell's theory of electromagnetism suggested a global world ether for the propagation of electromagnetic radiation. The 19th century experimental efforts, however, failed in showing the existence of the world ether. The approach adopted to solve the problem was the modification of the Galilean transformation, which meant rejection of absolute time and distance, and de-linearization of Newton's second law. The choice was encouraged by observations on the mass increase of electrons in fast motion.

In contemporary physics, the "relativistic" mass is a consequence of velocity. In the DU framework, the increased mass of a moving object is due to the mass contribution needed to obtain the motion at constant gravitational potential. In free fall in a gravitational field, the velocity is obtained against a reduction of local velocity of light, which means that there is no increase in mass in free fall.

In the DU framework, the effects of motion and gravitation on clocks, and the observed velocity of light are explained as consequences of the conservation of energy. This is a fundamental difference from the explanation given by the theory of relativity, which explains the effects in terms of modified coordinate quantities, time and distance.

A further difference between the theory of relativity and the Dynamic Universe is the nature of relativity. The theory of relativity does not recognize any absolute frame of reference; primarily, relativity appears as relativity between on object and the observer. In the DU, the state of rest in a hypothetical homogeneous space serves as the absolute reference for all energy states in space. Relativity in the DU can be characterized as relativity between the local and the whole.

The velocity of light

The constancy of the velocity of light is a cornerstone of the theory of relativity. In the theory of relativity, the constancy of the velocity of light is established by the postulated relativity principle, which fixes the velocity of light to any local frame of reference.

In the Dynamic Universe, the local velocity of light is linked to the local 4D velocity of space, which is a function of the local gravitational state. The local 4D velocity is not a function of the motion of a local energy frame.

Discontinuity and discreteness of physical systems

Since Newton's laws of mechanics, physical systems were seen as being driven by continuous processes, allowing any amount of mass or any energy state in a system. In spite of the early development of atomic theory as a tool for understanding chemical reactions, since John Dalton's discoveries in the early 19th century, the nature of atoms as discrete elements of matter was more or less untouched in physics for the whole century. A step towards understanding the nature of atoms as discrete elements in physical systems was taken with Ludwig Boltzmann's kinetic theory of gases and the statistical interpretation of the second law of thermodynamics, which he developed in the late 19th century and the beginning of the 20th century.

The next piece of the puzzle was Max Planck's solution of the spectrum of blackbody radiation discussed in Section 5.1.1. The excellent agreement between Planck's prediction and the observed blackbody spectra left no doubts about Planck's considerations on discrete emission–absorption processes behind the energy balance between radiation and the radiating body. Planck's findings in solving the blackbody radiation were extended to the solution of the specific heat of solids by Albert Einstein in 1906. In fact, inspired by statistical thermodynamics, Einstein had gone a step further in discreteness by proposing light quanta as the solution for the photoelectric effect in 1905. It took another 20 years for the formulation of quantum mechanics, which tied together the theoretical considerations and experimental findings. Quantum mechanics covers a wide range of phenomena and works well in practice in spite of the still disputable philosophical basis and limited compatibility of quantum mechanics with other branches of physics.

An essential part of the experimental work behind quantum mechanics comes from spectral analysis of matter–radiation interactions. Stationary quantum states are described in terms of resonant wave functions, like spherical harmonics in the case of electron states in an atom.

In the Dynamic Universe framework, a quantum of radiation is the energy injected into a cycle of radiation by a single electron transition in the emitter.

The solution of Planck's equation from Maxwell's equations is not tied to any assumption of the Dynamic Universe – the interpretation of quantum as the unit energy injected into radiation or absorbed by an absorber is equally valid in the contemporary physics framework. In fact, the energy injected by any antenna into a cycle of radiation is proportional to frequency.

Wavenumber, mass, and energy

The linkage of Planck's equation to Maxwell's equations links the Planck constant to primary electrical constants: unit charge, e, and vacuum permeability μ_0.

Further, it reveals the linkage of the Planck constant to the velocity of light and thereby the mass equivalence of electromagnetic radiation, and vice versa, the wavelength equivalence of a mass object.

In the DU framework, the conventional form of the Planck constant is replaced with the intrinsic Planck constant, where the linkage to the velocity of light is removed.

The intrinsic Planck constant, $h_0 = h/c$, has dimensions of mass-distance [kg·m], which establishes a direct linkage between mass and wavelength $m = h/\lambda$, or in terms of the wavenumber $m = h \cdot k$.

In the DU framework, mass appears as a wavelike substance for the expression of energy.

	Contemporary physics	Dynamic Universe
Wave number	k	k
Mass	$m = \dfrac{\hbar}{c} \cdot k$	$m = \hbar_0 k$
Rest energy	$E_0 = mc^2$	$E_0 = c_0 mc$
Critical radius	$r_c = 2\dfrac{Gm}{c^2}$	$r_c = \dfrac{G}{c_0 c_{0\delta}} m = \dfrac{R''}{M''} \cdot m$

Table 7.1.1-I. Linkage of mass, rest energy, and the critical radius to the Compton wave number k in contemporary physics and in the Dynamic Universe. The last form of the critical radius in the Dynamic Universe illustrates the linkage of the critical radius, r_c, related to the local mass m to the total mass M'' and the 4-radius of space R''. In contemporary physics, the critical radius is twice the critical radius in the DU. The difference comes from the way Newtonian gravitation is used as a boundary condition in the derivation of the critical radius in Schwarzschild's space.

Table 7.1.1-I illustrates the linkage of wavenumber, mass, momentum, rest energy, and the critical gravitational radius in contemporary physics and the Dynamic Universe. In the table, the DU expressions are reduced to the form of their counterparts in contemporary physics. In the DU framework, the expression for wavenumber, mass, and all forms of energy are primarily expressed as complex functions, which give more information about the physical nature of each (see Section 4.1).

In the DU, localized mass objects are described as resonant mass wave structures. The real component of the momentum of a mass object moving in a local frame in space can be described as a wave propagating in parallel with the object. The wave number of the momentum wave is created as the sum of the Doppler-shifted front and back waves in the moving resonator. Based on the parallel momentum wave, the DU explanation of the double slit experiment is given in Section 5.3.5.

The mass wave in the DU can be seen as a physical replacement of the wave function in quantum mechanics. The solution of the hydrogen atom in Section 5.1.4 demonstrates the use of the resonant mass wave concept.

7.2 Changes in paradigm

7.2.1 The basic postulates

We can summarize the main postulates of the Dynamic Universe as follows:

1. Space is defined as the three-dimensional surface of a four-dimensional sphere free to contract and expand in an infinite four-dimensional universe.
2. Time is a universal scalar. The fourth dimension is metric by its nature.
3. As the initial condition, all mass is homogeneously distributed in space. Total mass in space is conserved in all interactions in space.
4. The dynamics of space is determined by a zero-energy balance of motion and gravitation in the structure.
5. The inherent energy of gravitation is defined in hypothetical homogeneous space; the inherent energy of motion is defined in the hypothetical environment at rest.
6. The buildup of motion, electromagnetic energy, elementary particles, and mass centers *within space* conserves the total energy and the zero-energy balance created in the contraction–expansion process *of space*.

Classically, *force and force field* are primary quantities, and energy is derived by integration of force. The prevailing cosmological appearance and structure of space is an extrapolation of local spacetime based on the equivalence principle equating inertial and gravitational accelerations. The prevailing theory does not allow the determination of the total energy in space, nor does it rely on the conservation of total energy. In FLRW cosmology, due to the interpretation of Planck's equation as an inherent property of radiation, the redshift of radiation results in a disappearance of energy.

The postulates in the DU reflect the holistic approach. The total structure of space and the *inherent expressions of energy* are postulated; force is derived as the gradient of energy. Local structures are derived from the whole by conserving the total energy in space. Planck's equation is interpreted as the energy conversion at the emission of electromagnetic radiation. As a consequence, the energy of radiation is conserved in the redshift; the energy density of radiation is diluted.

The DU approach discards the central postulates of the prevailing theories:

1. The velocity of light is not postulated as a constant or an invariant.
2. Time is not regarded as a fourth dimension; the space-time concept is ignored.
3. There is no postulated equation of motion or force/acceleration.
4. There is no need or basis for the equivalence principle.
5. There is no need or basis for the Lorentz transformation.
6. There is no need or basis for the relativity principle.

7. There is no need or basis for dark energy or accelerating expansion.
8. There is no need to postulate the Planck equation; the Planck equation can be seen as a consequence of Maxwell's equations.
9. There is no need to postulate the Schrödinger equation or a wave function; a resonant mass wave structure can be used to describe localized energy structures.

In local considerations in DU space, it is useful to describe the fourth dimension as an imaginary direction; local momenta and energies are presented as complex functions with their imaginary parts showing the effects in the fourth dimension. As an example, the rest energy of matter appears as the imaginary component of the total energy of motion. A mass object moving in space has both imaginary and real components in the energy of motion. In the DU framework, the absolute values of the complex energies are equivalent to the corresponding energies in the prevailing theories, where energy is used as a scalar (real) function only.

The use of complex functions is a powerful tool for a detailed analysis of the energy structures in space. In spherically closed space, the imaginary direction links the global effect, the effect of the rest of space, to the local effects in space. The global gravitational energy arising equally from all space directions is equivalent to the gravitational energy arising from the barycentre of space in the fourth dimension. Also, local mass objects that are described as resonant mass wave structures with spherical symmetry in the three space directions have their rest momentum in the fourth dimension.

The system of nested energy frames is a central feature of the Dynamic Universe. The conservation of the overall zero-energy balance through the system of nested energy frames allows the understanding of the local state of rest and its relation to the state of rest in hypothetical homogeneous space, which serves as the universal reference for all energy states and the states of motion and gravitation in space.

The composition of the Planck constant and the identification of the intrinsic Planck constant are exceedingly important steps for the unified expression of energies and for understanding the wave-like nature of mass as the substance for all expressions of energy. Applying the intrinsic Planck constant, mass can be expressed in terms of a wavelength equivalence or wave number equivalence.

7.2.2 Natural constants

Gravitational constant

The gravitational constant, G, is considered a constant anywhere in space and throughout the contraction and expansion of space.

Total mass in space

Mass is defined as the substance for the expression of energy. The total mass in space is a constant determining, together with the gravitational constant, the dimensions and the dynamics of space. Estimates for the total mass M_Σ and the mass equivalence M'' in space are obtained from equation (3.3.2:4). Using the CODATA 2006 value ($G = 6.67428 \cdot 10^{-11}$

[Nm²/kg²]) for the gravitational constant and the Hubble constant $H_0 = 70.5 \pm 1.3$ [(km/s)/Mpc] [74,] we get

$$M" = \frac{c_0^2 R_4}{G} = 0.776 \cdot M_\Sigma \approx 1.78 \cdot 10^{53} \text{ [kg]} \quad ; \quad M_\Sigma \approx 2.30 \cdot 10^{53} \text{ [kg]}. \quad (7.2.2:1)$$

Substance, distance, and time are the fundamental quantities for human conception of observable reality by answering the questions *"what, where, and when?"*. Their primacy is reflected in the system of physical units, which relies on kilogram, meter, and second — or mass, distance, and time as the three primary quantities.

The velocity of light

Several quantities that traditionally are considered as physical constants appear in the DU as parameters related to the state of the Universe. The most important of these is the velocity of light, the cornerstone of the theory of relativity. As a consequence of the conservation of the zero-energy balance between motion and gravitation, the velocity of light is determined by the velocity of the expansion of space in the fourth dimension [equation (3.3.1:6)],

$$c_0 = c_4 = \sqrt{\frac{GM"}{R_4}}, \quad (7.2.2:2)$$

where the G is the gravitational constant, I_g a geometrical factor resulting from the integration of the total gravitational energy in space [equations (3.2.2:4 and (3.2.2:5))], M_Σ the total mass in space, and R_4 the 4-radius of spherically closed space.

Due to the expansion of space, the 4-radius increases with time. As a result, the velocity of light is also a function of time since the singularity [equation (3.3.3:8)]:

$$c_0 = c_4 = \left(\frac{2}{3}GM"\right)^{1/3} t^{-1/3}. \quad (7.2.2:3)$$

Velocity c_0 in equations (7.2.2:3) means the velocity of light in hypothetical homogeneous space where all mass is uniformly distributed. Because of the local geometry of space near mass centers, the local velocity of light is not only a function of the expansion of space but also a function of the local gravitational environment [equation (4.1.4:10)],

$$c = c_{\delta(n)} = c_0 \prod_{i=1}^{n}(1-\delta_i), \quad (7.2.2:4)$$

where δ_i means the gravitational factor [equation (4.1.1:30)] in the i:th gravitational frame under hypothetical homogeneous space

$$\delta_i = \frac{GM_i}{c_0 c_{0\delta(i)} r_{0\delta(i)}}, \quad (7.2.2:5)$$

where $c_{0\delta(i)}$ is the velocity of light in the apparent homogeneous space of the i:th gravitational frame and $r_{0\delta(i)}$ is the distance to the local mass center measured in the direction of the apparent homogeneous space of the frame.

Planck's constant

It has been shown that Planck's constant contains the velocity of light as a hidden factor. The energy of one cycle of radiation from a point emitter, when solved from Maxwell's equations as a dipole in the fourth dimension, is [equations (5.1.2:21)]

$$E_\lambda = N^2 X_\lambda \cdot 2\pi^3 e^2 \mu_0 c \, f = N^2 h_0 c \, f = N^2 \frac{h_0}{\lambda} c_0 c, \quad (7.2.2:6)$$

where N is the number of unit charges oscillating in the emitter, f is the frequency of radiation emitted, and h is Planck's constant

$$h = \chi_\lambda \cdot 2\pi^3 e^2 \mu_0 c = h_0 c, \quad (7.2.2:7)$$

where h_0 is referred to as the *intrinsic Planck's constant* independent of the velocity of light. The Maxwell frame conversion factor χ_λ in equation (7.2.2:7) has the value [see equation (5.1.2:9) and (5.1.3:17)]

$$\chi_\lambda = \left(\frac{z_0}{\lambda}\right)^2 \frac{c_0}{c} = 1.1049. \quad (7.2.2:8)$$

Based on the present knowledge of the gravitational frames in space, the ratio c_0/c is not high enough to explain the numerical value of χ_λ in the expected $z_0 = \lambda$ situation. The effective dipole length or the apparent effective length of a quantum emitter as a dipole may derive from four-dimensional geometry or the simplified way Maxwell's equations were applied in the fourth dimension (see Section 5.1.2).

The fine structure constant

By applying equation (7.2.2:7) for Planck's constant, the fine structure constant a obtains the form [equation (5.1.2:14)]

$$a \equiv \frac{e^2 \mu_0 c}{2h} = \frac{e^2 \mu_0 c}{2 \cdot \chi_\lambda \cdot 2\pi^3 e^2 \mu_0 c_0} = \frac{1}{1.1049 \cdot 4\pi^3} \approx \frac{1}{137.0360}, \quad (7.2.2:9)$$

which means that the fine structure constant is a pure numerical constant independent of any physical constant or the velocity of light, and accordingly independent of the expansion of space.

The Bohr radius

The Bohr radius of a hydrogen atom is a function of the rest mass of an electron in the nucleus frame. The rest mass of an electron in the nucleus frame is subject to reduction due to the motion of the atom in its local potential energy frame and the motions the of the local frame in the parent frames [equation (5.1.3:3)]

$$m_{e(Nucleus)} = m_{e,0} \prod_{i=1}^{n} \sqrt{1-\beta_i^2}, \quad (7.2.2:10)$$

where $m_{e,0}$ is the electron mass at rest in hypothetical homogeneous space, and velocities $\beta_i = v_i/c_\delta$ are velocities of the atom (the nucleus frame) in the local energy frame and the velocities of the local frame in the parent frames.

The Bohr radius obtains the form [see equation (5.1.3:13)]

$$a_{0(\beta_i)} = \frac{h_0^2}{\pi\mu_0 e^2 m_{e,(nucleus)}} = \frac{a_{0(0)}}{\prod_{i=0}^{n}\sqrt{1-\beta_i^2}}, \qquad (7.2.2{:}11)$$

where $a_{0(0)}$ is the Bohr radius at rest in hypothetical homogeneous space and velocities β_i are the velocities of the atom in the local frame and the relevant parent frames. The Bohr radius is a function of the velocity of the atom, but *not a function of the local velocity of light or the expansion of space*.

Vacuum permeability

As a consequence of the conservation of electromagnetic (Coulomb) energy in relation to the energy of electromagnetic radiation and the rest energy of matter, vacuum permeability μ_0 rather than vacuum permittivity ε_0 is a constant. Vacuum permittivity ε_0 is expressed in terms of μ_0 as [see equation (5.1.1:7)]

$$\varepsilon_0 = \frac{1}{c_0 c \cdot \mu_0} = 1 \Big/ \mu_0 c_0^2 \prod_{i=1}^{n}(1-\delta_i), \qquad (7.2.2{:}12)$$

which means that the value of ε_0 increases with the expansion of space. Vacuum permittivity is also a function of the local gravitational state due to the dependence of the local velocity of light on the local gravitational state.

Summary of natural constants

Table 7.2.2-I summarizes some fundamental physical constants and quantities in the theory of relativity, quantum mechanics, the standard cosmology model, and the Dynamic Universe.

The units of time and distance are constant, but instead of being classified as physical constants, they are coordinate quantities, measures used for the physical constants and derived quantities.

7.2.3 Energy and force

The Dynamic Universe concept is strongly based on energy. The energy available in local energy structures and energy objects in space originates from the energy of motion that matter in space possesses due to the expansion of space as the surface of a 4-sphere.

In planetary systems, the total energy is expressed in the form of Kepler's energy integral; in thermodynamic systems, it is referred to as the total internal energy of the system. In a quantum mechanical system, the total energy, as the sum of the kinetic energy and the potential energy, is expressed by the Hamiltonian in the Schrödinger equation.

Summary

Quantity		SR, GR, QM, SC	DU
gravitational constant	G	constant	constant
total mass in space	M_Σ	$M_\Sigma + E =$ constant	constant
electron (unit) charge	e	constant	constant
vacuum permeability	μ_0	constant	constant
velocity of light	c	constant	$c = c(G, M_\Sigma, R_4, \delta)$
vacuum permittivity	ε_0	constant	$\varepsilon_0 = f(c)$
Planck's constant	h	constant	$h_0 = h/c = C \cdot e^2 \mu_0 =$ constant
rest mass of a specific particle	m	constant	$m = m(\beta_i)$
Bohr radius	a_0	constant	$a_0 = f(\beta_i)$
Compton wavelength of a specific particle	λ_C	constant	$\lambda_C = f(\beta_i)$
characteristic emission/absorption wavelength of atomic objects	$\lambda_{n1,n2}$	constant	$\lambda_{n1,n2} = f(\beta_i)$
frequency of atomic oscillators	f	constant	$f = f(R_4, \delta_i, \beta_i)$
inertial force	F	$F = -m(\beta) \cdot a$	$F = -\chi m(\beta) \cdot a = f(\delta_i, \beta_i)$
planetary radii	r	constant, $r = r_{(0)}$	$r = r_{(0)} \cdot R_4 / R_{4(0)}$
dimensions of galaxies	d_G	constant, $d_G = d_{G(0)}$	$d_G = d_{G(0)} \cdot R_4 / R_{4(0)}$
unit of distance	dr	$dr = f(\delta, \beta)$; SR, GR constant; QM	constant
unit of time	dt	$dt = f(\delta, \beta)$; SR, GR constant; QM	constant

Table 7.2.2-I. Comparison of some fundamental physical quantities and the coordinate quantities in the theory of relativity (SR, GR), quantum mechanics (QM), standard cosmology model (SC), and the Dynamic Universe (DU).

Since Newton's revolutionary insights into the laws of physics, *force* has been used as a basic physical quantity, both for the postulates of gravitation and the laws of motion. The concept of kinetic energy was identified more than 100 years after Newton's laws of motion. Energy was derived as the integrated work done by force or a change in momentum, i.e., force was a postulated quantity, whereas energy became a derived quantity.

The concept of energy as a primary physical quantity was finally recognized in the 19th century. Conservation of energy and the natural trend to minimum potential energy were understood as basic laws of nature as a part of thermodynamics. Conservation of energy requires a closed system for a precise determination of the energy to be conserved. The closing of a system assumes the definition of the structure and the linkage of the potential energy of the system to the structure. The total energy of a system can be expressed as the sum of potential energy and the energy of motion.

The whole space in the Dynamic Universe is a Hamiltonian surface. More specifically, it is a zero-Hamiltonian surface in the 4-dimensional universe where the sum of the energies of motion and gravitation is zero. The natural trend towards minimum potential energy defines the concept of force and results in the dynamic behavior of the system.

Conservation of the total energy in space makes the locally available rest energy a function of the motion and gravitational state of the local system in space. Relativity in the DU is a consequence of the finiteness of total energy in space.

In the Dynamic Universe, the inherent forms of the energies of motion and gravitation are postulated in an "undisturbed" environment; gravitation for mass in homogeneous infinite space in equation (2.2.2:1) and motion in the environment at absolute rest in equation (2.2.2:4). The energies of motion and gravitation of mass at rest in spherically closed homogeneous space can be derived from the inherent forms of the energies due to the symmetry of spherically closed space. In real space, mass is not uniformly distributed but accumulated into mass centers in several steps. The environment within space is not "environment at rest" either. That is why both the energies of gravitation and motion are modified by the mass distribution and the motion of real space.

Force in the DU is a derived quantity. Force is an implication of the "natural trend" towards minimum potential energy; force is defined as the gradient of potential energy. The equations of motion are derived from the conservation of total energy of motion, comprising the energies due to the motion of space and the motion in space.

Unified expression of energy

Gravitation and motion are described as primary expressions of energy. Mass in space is energized through the primary energy buildup of space in contraction and expansion. The primary energy buildup creates the rest energy of matter as an excited state of motion and gravitation. In hypothetical homogeneous space, the rest energy and the balancing gravitational energy appear in their elemental forms [see equation (3.3.5:2)]

$$E_{(0,0)} = c_0 |\mathbf{p}_0| = c_0 m c_0 = c_0 \hbar_0 k_m c_0 = \frac{GM''}{R_4} m. \qquad (7.2.3:1)$$

The concept of mass as the substance for the expression of energy is reflected in equation (7.2.3:1) as equal first-order contributions of mass m both to the energy of motion and to the energy of gravitation.

Electromagnetic energy is described as a secondary form of the energy of motion comprising the expressions of electromagnetic radiation and Coulomb energy [see equations (5.1.2:4) and (5.1.1:11)]

$$E_{\mathrm{rad}} = N^2 \frac{h_0}{\lambda} c_0 c = m_{\mathrm{rad}} c_0 c, \qquad (7.2.3:2)$$

and

$$E_{\mathrm{EM}} = -\frac{q_1 q_2 \mu_0}{4\pi r} c_0 c = N^2 a \frac{h_0}{L} c_0 c = m_{\mathrm{EM}} c_0 c, \qquad (7.2.3:3)$$

where $h_0 = h/c$ is the intrinsic Planck constant. In the last form of Coulomb energy, a is the fine structure constant defined in equation (5.1.2:13) and $L = 2\pi r$ is the circumference of a circle with radius r. The substance of electromagnetic energies in equations (7.2.3:2) and (7.2.3:3) can be expressed as mass equivalences

$$m_{rad} = N^2 \frac{h_0}{\lambda} \quad [kg], \tag{7.2.3:4}$$

and

$$m_{EM} = N^2 a \frac{h_0}{L} \quad [kg]. \tag{7.2.3:5}$$

In radiation, the elemental quantity of substance, a quantum, is h_0/λ ($N = 1$), which describes a quantum as one wavelength of radiation related to emission from a transition of a unit charge (see Section 5.1.2).

The rest mass of a mass object can be expressed in terms of wavelength equivalence

$$\lambda_{Compton(\beta)} = \frac{h_0}{m_{rest(\beta)}} \quad (\text{Compton-wavelength}), \tag{7.2.3:6}$$

and the relativistic mass $m_{(\beta)} = (m + \Delta m)$ in terms of wavelength equivalence

$$\lambda_{(\beta)} = \frac{h_0}{m_{(\beta)}} = \frac{h_0}{m_{(0)}} \sqrt{1-\beta^2}. \tag{7.2.3:7}$$

Combined with the motion of space in the fourth dimension, the rest momentum of a mass object is

$$\mathbf{P}_{rest} = \mathbf{i} \frac{h_0}{m_{rest(\beta)}} c = \mathbf{i} \hbar_0 k_{rest(\beta)} c = \mathbf{i} m_{rest(\beta)} c. \tag{7.2.3:8}$$

The rest energy of a mass object appears as the energy of motion in the fourth dimension. A mass object can be described as a standing wave structure with a wavelength equal to the wavelength equivalence of the rest mass.

The momentum of an object moving at velocity β in a local frame is expressed by a wave front with wavelength $\lambda_{(\beta)}$ propagating in parallel with the object at velocity β

$$\mathbf{P}_\beta = m_{(\beta)} \mathbf{v} = \frac{h_0}{\lambda_{(\beta)}} \mathbf{v} = \hbar_0 k_{(\beta)} \mathbf{v} = \hbar_0 k_{deBroglie} c. \tag{7.2.3:9}$$

In the DU framework, the total momentum of an object can be expressed in complex form as

$$\hbar_0 k_{(\beta)} c = \hbar_0 k_{(\beta)} \beta c + \mathbf{i} \hbar_0 k_{(0)} c, \tag{7.2.3:10}$$

which, by multiplying by c_0, returns the complex energy of motion

$$c_0 \hbar_0 k_{(\beta)} c = c_0 \hbar_0 k_{(\beta)} \beta c + \mathbf{i} c_0 \hbar_0 k_{(0)} c, \tag{7.2.3:11}$$

equal to

$$c_0 \hbar_0 k_{(\beta)} c = c_0 \hbar_0 k_{(\beta)} \beta c + \mathbf{i} c_0 \hbar_0 k_{(0)} c. \tag{7.2.3:12}$$

Solving for the scalar value of the total energy squared (7.2.3:11) returns the energy-momentum four vector in the form

$$E_{tot}^2 = c_0^2 p^2 + c_0^2 (mc)^2, \qquad (7.2.3:13)$$

and by setting $c_0 = c$, we get the traditional form of energy-momentum four vector

$$E_{tot}^2 = c^2 p^2 + c^2 (mc)^2. \qquad (7.2.3:14)$$

7.3 Comparison of DU, SR, GR, QM, and FLRW cosmology

Philosophical basis

The Theory of Relativity and standard cosmology model	The Dynamic Universe
Spacetime is a four-dimensional continuum with a local structure dependent on the distribution of mass density.	Space is a dynamic spherical structure closed through the fourth dimension.
Time is considered the fourth dimension. Time and distance are attributes of local gravitation and relative velocities.	The fourth dimension is geometrical in nature, but inaccessible from space. Time and distance are universal and absolute.
On a cosmological scale, space is homogeneous, i.e., it looks essentially the same at any location (the cosmological principle).	On a cosmological scale, mass is uniformly distributed in space, which, together with the spherical symmetry, makes space look essentially the same at any location.
The velocity of light is a physical constant and the maximum velocity by definition. Mathematically, the velocity of light is made the maximum velocity in space through the Lorentz transformation.	The velocity of light is determined by the velocity of space in the fourth dimension. The velocity of light is affected by the local gravitational state. The maximum velocity of light, which occurs in hypothetical homogeneous space, is determined by the state of the Universe.
Lorentz covariance is a property of spacetime required by the special theory of relativity. The principle is postulated to make the law of nature look the same for any observer. The equivalence principle is postulated.	The laws of nature are the same anywhere in space. No relativity principle, equivalence principle, or Lorentz covariance is postulated.
Relativity means relativity between an observer and an object.	Relativity means the relativity between the local and the whole.
Total energy in space cannot be defined.	Total energy in space is conserved in all interactions in space.
Mass is a form of energy.	Mass is the substance for the expression of energy. Total mass in space is conserved throughout the contraction–expansion process of space.
Gravitational interaction between masses is actuated by gravitons propagating at the velocity of light.	Gravitational interaction is instantaneous. It can be described as sensing of the local potential energy and its gradient.

The energies of matter and radiation are assumed to appear instantaneously in the Big Bang. Gravitational energy and the possible dark energy cannot be defined quantitatively.	The rest energy of matter is the energy of motion mass possesses due to the motion of space in the fourth dimension. The rest energy of matter is obtained in a zero-energy process against the release of gravitational energy in the contraction of space preceding the ongoing expansion phase.
The flow of time started at the Big Bang. Locally, the flow of time (proper time) depends on the state of motion and gravitation of the object, relative to an observer.	Time is eternal and absolute in nature, but the frequencies of oscillations (like the frequency of an atomic clock) depend on the state of gravitation and motion of the object.

Physics

The theory of relativity and quantum mechanics	The Dynamic Universe
Rest energy is a property of mass. The rest energy of matter is expressed $E = mc^2$, where both m and c are independent of the state of motion and gravitation.	Rest energy is the energy of a mass object due to the motion of space in the fourth dimension. The rest energy of matter is an attribute of the local gravitational state and the motion of the object in the local frame, and the gravitational state and the motion of the local frame in its parent frames $$E_{rest(n)} = c_0 mc = m_0 c_0^2 \prod_{i=1}^{n}(1-\delta_i)\sqrt{1-\beta_i^2}$$
Any state free of acceleration can be defined as a state of rest.	The state of rest is an attribute of an energy frame.
Proper time and unit length in a local frame are dependent on the state of motion and gravitation of the frame relative to an observer.	The units of time and length are absolute and universal, and independent of each other. The rate of physical processes is dependent on the local energy state.
The total energy-momentum four vector is expressed as $E_{tot}^2 = c^2 p^2 + c^2(mc)^2$.	The energy-momentum four vector results from the complex nature of the energy of motion $$E_{(m),tot}^\square = c_0 p^\square = c_0(p + \mathrm{i}\,mc),$$ resulting in $$E_{tot}^2 = c_0^2 p^2 + c_0^2(mc)^2.$$

Summary

Planck's equation, postulated as $$E_{rad} = hf,$$ means the energy of a quantum of electromagnetic radiation.	The energy of a quantum in Planck's equation $$E_{rad}(=hf) = h_0 c_0 f = \frac{h_0}{\lambda} c_0 c$$ is linked to Maxwell's equations as the energy emitted by a single electron transition into one cycle of radiation by a point source as a one-wavelength dipole in the fourth dimension.
The quantum of action, Planck constant h, has the dimensions of momentum-length.	The intrinsic Planck's constant, $$h_0 = \frac{h}{c},$$ has the dimensions of mass-length.
Inertia is a property of mass. It is the same everywhere in space.	Inertia is related to the work done in reducing the rest mass of an object moving in space and hence the gravitational effect of the rest of the mass in space on the object in motion.

Cosmology

FLRW cosmology	The Dynamic Universe
The Universe came into being in a "Big Bang" about 14 billion years ago and obtained its energy instantaneously as the energy of radiation in a quantum jump. The energy created in the Big Bang has been conserved as a constant.	The energy excitation of the Universe was built up in a continuous contraction – expansion cycle. The energy buildup culminated in a singularity about 9.3 billion years ago when the contraction of space reverted into expansion. As space expands, the energy excitation built up in the contraction phase is gradually released. The energy of the motion of space appears as the rest energy of matter.
The future of the Universe is unclear. The expansion of space continues forever if the density of mass in space is less than or equal to the Friedman critical mass density. Alternatively, it collapses in a "Big Crunch".	The expansion continues to infinity. In the process, the energies of motion and gravitation are consumed, and all expression of matter and radiation terminates. The DU concept does not exclude repeated contraction – expansion cycles.
The expansion of space occurs in free space between gravitationally bound local systems only. The dimensions of galaxies and the radii of orbiting stellar systems are conserved in the course of the expansion of space.	The expansion of space occurs uniformly everywhere in space; the dimensions of galaxies and the orbital radii of stellar systems are also subject to expansion. However, the radii of solid objects conserve their dimensions because atomic radii remain unchanged in the course of the expansion of space.

Light propagates along geodetic lines of spacetime.	Light follows the shape of space. At cosmological distances, the propagation path of light is a spiral in the fourth dimension.
The original (linear) Hubble law for redshift is $$z = \frac{D}{R_H} = \frac{H_0 D}{c},$$ where R_H is the Hubble radius, D is the distance of the object, and H_0 is the Hubble constant. The standard model does not provide a general prediction for the Hubble law due to the lack of an unambiguous prediction of the development of the expansion of space.	The Hubble law for redshift has the form $$z = \frac{D/R_4}{1 - D/R_4} = e^{\theta} - 1,$$ where R_4 is the 4-radius of space, D is the optical distance of the object, and θ is the distance angle of the object in the universal coordinate system.
The prediction of the angular size of a standard rod and the size of objects shows a minimum at about $z \approx 2$, which means a radical deviation from a Euclidean view.	Due to the spherical geometry and the expansion of local systems, the angular sizes of distant galaxies and quasars are observed in Euclidean geometry $$\psi = \frac{\theta_d}{z},$$ where z is the redshift observed, and θ_d is the angular size of the object in the universal coordinate system with origin at the 4-center of space.
Recent observations of the magnitude/redshift relationship in distant supernova explosions mean the acceleration of the expansion of space. Acceleration is assumed to be caused by "dark energy" working against gravitation between galaxies.	Recent observations of magnitude/red-shift in distant supernova explosions follow the DU-prediction without any additional assumptions or free parameters.
Matter identified through its gravitational effect alone is referred to as dark matter. There is no theoretical description of dark matter.	Unstructured matter is the initial form of energized mass. A certain share of unstructured matter is converted into visible, structured material.

7.4 Conclusions

The Dynamic Universe model provides a comprehensive description of observable physical reality. It is based on a few rational assumptions and straightforward mathematics and produces precise predictions that are in excellent agreement with observations.

The Dynamic Universe model means a major restructuring of the picture of reality. Space is seen as a highly ordered entity where local structures are results of diversification of the whole in a process directed by an overall zero-energy balance.

The Dynamic Universe model re-establishes the concepts of absolute time and distance. It is seen that the essence of relativity is in the finiteness of total resources in space. Relativity in the DU is not described by distorting the coordinate quantities, time and distance, but by showing the effects of motion and gravitation on the locally available share of the total energy.

Reflecting on the history of the theory of relativity, in 1948, Einstein asked himself, *"… why was another seven years required for the construction of the general theory of relativity? The main reason lies in the fact that it is not so easy to free oneself from the idea that coordinates must have an immediate metrical meaning."* [89]

To now free ourselves from the idea that coordinates are used as parameters to explain observations and to return to coordinates with direct metrical meaning may require an even greater shift in thinking.

Description of relativity as a consequence of finite total energy in space conserves absolute coordinate quantities but makes the locally observed rest energy of matter a function of the velocity and gravitational environment of the object studied. A local state of rest is linked to the universal state of rest, and relativity is inherently brought into quantum mechanical considerations, where the energy states are related to the rest energy of the objects in the system. In the Dynamic Universe, any local object or event is linked to the rest of space.

The Dynamic University approach means a major step in the unification of quantum mechanics, relativistic physics, and cosmology. Unification comes from the primary postulate basis and unified expressions of energy that allow an unbreakable study of the conservation of energy in all interactions in space.

8. Index

	symbol	Section
acceleration	a	
central acceleration (in the imaginary direction)		4.1.8
gravitational acceleration		4.1.6, 4.2.2
inertial acceleration		4.1.2, 4.1.7
angle between locations in space, see distance angle	θ	2.1.1, 3.3.4
angular size (observed)		6.2.2, 6.2.3'
of standard rod		6.2.2
of expanding objects		6.2.3
anti-matter		3.3.5
Bohr radius		5.1.4
bolometric energy flux		1.3.3
bending of light path	ψ	5.4.3
black hole (local singularity in space)		1.2.6, 4.2.8
complex function		2.1.4
notation of complex quantities		2.1.4
Compton wavelength, wave number		1.1.5, 1.2.1, 5.1.3
conservation laws		2.2.2
of energy		2.2.2
of momentum		2.2.2
of mass		2.2.2
of phase velocity		5.2.3
contraction of space		3.3.1
Coulomb energy		5.1.2
mass equivalence of Coulomb energy		5.1.2
critical mass density (FLRW, Friedman)	ϱ_c	3.3.2
critical radius	r_c	1.2.6, 4.1.6, 4.1.9, 4.2
critical radius in Schwarzschild space	$r_{c(Schw)}$	1.2.6,
critical radius in DU space	$r_{c(DU)}$	1.2.6, 4.1.6, 4.1.8
curvature of space		4.1.9
dark energy (FLRW)		6.2.3, 6.3.4
dark matter		3.3.2, 3.3.5,
de Broglie wavelength, wavenumber	λ_{dB}, k_{dB}	1.1.5, 5.1.3
distance in space, in the direction of	$s = \theta \cdot R_4$	2.1.1, 3.3.4
apparent homogeneous space (flat space distance)	$r_{0\delta}$	1.2.2, 4.1.1
local space (tilted space)	r_δ	4.1.1, 4.2.1,
distances (cosmological definitions)		
optical distance (DU)	D	1.3.3, 3.3.4, 6.2.1
physical distance (DU)	D_n	1.3.3, 3.3.4
co-moving distance (FLRW)	D_C	1.3.2
coordinate distance (FLRW)	D_C	1.3.2
angular diameter distance (FLRW)	D_A	1.3.2,
luminosity distance (FLRW)	D_L	6.3.1
Doppler effect		1.1.5, 1.3.2, 5.2.3
double slit experiment		5.3.5
energy buildup in space		3.3.1
energy object		4.1.4, 5.3.2, 5.3.4

energy of gravitation	E_g	2.2.2, 3.2.2, 3.3.2
complex presentation of the energy of gravitation	$E^\square{}_g=E_g'+iE''_g$	3.3.1
global gravitational energy	$E_{g(total)}$	3.2.2, 4.1.1
imaginary energy of gravitation	E''_g	4.1.1
inherent energy of gravitation	$E_{g(0)}$	2.2.2
local gravitational energy	$E_G=\Delta E_{g(\delta)}$	4.1.1,
total energy of gravitation	$E_{g(tot)}$	3.3.2
energy of motion	E_m	2.2.2
complex presentation of energy	$E^\square{}_m=E_m'+iE''_m$	2.1.4, 4.1.1
inherent energy of motion	$E_{m(0)}$	2.2.2
internal energy of motion	E_I	4.1.3
kinetic energy	E_{kin}	4.1.2,
rest energy	E_{rest}	4.1.4, 4.1.5
imaginary energy of motion	E''_m	4.1.1
total energy of motion	$E_{m(tot)}$	4.1.2
expansion of space		3.3.1
fine structure constant	a	5.1.2
force	**F**, F	2.2.2
central force	**F**$_C$, F_C	4.1.3, 4.1.8
gravitational force	**F**$_g$, F_g	3.2.2
inertial force	**F**$_i$, F_i	4.1.7
fourth dimension		2.1.1, 3.3.1
in hypothetical homogeneous space		2.1.1, 4.2.1
in local space		4.2.1
four-radius of hypothetical homogeneous space	R_4 or R_0	3.3.3
apparent 4-radius of local space	R''	4.1.1
frame conversion factor	$\chi = c_0/c$	4.1.4
frame dragging (by optically dense medium)		5.4.6
frame of reference, see system of energy frames		4.1.4
free fall (gravitational)	v_{ff}	4.1.1
gravitational constant	G	3.3.2
gravitational energy, see energy of gravitation		2.2.2, 3.2.2, 3.3.2
gravitational factor	δ	4.1.1, 4.4.4,
homogeneous space		4.1.1
apparent homogeneous space		4.1.1
hypothetical homogeneous space		4.1.1
Hubble constant	H_0	3.3.2
Hubble flow		1.3.2
Hubble law		6.1.1, 6.1.2
Hubble radius	R_H	3.3.2, 6.1.2
hypothetical homogeneous space		4.1.1
imaginary direction (direction of imaginary axis)	Im	2.1.1, 2.1.4
in apparent homogeneous space	Im$_{0\delta}$	4.1.1
in hypothetical homogeneous space	Im$_0$	4.1.1
in local space	Im$_\delta$	4.1.1
imaginary energy	iE''	2.1.4, 3.3.1, 4.1.1
imaginary momentum	**ip**'', ip''	2.1.4, 4.1.1
inertia		2.2.3, 4.1.3, 4.1.7
intensity factor	I_λ	5.1.1
internal energy	E_I	4.1.3
internal momentum	p_I	4.1.3
inherent energy of gravitation	$E_{g(0)}$	2.2.2
inherent energy of motion	$E_{m(0)}$	2.2.2
K-correction	K	6.3.4
kinetic energy, see energy of motion	E_{kin}	4.1.2

Index

local space		4.1.4
local frame of reference		4.1.4
local velocity of light		2.1.4,
local gravitational state		4.1.4
local rest energy		4.1.4
Mach's principle		1.2.2, 4.1.2, 4.1.3
magnitude (absolute)	M	6.3.1
apparent magnitude	m	6.3.1
distance modulus	μ	6.3.1
mass	m	2.2.1
definition	M	2.2.1
total mass in space	M_Σ	3.3.1, 3.3.2
mass equivalence of the total mass	M''	3.2.2
mass equivalence of radiation	m_λ	2.1.1, 5.1.1
mass equivalence of Coulomb energy	m_{EM}	5.1.2
mass object		5.3.4
relativistic mass (effective mass) (SR)	m_{rel}	1.2.2, 1.2.6, 4.1.2
rest mass	m	4.1.4
matter		3.3.5
anti-matter		3.3.5
baryonic matter		3.3.5
dark matter		3.3.5
Maxwell's equations	χ_λ	1.1.5, 1.4.2, 5.1.1
momentum	p	2.2.2
complex presentation of momentum	$p^\square = p' + ip''$	2.1.4
in space	$\mathbf{p} = \mathbf{p}'$	2.1.4
in the fourth dimension	$\mathbf{p}'' = \mathbf{i}p''$	2.1.4
internal momentum	$p_I^\square$	2.2.2
rest momentum	$p_{rest} = |\mathbf{p}''|$	2.2.2, 4.1.1, 4.1.3
Planck constant	h	1.1.5, 5.1.1
intrinsic Planck constant	$h_0 = h/c_0$	5.1.1
intrinsic reduced Planck constant	$\hbar_0 = h_0/2\pi$	5.1.3
Planck equation		1.1.5, 5.1.1
Planck units		5.3.6
Planck distance	r_0	5.3.6
Planck mass	m_0	5.3.6
phase velocity (of electromagnetic radiation)	c	5.3.2
primary energy buildup		3.3.1
radius (4-radius) of space		1.1.2, 1.1.3, 2.1.1
recession velocity		3.3.4, 6.1.2
physical (of distant objects)	v_n	3.3.4, 6.1.2
optical		6.1.2
redshift of radiation (cosmological)	z	6.1.2
reference at rest		2.1.3, 4.1.4
absolute reference at rest		2.1.3
the state of rest in hypothetical homogeneous space		1.3.1, 3.3.4
local state at rest		4.1.3
refractive index	n	5.4.6
rest energy	E_{rest}	4.1.4, 4.1.5
rest mass	m_{rest}	4.1.4, 4.1.5
rest momentum	p_{rest}	4.1.4, 4.1.5
Sagnac effect		5.4.5, 7.2.2, 7.3.2
Shapiro delay		5.4.1, 5.1.1, 7.3.4
space		1.1.1
spherically closed		1.1.1, 1.2.1
flat space (FLRW)		1.1.3, 3.3.2

system of energy frames		4.1.4
tilting angle of space	ψ	2.1.4, 4.1.1, 4.1.9
time		1.1.1, 2.1.2
SI second		5.7.1, 5.7.2
unit charge	e	1.1.5, 5.1.1, 7.2.2
vacuum permeability (electric constant)	ε_0	5.1.1, 5.1.2, 5.4.6
vacuum permittivity (magnetic constant)	μ_0	5.1.1, 5.1.2, 5.4.6
wave number	k	2.2.1, 5.1.3
of mass wave	k_m	5.1.3,
Compton wave number	$k_{Compton}$	1.1.5, 5.1.1, 5.1.3
de Broglie wave number	k_{dB}	5.1.3
velocity of light	c	3.3.2, 4.1.4
dependence of the expansion of space	$c_0 = 2/3 \cdot R_4 / t$	3.3.3
in hypothetical homogeneous space	c_0	3.3.1
in apparent homogeneous space	$c_{0\delta}$	4.1.4
in local space	c_δ	4.1.1, 4.1.4
phase velocity		5.3.2, 5.4.5
present definition		3.3.2
velocity of space	$c_4 = c_0$	3.3.3
wavelength equivalence of		
Coulomb energy	λ_{EM}	5.1.2
mass	λ_m	2.2.1, 5.1.3
temperature	λ_T	Appendix 1
zero-energy principle		2.2.2

Appendix 1, Blackbody radiation

Energy density of radiation in a blackbody cavity

The wavelength equivalence λ_T, and the frequency equivalence f_T of temperature T, respectively, are defined as

$$\lambda_T = \frac{hc}{kT} = \frac{h_0 c_0 c}{kT} = \frac{h_0}{k_0 T}, \qquad (A1:1)$$

and

$$f_T = \frac{c}{\lambda_T}, \qquad (A1:2)$$

where k is the Boltzmann constant. In the DU framework, the product kT is proportional to the velocity of light squared, or $kT \sim c_0 c$. In order to keep the temperature T independent of the velocity of light, it is useful to define the *intrinsic Boltzmann constant* k_0, defined as

$$k_0 \equiv \frac{k}{c_0 c} \quad ; \quad E_{\lambda_T} = \frac{h_0}{\lambda_T} c_0 c = k_0 T \cdot c_0 c \quad (=kT). \qquad (A1:3)$$

In the DU framework, k_0 is constant in the course of the expansion of space, and allows the use of temperature T, in Kelvins, as a parameter independent of the expansion of space and the declining velocity of light.

By applying the wavelength equivalence λ_T, the energy of radiation in a wavelength differential $d\lambda$ in a blackbody cavity is

$$dE_\lambda = E(\lambda) d\lambda = \frac{8\pi h_0 c^2}{\lambda_T^5} \frac{(\lambda_T/\lambda)^5}{(e^{\lambda_T/\lambda}-1)} d\lambda \qquad \left[\frac{J}{m^3}\right], \qquad (A1:4)$$

and by applying the frequency equivalence f_T, the energy of radiation in a frequency differential df is

$$dE_f = E(f) df = \frac{8\pi f_T^3 h_0}{c^2} \frac{(f/f_T)^3}{(e^{f/f_T}-1)} \cdot df \qquad \left[\frac{J}{m^3}\right]. \qquad (A1:5)$$

The total energy density in a blackbody cavity is obtained by integrating equation (A1:5)

$$E_{total} = \frac{8\pi f_T^3 h_0}{c^2} f_T \int_0^\infty \frac{(f/f_T)^3}{(e^{f/f_T}-1)} d\left(\frac{f}{f_T}\right) = \frac{8\pi f_T^3 h_0}{c^2} f_T \frac{\pi^4}{15}, \qquad (A1:6)$$

where the factor $\pi^4/15$ comes from the definite integral. In terms of λ_T the total energy in equation (A1:6) obtains the form

$$E_{total} = \frac{8\pi^5}{15\lambda_T^3} \cdot \frac{h_0}{\lambda_T} c^2 = \frac{8\pi^5}{15\lambda_T^3} E_{\lambda_T} = I_T \cdot E_{\lambda_T}. \tag{A1:7}$$

In the last two forms of (A1:7), the energy density in the cavity is related to the energy of a cycle of radiation emitted at wavelength λ_T. The factor I_T is the intensity factor characteristic to blackbody temperature T or the corresponding wavelength equivalence λ_T as

$$I_T = \frac{8\pi^5}{15\lambda_T^3}. \tag{A1:8}$$

Radiation emittance

The radiation emittance, i.e., the total bolometric energy flux emitted by a blackbody to the surrounding space, is obtained from equation (A1: 7) by multiplying the energy density in the cavity by the Stefan-Boltzmann factor $c/4$

$$F_{bol} = \frac{c}{4} \cdot I_T \cdot E_{\lambda_T} = \frac{2\pi^5}{15\lambda_T^2} \cdot \frac{c}{\lambda_T} \cdot \frac{h_0}{\lambda_T} c^2 = \frac{2\pi^5}{15\lambda_T^2} f_T \cdot \frac{h_0}{\lambda_T} c^2 \quad \left[\frac{W}{m^2}\right], \tag{A1:9}$$

which relates the total energy flux to the unit energy flux carried by one cycle of radiation at wavelength λ_T. Substitution of (A1:1) for λ_T in (A1:9) returns the classical expression of radiation emittance

$$F_{bol} = \frac{2\pi^5 k^4}{15 h_0^3 c^5} \cdot T^4 = \sigma T^4 \quad \left[\frac{W}{m^2}\right], \tag{A1:10}$$

where the constant σ is referred to as the Stefan-Boltzmann constant, $\sigma = 5.6693 \cdot 10^{-8}$ [W/m² /°K⁴].

Spectral distribution of blackbody radiation

The energy flux emitted in a wavelength differential $d\lambda$ to a steradian is obtained by applying the Stefan-Boltzmann factor $c/4$, and factor $1/4\pi$ for the flux density per steradian, to (A1:5)

$$dF_\lambda = \frac{c}{4 \cdot 4\pi} dE_\lambda = \frac{hc^2}{2\lambda_T^5} \frac{(\lambda_T/\lambda)^5}{(e^{\lambda_T/\lambda} - 1)} d\lambda = F(\lambda) d\lambda \quad \left[\frac{W}{m^2 sr}\right], \tag{A1:11}$$

where $F(\lambda)$ is

$$F(\lambda) = \frac{hc^2}{2\lambda_T^5} \frac{(\lambda_T/\lambda)^5}{(e^{\lambda_T/\lambda} - 1)} = \frac{F_0}{\lambda_T} \frac{(\lambda_T/\lambda)^5}{(e^{\lambda_T/\lambda} - 1)} \quad \left[\frac{W}{m^2 sr}/m\right]. \tag{A1:12}$$

In a frequency differential df the corresponding energy flux is

$$dF_f = \frac{c}{4 \cdot 4\pi} dE_f = \frac{f_T^3 h}{2c^2} \frac{(f/f_T)^3}{(e^{f/f_T} - 1)} df = F(f) df \quad \left[\frac{W}{m^2 sr}\right], \tag{A1:13}$$

where $F(f)$ is

$$F(f) = \frac{f_T^3 h}{2c^2} \frac{(f/f_T)^3}{(e^{f/f_T} - 1)} = \frac{F_0}{f_T} \frac{(f/f_T)^3}{(e^{f/f_T} - 1)} \qquad \left[\frac{W}{m^2 sr\, Hz}\right]. \qquad (A1:14)$$

The factor F_0 in equations (A1:12) and (A1:14) is

$$F_0 = \frac{1}{2\lambda_T^2} f_T \frac{h_0}{\lambda_T} c^2 = \frac{15}{\pi^4} F_{bol(sr)} \qquad \left[\frac{W}{m^2 sr}\right], \qquad (A1:15)$$

where $F_{bol(sr)}$ is the bolometric flux of (A1:10) per steradian $F_{bol(sr)} = F_{bol}/4\pi$.

The energy flux emitted in the wavelength or frequency range of a narrowband filter with relative width $W = W_\lambda = \Delta\lambda/\lambda = W_f = \Delta f/f$ is obtained from equations (A1:12) and (A1:14), respectively, as

$$F_{W(\lambda)} = F_0 \frac{(\lambda_T/\lambda)^5}{(e^{\lambda_T/\lambda} - 1)} \frac{\Delta\lambda}{\lambda} \cdot \frac{\lambda}{\lambda_T} = F_0 \frac{(\lambda_T/\lambda)^4}{(e^{\lambda_T/\lambda} - 1)} W \qquad \left[\frac{W}{m^2 sr}\right] \qquad (A1:16)$$

$$F_{W(f)} = F_0 \frac{(f/f_T)^3}{(e^{f/f_T} - 1)} \frac{\Delta f}{f} \frac{f}{f_T} = F_0 \frac{(f/f_T)^4}{(e^{f/f_T} - 1)} W \qquad \left[\frac{W}{m^2 sr}\right], \qquad (A1:17)$$

or by relating the narrow band power density to the bolometric flux density by substituting (A1:15) for F_0 in (A1:16) and (A1:17) as

$$F_{W(f,\lambda)} = \frac{15}{\pi^4} \frac{(f/f_T)^4}{(e^{f/f_T} - 1)} W \cdot F_{bol(sr)} = \frac{15}{\pi^4} \frac{(\lambda_T/\lambda)^4}{(e^{\lambda_T/\lambda} - 1)} W \cdot F_{bol(sr)} \qquad \left[\frac{W}{m^2 sr}\right]. \qquad (A1:18)$$

The distribution function $D = x^4/(e^x - 1)$ obtains its maximum at $x = 3.9207$

$$D_{max} = \left(\frac{x^4}{e^x - 1}\right)_{max} = D_{(x=3.9207)} = 4.780. \qquad (A1:19)$$

At a fixed relative bandwidth W the maximum flux occurs when the nominal frequency or wavelength of the filter is $f_W/f_T = \lambda_T/\lambda_W = 3.9207$ ($f_W = c/\lambda_W$)

$$F_{W(f,\lambda)} = \frac{15}{\pi^4} \cdot D_{max} \cdot W \cdot F_{bol(sr)} \qquad \left[\frac{W}{m^2 sr}\right], \qquad (A1:20)$$

which relates the energy flux through an ideal narrow band filter to the bolometric energy flux of the radiation. The nominal frequency of the filter is matched to the maximum power throughput of blackbody radiation by setting $f_W = 3.9207 \cdot f_T$.

Table A1-I lists the frequencies and wavelengths corresponding to the maximum power density obtained from equations (A1:12), (A1:19), and (A1:14) for the cosmic microwave background. The blackbody temperature of the CMB is 2.725 °K, and the corresponding frequency and wavelength equivalences are $f_T = 56.8$ GHz and $\lambda_T = 5.28$ mm, respectively.

Figure A1-1 illustrates the energy spectra calculated for the cosmic microwave background as 2.725 °K blackbody radiation.

Equation	Distribution function D	x for D maximum	Unit	f (F_{max}), CMB [GHz]	λ (F_{max}) CMB [mm]
(A1:12)	$\dfrac{x^5}{(e^x-1)}$	4.9651	$\dfrac{W}{m^2 \cdot sr}\Big/m$	282	1.06
(A1:19)	$\dfrac{x^4}{(e^x-1)}$	3.9207	$\dfrac{W}{m^2 \cdot sr}$	223	1.35
(A1:14)	$\dfrac{x^3}{(e^x-1)}$	2.8214	$\dfrac{W}{m^2 \cdot sr}\Big/Hz$	160	1.87

Table A1-I. Application of equations (A1:12), (A1:19), and (A1:14) for determining the frequencies and wavelengths of the maximum power density of the cosmic microwave background in units of [Wm^{-2}sr^{-1}m^{-1}], [Wm^{-2}sr^{-1}], and [Wm^{-2}sr^{-1}], respectively. See also Figure A1-1.

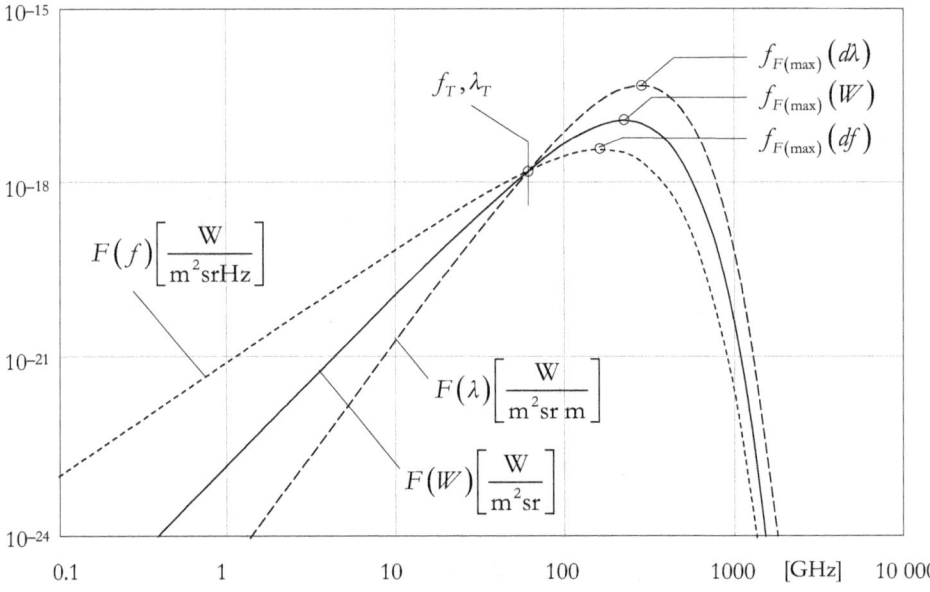

Figure A1-1. The energy flux density of the cosmic microwave background (CMB) in terms of $F(\lambda)$ [Wm^{-2}sr^{-1}m^{-1}] (A1:12), F_W [Wm^{-2}sr^{-1}] (A1:17), and $F(f)$ [Wm^{-2}sr^{-1}Hz^{-1}] (A1:14) in the frequency range from 100 MHz to 10 THz. The wavelength of the observed maximum power density in terms of $F(f)$ [Wm^{-2}sr^{-1}Hz^{-1}] is 1.87 mm. In terms of $F(\lambda)$ [Wm^{-2}sr^{-1}m^{-1}], the maximum occurs at wavelength 1.06 mm. The integrated total energy is equal for each flux density function. Curve F_W [Wm^{-2}sr^{-1}] shows the shape of the flux density function observed in narrow band filters with $W = \Delta\lambda/\lambda = \Delta f/f$.

References

1. A. Einstein, *Kosmologische Betrachtungen zur allgemeinen Relativitätstheorie*, Sitzungsberichte der Preussischen Akad. d. Wissenschaften (1917)
2. R. Feynman, W. Morinigo, and W. Wagner, *Feynman Lectures on Gravitation (during the academic year 1962-63)*, Addison-Wesley Publishing Company, p. 164 (1995)
3. R. Feynman, W. Morinigo, and W. Wagner, *Feynman Lectures on Gravitation (during the academic year 1962-63)*, Addison-Wesley Publishing Company, p. 10 (1995)
4. G. Leibniz, *Essay in Dynamics, Part 2, The Laws of Nature*, http://www.earlymodern-texts.com/assets/pdfs/leibniz1695b.pdf
5. D.W. Sciama, *On the origin of inertia*, Monthly Notices of the Royal Astronomical Society, Vol. 113, p.34. (1953) http://adsabs.harvard.edu/full/1953MNRAS.113...34S
6. A. Einstein, The Berlin Years: Writings, 1914-1917 Volume 6, p 370: *The Structure of Space According to the General Theory of Relativity*, http://einsteinpapers.press.princeton.edu/vol6-trans/382
7. Planck, Max (1901), "*Ueber das Gesetz der Energieverteilung im Normalspectrum*", Ann. Phys. 309 (3): 553–63
8. G.W. Leibniz, Mathematischer Naturwissenschaftlicher und Technischer Briefwechsel, Sechster Band (1694)
9. G.W. Leibniz, Principles of Nature and Grace Based on Reason, http://www.earlymodern-texts.com/pdf/leibprin.pdf
10. H.E. Ives and G.R. Stilwell, J. Opt. Soc. Am. **28** (1938) 215
11. H.E. Ives and G.R. Stilwell, J. Opt. Soc. Am. **31** (1941) 369
12. H.I. Mandelberg and L. Witten, J. Opt. Soc. Am. **52**, 5 (1962) 529
13. H.J. Hay, J.P. Schiffer, T.E. Cranshaw, and P.A. Egelstaff, Phys. Rev. Letters **4**, 4 (1960) 165
14. D.C. Champeney, G.R. Isaak, and A.M. Khan, Nature **198,** 4886 (1963) 1186
15. W. Kundig, Phys. Rev. **129** (1963) 2371
16. Turner, K.C., and Hill, H.A., Phys. Rev. B, **134** (1964) 252
17. J.C. Hafele and R.E. Keating, Science **177** (1972) 166
18. J.C. Hafele, Nature Phys. Sci. **229** (1971) 238
19. D. Kleppner, R.F.C. Vessot, and N.F. Ramsey, Astrophysics and Space Science **6** (1970) 13
20. R.F.C. Vessot et al., Phys. Rev. Letters, **45**, 26 (1980) 2081
21. T.E. Cranshaw, J.P. Schiffer, and A.B. Whitehead, Phys. Rev. Letters **4**, 4 (1960), 163
22. R.V. Pound and G.A. Rebka Jr., Phys. Rev. Letters **4** (1960) 337
23. R.V. Pound and J.L. Snider, Phys. Review **140,** 3B (1965) B788
24. R. Schlegel, Nature Phys. Sci. **229** (1971) 237
25. T. Suntola, Galilean Electrodynamics, 14, No.4 (2003) http://physicsfoundations.org/data/documents/2003_TS_GE_Re-evaluation_of_Scout_D.pdf
26. J.C. Maxwell, *A treatise on electricity and magnetism*, Chapter XX, (1873), openlibrary.org ja wikisource.org

27. A.A. Michelson, *The Relative Motion of the Earth and the Luminiferous Ether* (1881), wikisource.org
28. Albert A. Michelson and Edward W. Morley, *Influence of Motion of the Medium on the Velocity of Light,* American Journal of Science, 1886, Ser. 3, Vol. 31, Nr. 185: 377-386, openlibrary.org, wikisource.org
29. A.A. Michelson et al., *Conference on the Michelson–Morley Experiment, Mount Wilson, February, 1927*, Astrophysical Journal 68, 341 (1928)
30. G. Joos, Annalen der Physik. **7**, Heft 4 (1930) 385
31. R.J. Kennedy and E.M. Thorndike, Phys.Rev. **42** (1932) 400
32. E.W. Silvertooth, J. Opt. Soc. Amer. **62** (1972) 1330
33. L. Essen, Nature **175** (1955) 793
34. T.S. Jaseda et al., Phys. Rev. A **133** (1964) 1221
35. A. Brillet and J.L. Hall, Phys. Rev. Lett. **42**, 9 (1979) 549
36. G. Sagnac, *Comptes Rendus* **157**, 708 (1913)
37. G. Sagnac, *Comptes Rendus* **157**, 1410 (1913)
38. A.A. Michelson and H.G. Gale, Astrophys. J., **61**, 7 (1925) 140
39. J. Weber, General Relativity and Gravitational Waves, Interscience Publishers, Inc. pp. 64-67 (1961)
40. M. Berry, Principles of cosmology and gravitation, Cambridge University Press, p.83 (1989)
41. J. Foster and J.D. Nightingale, A Short Course in General Relativity, Springer, Second Edition, p. 147 (2001)
42. R. Genzel et al., Nature **425** (2003) 934
43. J.H. Taylor, R.A. Hulse, L.A. Fowler, G.E. Gullahorn, and J.M. Rankin, The Astrophysical Journal, 206:L53-L58 (1976)
44. C.M. Will, Was Einstein Right? Oxford University Press (1993)
45. Carl Sagan and George Mullen, Science 177 (4043), 52-56 (1972)
46. W. de Sitter, *Do the galaxies expand with the universe*, BAN 6, 146D (1931)
47. R.C. Tolman, PNAS 16, 511, 1930
48. E. Hubble and R. C. Tolman, ApJ, 82, 302 (1935)
49. E. Hubble, M. L. Humason, Astrophys.J., 74, 43 (1931)
50. W. de Sitter, B.A.N., 7, No 261, 205 (1934)
51. H.P. Robertson, Zs.f.Ap., 15, 69 (1938)
52. E. Hubble, M. L. Humason, Astrophys.J., 74, 43 (1931)
53. A. Sandage, J-M. Perlmuter, ApJ, 370, 455 (1991)
54. K. Nilsson et al., Astrophys. J., 413, 453, (1993)
55. J.T. Tonry et al., ApJ, 594, 1 (2003)
56. J.W. Wells, Nature 197, (1963) 948
57. J.W. Wells in Paleogeophysics, Edited by S.K. Runcorn, Academic Press, London (1970)
58. D.L. Eicher, Geologic Time, 2nd edition, Prentice/Hall International Inc., London 117 (1976)
59. Stephenson, F.R.; Morrison, L.V., Philosophical Transactions: Physical Sciences and Engineering, Volume 351, Issue 1695, pp. 165-202 (1995)

60. F.R. Stephenson, L.V. Morrison, C.Y. Hohenkerk, *Measurement of the Earth's rotation: 720 BC to AD 2015*, Proceedings of the Royal Society A, 7 December 2016. DOI: 10.1098/rspa.2016.0404, http://rspa.royalsocietypublishing.org/content/472/2196/20160404
61. H. Schuh, H. Schmitz-Hübsch, Surveys in Geophysics **21**, 499–520 (2000)
62. J.P. Vanyo and S.M. Awramik, Precambrian Research **29**, 121 (1985)
63. I. Newton, The Principia, Translation by J. Budenz and A. Whitman, Univ. of California Press (1999)
64. A.H. Guth, Proc. Natl. Acad. Sci. U.S.A., 90, **11**, 4871 (1993)
65. P.C. Peters and J. Mathews, Phys. Rev. 131, 435 (1963)
66. J.M. Weisberg, J.H. Taylor, Binary Radio Pulsars, ASP Conference Series, Vol. 328, **pp. 25-31,** (2005)
67. Citation from Max Planck's Nobel lecture, June 2, (1920)
68. A. Einstein, *"Über die Möglichkeit einer neuen Prüfung des Relativitätsprinzips"*, Annalen der Physik SER.4, no.23 (1907)
69. M.L. Baker Jr., Astrodynamics, Academic, New York, 218 (1967)
70. C. Jönsson, American Journal of Physics **4**, 4 (1974)
71. A. Lehto, Nonlinear Dynamics 55, 279 (2009)
72. J.O. Dikey, et.al. Science **265**, 482 (1994)
73. National Institute of Standards and Technology, http://www.nist.gov/
74. E. Komatsu, et al., arXiv:astro-ph/0803.0547v2 (2008)
75. S.M. Carroll, W.H. Press, and E.L. Turner, ARA&A, 30, 499 (1992)
76. J.M.H. Etherington, Phil. Mag. 15, 761 (1933)
77. A. Kim, A. Goobar, & S. Perlmutter, PASP, 190–201 (1996)
78. A.G. Riess et al., [Supernova Search Team Collaboration], Astron. J. 116, 1009 (1998)
79. A.G. Riess et al., Astrophys. J., 607, 665 (2004)
80. A.G. Riess et al., [arXiv:astro-ph/0611572] (2006)
81. S. Perlmutter et al. [Supernova Cosmology Project Collaboration], Astrophys. J. 517, 565 (1999)
82. R. Knop et al., ApJ, 598, 102 (2003)
83. B. Barris et al., ApJ, 602, 571(2004)
84. P. Astier et al. [Supernova Legacy Survey, SNLS]: A&A 447, 31–48 (2006)
85. A. Conley et al., ApJ, 644, 1 (2006)
86. T. Suntola and Robert Day, arXiv/astro-ph/0412701 (2004)
87. G. Goldhaber et.al., Astrophys. J., **558**, Issue 1, 359 (2001)
88. J.P. Norris, Astrophys. J., **424**, Part 1, no. 2, 540 (1994)
89. A. Einstein, statement (1949) – citation in C.W. Misner, K.S. Thorne, and J.A. Wheeler, *Gravitation*, W.H. Freeman and Company, San Francisco, p. 5 (1973)